THE TECHNIQUES OF
MODERN
STRUCTURAL
GEOLOGY

Volume 1: Strain Analysis

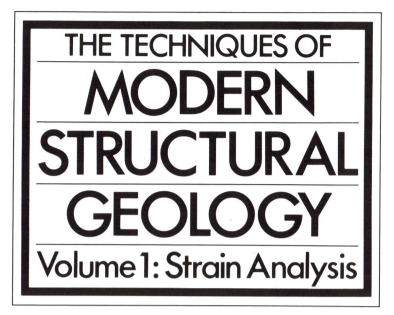

THE TECHNIQUES OF MODERN STRUCTURAL GEOLOGY

Volume 1: Strain Analysis

JOHN G. RAMSAY
MARTIN I. HUBER

ACADEMIC PRESS

Harcourt Brace & Company, Publishers

London   San Diego   New York
Boston   Sydney   Tokyo   Toronto

ACADEMIC PRESS LIMITED
24/28 Oval Road
London NW1 7DX

United States Edition published by
ACADEMIC PRESS, INC.
San Diego, CA 92101

British Library Cataloguing in Publication Data
Ramsay, J. G.
    The techniques of modern structural geology.
    Vol. 1
    1. Geology, Structural
    I. Title   II. Huber, M.
    551.8   QE601
    ISBN 0-12-576901-6
    ISBN 0-12-576921-0 Pbk

LCCCN 82-074569

Printed in Great Britain at The Alden Press, Oxford

Preface

In the case of nearly all branches of science a great advance was made
when accurate quantitative methods were used instead of more qualitative.
One great advantage of this is that it necessitates more accurate thought,
points out what remains to be learned, and sometimes small residual
quantities, which otherwise would escape attention, indicate important
facts.

J. C. Sorby, 1908

Structural geology, a branch of earth science, has leapt ahead over the past decade. This rapid development has come about for several reasons. First has been the realization that naturally deformed rocks seen in field outcrops contain a fantastic range of small scale features which relate directly to large-scale orogenic deformation systems. No longer is it wise for the geologist interested in large-scale crustal structure to write off such small-scale features as trivia. It has become clear today that the understanding and interpretation of these geometric features enables the investigator to build up a much more comprehensive understanding of the subtle but very important features of an orogenic zone. The field structural geologist with a developed "eye" for these features can often work over large areas to much more effect than was possible in the past. No longer is it adequate to determine the fold trends, or to map a structural lineament, or even to realize that here we have a suture between two moving continental masses. Today we can and should go much further—we can use our studies of strain state to say exactly how much deformation and how much displacement have occurred. When we have made an assessment of the overall strain pattern of a region we are in a good position to say what the relative movement directions between large and small sections of the crust were, whether the great bends in the major orogenic zones of the earth for example are primary features, or result from secondary bowing of a previously rectilinear zone.

The great interest in structural geology in many earth science institutes nowadays can be attributed at least in part to the practicality of the subject. We have found that students become very involved in the subject when they go into the field, which may be a well known locality, and find that the tectonic structures may have been neglected in previous publications. They often find that they can make significant original observations which can lead to extremely stimulating and exciting interpretations of the tectonics, interpretations which can often be tested by extending the investigations into adjacent terrain.

An Instructors' Manual is available for this book. This can be obtained from Academic Press by recognised teachers.

The second reason for the recent advances in structural geology has been the realization that rock deformation, like any other form of deformation, obeys well-defined physical and chemical laws. The concepts of material science have therefore been applied to the study of deformed rocks with great effect. Particularly important have been the use of mathematical methods and the application of the concepts of continuum mechanics. These numerical approaches have revolutionized the analysis of the geometry of naturally deformed rocks and both theoretical and practical interpretations of what these structures signify.

These books have grown out of a need to teach fundamental, practical aspects of structural geology to undergraduate and postgraduate students in the earth sciences and they have been written to provide a basic text at undergraduate university level. We have tried to assemble a comprehensive account of such basic techniques as could be the foundation of a practical and theoretical course in the analysis of tectonic structures, stress and strain. Volume 1 covers the principles of deformation, and volume 2 applies these principles specifically to the analysis of folds and fractures. The material covered in these first two volumes has formed the core of various university courses given by us in Britain and at the ETH Zürich, and the two volumes have been designed to provide material for two successive 15-week teaching semesters. The material of the planned third volume would be appropriate for advanced specialist undergraduate or postgraduate courses.

We have tried to encourage independent thinking by the student by leading him through progressive ways of thought that we ourselves have found helpful in building an understanding of the subject. Our approach has been to help the student to develop his own methods and to solve problems by logical thinking, rather than to rely on the routine application of a previously described and well-worn "magic" formula. We have assembled the various sessions so that they build logically one on the other from two-dimensional to three-dimensional analysis, and from simple to complex analytical methods.

Each "session" starts with the formulation of a specific

problem and presentation of any essential background or necessary mathematical techniques. This is followed by a practical part in which we outline experiments, pose problems and set up a number of specific questions aimed at focusing attention on key points, perhaps assisting the solutions by giving a few hints on how to get started. After the questions have been answered or a technique has been applied to solve a particular problem we provide a section entitled "Answers and Comments". Here we set out the key features of the solutions, give numerical answers, and comment on why one technique may be preferable to another. We also develop a commentary on the geological significance of the results. We are of the opinion that such a commentary will link the sessions and, by providing an important background, enable the student to see how wider applications of the techniques can lead to deeper understanding of tectonic studies.

The problems we present are based on actual geological examples or on experiments and methods which we have found particularly useful in understanding rock strain. We are a little critical of some previously published collections of structural geology problems and maps in that, in our experience, they often do not relate to what occurs in nature. It also seems to us that the complexities of real geological systems are much more stimulating than carefully selected or "invented" problems which abstract too cleanly one particular aspect of a problem. We have also strived to produce new and relevant illustrative material, and we hope that the many photographs and diagrams will assist the student to understand the geometric complexities that sometimes arise and to relate more clearly the theoretical and practical aspects of the subject.

For each practical session we have set out material that the student should be able to cover in a three-hour work period. We are aware that earth science students have dissimilar backgrounds, particularly in mathematics, and that they may, therefore, find too much or too little to cover in a three-hour period. The questions we pose have been designed to develop a thorough understanding of the geometric features of deformation, because we are convinced that such a background is critical for the future development of the subject. We have also aimed at the development of mathematically-based thought processes and analyses. The concepts of displacement and strain, which form the basis of so much modern analysis, are most naturally developed with some mathematical formulation. Therefore, we have designed mathematical problems of complexity increasing through the sessions to build a strong background in mathematical theory. The mathematical techniques are generally straightforward, being based on simple algebraic methods, two- and three-dimensional co-ordinate geometry and some elementary calculus. Although the specialist may see more elegant ways of arriving at our results, we strongly hold the view that it is better to get a student excited and interested in discovering geometric principles with the basic mathematics he already possesses, rather than insist on more sophisticated techniques which might leave him struggling. From our experience we have also found that an initial mastery and understanding of geometric principles in two dimensions is an absolute necessity. When these principles are grasped, it is generally a relatively easy matter to extend them into the third dimension.

Because of the range of talent and the differences in student background we have developed a double system of problems. The first set is basic. It should be completed in the three-hour session and it contains fundamental material that will be essential before the following sessions can be tackled. A second set, marked with starred numbers, is more advanced. It has been designed to lead to a deepening of the understanding of the topic in hand, or to develop stronger mathematical techniques. It is not essential to complete this second set of problems before passing to the next session. They can either be tackled directly after the basic set, or be used for a "second round" or revision approach.

We have developed a number of appendices separate from the sessions, in which we set out the most important mathematical proofs required to establish the basic formulae used in the sessions. We have done this because we thought that an overloading of the sessions themselves with somewhat dry basic mathematics (important as it is) might tend to inhibit the flow of problem solving. In the session questions we have suggested particular mathematical exercises which relate directly to the specific problems in hand, whereas in the appendices we have developed more general solutions of displacement and strain covering a much wider and less specific field. We also think that this general and basic information is best concentrated in appendices for easy access when required at a later stage.

In producing these books we have relied heavily on help from many people and in conclusion we would like to thank most sincerely all those who have assisted us to put these volumes together. In particular we wish to convey our thanks to Urs Gerber for producing so expertly black and white photographic material from our colour diapositives, and for assisting in many ways the production of the illustrations. We thank our secretary, Barbara Das Gupta, for the long hours she has put into this work and the expert production of the manuscript. Many course students and assistants have battled with the session questions, discovering our mistakes and helping us to select from a range of material what would work well. In particular, Roy Kligfield provided us with a very full and useful commentary on the value of the practical material. Special thanks go to David Durney for also providing us with an appraisal of the text. We also have drawn extensively on his ideas relating to the effects of volumetric dilatation in shear zones. Some data has been drawn from his unpublished Ph.D. thesis to help produce the profiles of the Helvetic nappes (Fig. 11.10), and the authors and he seem to have discovered at the same time the effects of inequantly shaped objects on calculations of incremental strains from pressure shadows. We would like to thank Dorothee Dietrich, David Durney and Andrew Siddans for providing unpublished data that went into the production of the strain maps of the Helvetic nappes, to Pierre Choukroune for generously providing rock specimens and photographs of pressure shadows, and to Shankar Mitra for providing photographs we used in Session 7. Final thanks go to Ishbel Ramsay for drawing the deformed version of the authors and to Adrianna Huber and Chris Ramsay for keeping the authors more or less along the right displacement paths.

September 1983
JOHN RAMSAY
MARTIN HUBER

Contents

Contents of Volume 2: Folds and Fractures

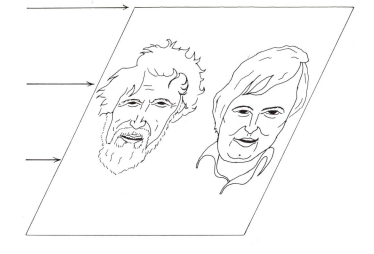

SESSION 1

Displacement: Changes in Length and Angles

Simple experiments with a card deck model are used to establish the concepts of longitudinal strain and shear strain during progressive simple shear. The methods for mathematically describing the displacements in the form of displacement equations and coordinate transformation equations are developed. Some simple geological implications of length changes are discussed and the development of boudin structures and buckle folds are related to stretching and shortening taking place in competent rock layers embedded in a less competent matrix.

INTRODUCTION

A study of the geometric changes that take place during deformation is probably the most important basic necessity for an understanding of deformed rocks and for the interpretation of the geological significance of tectonic structures. The Earth during its evolution has been acted on by a constantly changing force field which has led to variations in the state of stress of its various rock constituents. The application of force to a material causes the mass to change its position and often its shape. The changes of the positions of points in a body is known as **displacement** and any resulting change of shape as **deformation** or **strain**. The strains set up may be small and reversible, that is to say the material may return to its original shape on removal of the applied stresses. Such deformations are characteristic of solid bodies deformed in their **elastic** range. At a certain state of **elastic deformation**, at the elastic limit, the solid rocks lose their internal cohesion on certain surfaces and undergo **brittle deformation**. Observations on many naturally deformed rocks indicate that the rocks have undergone much larger deformations than those characteristic of elastic bodies. These deformations developed large permanent strain, often without the development of specific surfaces of rupture. Such rocks are said to have undergone **ductile flow**. The aim of the first few Sessions of our work is to investigate in detail the geometric features of deformed materials so that we can start to appreciate the reasons for the wide variety of structures seen in naturally deformed rocks.

The relationship between displacement and the internal shape changes, known as strain, is generally rather complicated. One of the practical ways of getting insight into these relationships is by performing experiments with simple models. It is possible to deform models of ductile materials such as clay, putty and Plasticene and to study the geometric changes made by lines, grids or circular markers inscribed on the model surface. To begin our study, however, we will study strain with some more easily worked material: paper! In this age of computers (when no scientist is considered a Scientist unless he has a print out under his arm or a pack of computer cards in his hand) we have easy access to a very simple modelling method—the *card deck*. Simple experiments made by shearing a stack of computer cards and observing the changes in geometric form of various grids and circular markers drawn on the edges of the card deck provide an extremely convenient way of helping us understand some of the key features of deformation geometry. Although the experiments proposed below will provide excellent visual aids for helping us appreciate the principles of deformation, we should straight away stress the limitations of this model technique as applied to natural deformations. Rocks are not packs of gliding cards and their internal deformations are generally much more varied and complex than those we will produce in our experiments. The student should avoid the idea that our experiments provide a universal model for the behaviour of natural systems. Sometimes natural deformations do take place by shearing along narrow, planar, card-like elements; but generally they do not. Sometimes they produce geometric features like those we will see in our model; at other times they do not. So long as we are aware of these limitations we should be safe, and we will find, in fact, that our card model is able to establish practically all the important fundamental features of displacement and strain.

Principles of displacement and strain using simple shear of a card deck

Practical details

The student is recommended to set up the model, perform the experiments himself and answer the questions posed.

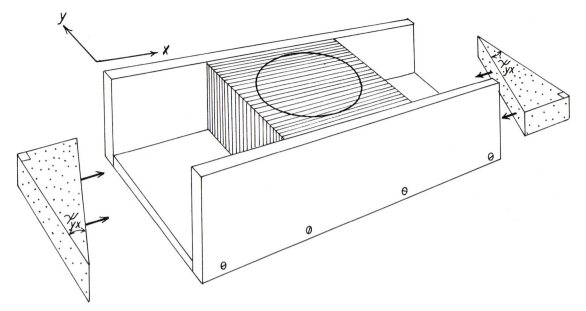

Figure 1.1. *Box used for holding cards together for simple shear experiments.*

When he has answered these questions he should then check his results with the answers (pp. 5–10) and carefully go through the notes provided on the significance of the results.

The experiments set out below can be carried out with any set of cards, but the most suitable are standard IBM computer cards which are thin and slip easily one on the other. The cards are best held in a box, such as that illustrated in Figure 1.1, so that they will not splay out sideways. The box should have sides which are at least three times the length of the cards, and the height of the walls should be the same as the width of the cards. The card displacement can be done very simply by hand, but best results will be obtained using pairs of wooden or plastic shaped endpieces or formers as illustrated in Figures 1.1 and 1.2. Using these formers, known values of shear can be induced into the card deck simply and accurately, while the free ends of the cards are held closely by the wedges, eliminating their tendency to open out one from the other. The student needs sets of paired triangular wedges with different apical angles (ψ). Values of ψ over the following ranges will be found most useful: 11·3°, 21·8°, 31°, 38·7°, 45°, 50·2°, 54·5°, 58°, 61°, 63·5°. The amount of shear induced into the card deck by formers of this type is best expressed in terms of the shear strain γ parallel to the walls of the shear box, defined as Greek "gamma":

$$\gamma = \tan \psi = a/b$$

Formers with the angles given above will have shear

strain values from 0·2 to 2·0 at increments of 0·2 γ. It will be convenient to use γ as an absolute measure of the displacement because if, for example, we double the displacement across the box, that is equivalent to doubling the gamma value.

Formers with more elaborately curved edges should also be made (see Figure 1.2) so that complex displacements and deformations can be induced into the card deck, and pairs with sinusoidal edges will also be useful for the more complicated experiments of Sessions 2 and 3.

The mathematical name for the type of displacement we will induce by sliding the cards one against the other is **simple shear**. In this and later Sessions we will want to have some reference axes so that we can refer more exactly to the displacement and resulting geometric modifications that take place on our deck surface. Let us therefore, choose an orthogonal coordinate reference system with x parallel to the walls of our shear box and y perpendicular to these walls (Figures 1.1 and 1.3). We will also need to define what we mean by shear strain with reference to

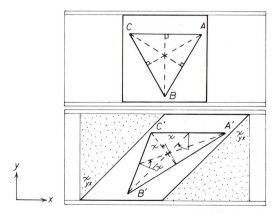

Figure 1.3. *Constructions to be made on the top of the cards for the experiments of Questions 1.1, 1.2 and 1.3. The lower diagram shows the configuration of the original equilateral triangle after displacement (A'B'C').*

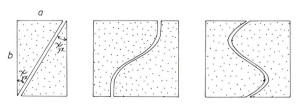

Figure 1.2. *Shaped endpieces or formers used to deform cards in model experiments.*

these coordinates (see above, γ). We use subscripts for shear with reference to our chosen coordinate frame as follows: the first subscript refers to the perpendicular to our shear surface (cards, shear box walls) and is the direction y, and the second refers to the shear direction, i.e. the x-direction—hence subscripts yx in γ_{yx}.

Changes in lengths and angles

Draw in black ink on the surface of the cards an equilateral triangle ABC (Figure 1.3, equal side lengths) so that AC is parallel to the card direction and AB and BC make angles of 60° and 120° respectively to the cards. In ink of some other colour draw perpendicular lines from A to side BC, B to side AC and C to side AB.

Shear the card deck with different pairs of formers so that the points A, B and C are displaced to new positions A', B' and C' (Figure 1.3).

Question 1.1

Do the lengths of the sides of the original triangle change? Measure them and record in tabulated form these data (see below).

Do the initially perpendicular lines remain perpendicular? Measure the angular deflection from the initial perpendicular (ψ, see Figure 1.3). Record the following data in column form for various values of simple shear displacement, $\gamma_{yx} = \tan \psi$.

γ_{yx}	$A'B'$	$B'C'$	$C'A'$	Direction $A'B'$ ψ	$B'C'$ ψ	$C'A'$ ψ
$\gamma_{yx} = 0{\cdot}2$ $\gamma_{yx} = 0{\cdot}4$						

In these experiments it should be clear that simple shear displacement does *generally* lead to changes in length of lines, and changes of angles between initially perpendicular lines. These two types of geometric effect are used to define strain parameters.

Definition: extension e

Changes in length are known as **longitudinal strains** and are recorded by a measure known as **extension** always referred to by the lower case letter e. For direction $A'B'$ this would be $e_{A'B'}$. Extension is defined as the change in length divided by the original length, i.e.

$$e_{A'B'} = \frac{A'B' - AB}{AB}$$

Extension can be either positive (when the length increases) or negative (when the length decreases), and its value lies between minus one and plus infinity ($-1 < e < +\infty$). Why are values of e less than -1 impossible? If the initial line length was unity, then e would record an actual measure of the change of length of this line; clearly

a value for e of -1 would imply that the line had no existence.

Definitions: angular shear strain ψ, shear strain γ

Angular changes for any specific direction are defined by measuring the change in angle between a line in that direction and another that *was originally perpendicular* to it. This angle is known as the **angular shear strain** and is always referred to by the Greek letter "psi" (ψ) (Figure 1.3). The deflection of the perpendicular to our reference direction can be either clockwise or anticlockwise and we differentiate these by using a sign convention. If the original perpendicular is rotated *clockwise* relative to the specific direction, the angular shear strain is *negative*; if rotated *anticlockwise* then *positive*. For example, in Figure 1.3, the signs of the angular shear strains with reference directions $A'B'$, $B'C'$ and $C'A'$ are +ve, +ve and −ve respectively. Go back to your tabulated results and indicate the signs of ψ. For purposes of convenience in later mathematical analyses we take the tangent of the angular shear strain; this defines the **shear strain**, and is referred to by the Greek letter "gamma" (γ), e.g.

$$\gamma = \tan \psi$$

The geometrical significance of γ is seen if we consider the displacement of the perpendicular line from the reference line at unit distance from the reference line after deformation (Fig. 1.4).

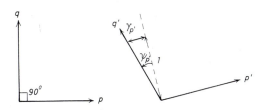

Figure 1.4. Relationships of ψ and γ. Two initially perpendicular lines p and q change orientation to take up positions p' and q'.

Question 1.2

Now extend your tabulated data, calculating the value of the extension and shear strain for each direction. The subscripts for e and γ refer to the direction of the side of the deformed triangle $A'B'C'$.

γ_{yx}	$e_{A'B'}$	$e_{B'C'}$	$e_{C'A'}$	$\gamma_{A'B'}$	$\gamma_{B'C'}$	$\gamma_{C'A'}$
$-0{\cdot}2$ $-0{\cdot}4$						

Graphically record these changes in extension (abscissa γ_{yx}, ordinate $e_{A'B'}$, etc.) and shear strain (abscissa γ_{yx}, ordinate $\gamma_{A'B'}$, etc.).

Question 1.3

Describe in words the changes in lengths of lines AB, BC and CA with increasing shear of the card deck. What is the significance of the minimum on the curve for changes in $e_{A'B'}$? What is the orientation of the line $A'B'$ at this minimum point? What is the significance of the point where the $e_{A'B'}$ curve crosses with γ axis? Are any of the variations in e linear with γ?

Could we describe the overall features of the deformation by the changes of extension in any one direction?

Question 1.4

Describe the changes that take place in the shear strains along lines AB, BC and CA with increase in γ_{yx} and again discuss the significance of these variations.

Question 1.5

Write two equations which express the displacement of *any initial point* in the card deck having coordinates (x, y) (x parallel to the shear box walls, y perpendicular to the cards) *to its final position* (x', y') after shear γ with reference to an origin of the coordinate system situated at one of the corners of the card deck (Figure 1.5). Put these in the form

$$x' = f_1(x, y)$$

$$y' = f_2(x, y)$$

where $f_1(x, y)$ and $f_2(x, y)$ are some mathematical functions of the initial position of the point. These equations are known as **coordinate transformation equations** because they relate the coordinates of points before and after displacement. Now check your answers and read the commentary on the significance of the results. When this has been done either tackle the more advanced starred questions below *or* move on to Session 2.

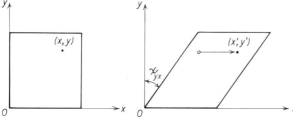

Figure 1.5. *Displacement of a point (x, y) to new position (x', y') during simple shear. Note that ψ_{yx} is negative.*

Advanced Problems (all denoted with stars★)

The problems set out below should lead to a furthering of a general understanding of the investigation carried out previously. They are designed to extend the student in various ways—to increase the depth of his understanding of the geometry of the experiments, to develop mathematical methods of thinking, to encourage his formulation of various aspects of the geometrical problems more exactly than has been possible in the questions posed earlier, and to develop a deeper insight into the geological potential of the material.

It is not necessary for the student to complete these problems before he moves on to Session 2, although clearly their completion is likely to enhance his understanding of the later Session work. They could be tackled much later as a revision exercise, or as a second round of work. Answers and comments to these starred questions will be found at the end of the Session.

Question 1.6★

Develop a mathematical formula that expresses the values of extension e in terms of the *initial orientation* of a specific line (angle α with the positive x-axis direction) before shear and the simple shear displacement γ_{yx}. Hint: choose a line with initial unit length, and describe how this length is modified as γ_{yx} changes.

Question 1.7★

Find a mathematical expression which gives the *final orientation* α' of a line in terms of its *initial orientation* α and the shear strain γ_{yx}.

Question 1.8★

Find a formula that expresses the values of extension e in terms of the final orientation α' of a line after shear and the simple shear displacement γ_{yx}.

Question 1.9★

Find an expression which mathematically describes the shear strain γ along any initial direction α in terms of α and γ_{yx}. Hint: if a line has an orientation described by its slope $m = \tan \alpha$, then the slope of a perpendicular is $-1/m$.

Question 1.10★

Define mathematically the relationships of shear strain variation along any direction α' after deformation in terms of the amount of shear γ_{yx}.

Question 1.11★

If you have a programmable calculator graph the functions resulting from Questions 1.6★ (plotting γ_{yx} as abscissa, e as ordinate) and 1.9★ (plotting γ_{yx} as abscissa and γ_α as ordinate) over ranges of value $\alpha = 20°$, $40°$, $60°$, $80°$, $100°$, $120°$, $140°$, $160°$ and $\gamma_{yx} = 0$ to $2\cdot0$, so as to realize fully the changes in extension and shear strain during progressive, simple shear. Discuss the significance of the curves (changes from positive to negative values, changes of slopes etc.).

ANSWERS AND COMMENTS

We have investigated in these first experiments a process known as *progressive displacement by simple shear*. The progression of the displacement has been recorded by measuring the changes in the shear strain γ_{yx}.

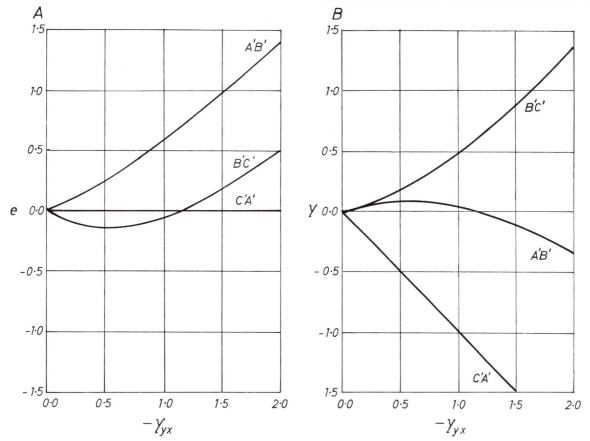

Figure 1.6. A shows the variations in extension along three line directions as a result of simple shear γ_{yx}, and B shows the variations in shear strain γ along the same directions.

Changes in lengths

Answers 1.1, 1.2, 1.3

The process of simple shear leads to quite complex changes in lengths and angles on the surface of the card deck. Lengths are generally changed, although those parallel to the cards (like CA) remain unaltered for all amounts of shear. If, after a shear displacement, a line is of unchanged length it is known as a direction of **no finite longitudinal strain** ($e_f = 0$). Direction CA is a line of no finite longitudinal strain for all values of simple shear displacement (Figure 1.6A). All other directions in our card surface show *progressive changes in extension* with displacement. *None of these length changes*, as recorded by values of e, *is linear* with increase in amount of shear as measured by γ_{yx}. We will see later that this is a general feature of systems undergoing displacement and internal strain. Doubling the amount of displacement does not lead to a simple doubling of the longitudinal strain.

Some directions in our model surface (line AB, Figure 1.6A) show progressive increases in extension with increase in shear, whereas others (e.g. BC with initial orientation $\alpha = 120°$) show a *complex sequence of changes, first shortening, then stretching*. During the early stages of simple shear, line BC becomes progressively shorter. During these early stages we say that the *longitudinal strain e_f* recording the *total* deformation is negative (e_f − ve) and that the **incremental longitudinal strain e_i** recording the changes taking place are also negative (e_i − ve) (Figure 1.7, step 1). As the experiment proceeds the rate of

shortening with shear becomes less and less, and when $B'C'$ is aligned perpendicular to the card surfaces it ceases to become shorter (step 2). At this position the *incremental longitudinal strain* is zero ($e_i = 0$). As we again increase the shear, the line starts to increase its length, although its *total* length is still shorter than it was originally (step 3 e_f − ve, e_i + ve). The graph of the finite elongation along $B'C'$ (Figure 1.7) eventually crosses the $e = 0$ abscissa line with a positive slope (step 4). At this stage in our experiment the later positive incremental elongations have accumulated to such an extent that they have exactly compensated for the earlier history of shortening and it is clear that where this happens we must have at least *two directions of no finite longitudinal strain*: $C'A'$, parallel to the cards (x-direction); and $B'C'$, making an angle of 60° to the x-direction. We will see later that the presence of two (and not more than two) directions of no finite longitudinal strain is a characteristic feature of many types of two-dimensional displacement processes (not only simple shear) and that the angle between these two directions has some special functional relationship to the strain state. With further shear, so that $B'C'$ makes angles of less than 60° with the x-direction, finite elongations and incremental elongations are always positive (Figure 1.7, step 5).

From this simple experiment it is clear that some rather remarkable and perhaps initially unexpected geometric happenings take place as a result of a very simple displacement process. If we had seen only the end of our experiment with a total shear $-\gamma_{yx} = 2.0$, we might never have guessed that line $B'C'$ had undergone such a complicated series

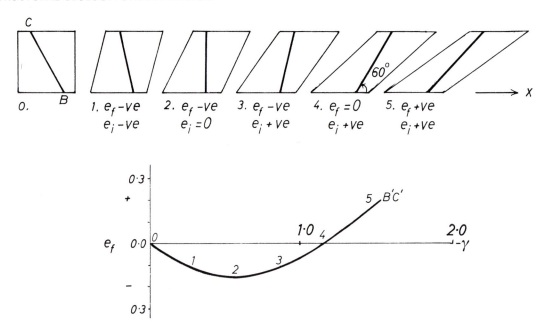

Figure 1.7. *Geometrical representation of changes in length of a line* BC *during simple shear.* e_f *refers to the finite longitudinal strain at successive times 1–5 and* e_i *refers to the incremental longitudinal strain.*

of length changes. Geologically, these complex series of length changes that occur in practically all progressive deformation sequences have most important implications and the small scale structures that the field geologist observes in naturally deformed rocks often show most interesting traces of these changing geometric events. We shall study later further properties of progressive deformation and the fascinating possibilities whereby, for example, folded strata might become unfolded or stretched, and mineral alignments produced in an early stage of a deformation might be severely modified during the later stages of a deformation. Although at this early stage in our study these prospects might appear alarming, perhaps it should be emphasized straight away that such geometric changes are neither haphazard nor chaotic, and that the changes all obey well defined rules. The important lesson for us to learn at this stage is that we must work to establish what these rules are so that we are in a good position to interpret sensibly the structures seen in naturally deformed rocks.

At the end of Question 1.3 we asked if the overall features of the deformation could be described by the changes of elongation in any *one* direction. Clearly, strain is not like many of our everyday measurement parameters such as length or angle; it cannot be defined by just one number. Within the surface of our card deck model, each direction has its own characteristic number which defines what changes have gone on in that direction. Strain is a so-called *tensor* quantity, and requires *several numbers* to define its "state". One of our problems in the next Session will be to look into the problem of variations in longitudinal strain in different directions to discover practical ways of defining features of the **finite or total strain tensor**.

Changes in angles

Answer 1.4

The variations in shear strains with change in initial orien-

tation α are complicated, as are those for longitudinal strains. We have already seen that the longitudinal strains along the direction $A'C'$ remain constant and zero. However, the shear strain along this line does change progressively as we increase the displacement. It should be clear from the way we defined γ_{yx} that the changes in $\gamma_{A'C'}$ must be the same as those of γ_{yx} and that the scale along the abscissa axis records negative values of γ_{yx}.

For the initial direction AB with $\alpha = 60°$ we have seen that the extension e increases progressively in a positive sense with increase in γ_{yx}, although it does so in a non-linear manner. However, the changes in shear strain along this line $A'B'$ are rather more complex (Figure 1.6B). During the first displacement stages the material line with direction AB changes orientation more rapidly than a line initially perpendicular to AB. This implies that we must develop a shear strain along $A'B'$, and that, because of the differences in rates of change of orientation, the measured shear strain deflection relative to $A'B'$ will be anticlockwise. The shear strain $\gamma_{A'B'}$, therefore, has a positive sign. The rate of change of this shear strain first decreases and later increases in a negative sense, so that at one particular stage in the deformation the positive anticlockwise shear effects become exactly compensated by later negative clockwise shear effects. On our graph this occurs where the shear strain line crosses the abscissa axis ($\gamma = 0$). Where this occurs these lines have very special properties. The two initially perpendicular directions which early in the experiment (between γ_{yx} values of 0·00 and $-1·15$) lost their perpendicularity *now return to a perpendicular relationship*. One can prove mathematically (see Question 1.9★) that these are the *only* two directions in the surface that were initially perpendicular and which at that particular stage of the strain history remain perpendicular. We will be able to prove later that where this happens the two directions are *parallel to the directions of maximum and minimum elongation* (see p. 27). As the experiment proceeds, the deflection of the initial perpendicular to $A'B'$ becomes clockwise with reference to the direction of $A'B'$, and the shear strain takes up negative values.

Finally, over the range of shear values in our experiment, the shear strain parallel to $B'C'$ is always positive, and it progressively increases in a non-linear fashion.

Answer 1.5

Displacement in two dimensions is defined as the vector joining an initial point (x, y) and its final position (x', y'). Displacement has two components: u parallel to the x-axis, and v parallel to the y-axis so that

$$u = x' - x$$
$$v = y' - y,$$

In simple shear, because of the parallel sliding of the cards, points are only displaced parallel to the x-axis, and the y coordinate value of any point does not change. The distance moved parallel to the x-axis increases linearly with distance from this axis. In simple shear we can therefore write two **displacement equations**:

$$u = -\gamma y$$
$$v = 0 \tag{1.1}$$

The general equations relating every initial and final point, known as the **coordinate transformation equations**

$$x' = x - \gamma y$$
$$y' = y \tag{1.2}$$

are related to the displacement equations. These equations are linear in that they do not involve complex functions of x and y. In their most complete form they involve four terms,

$$x' = x - \gamma y$$
$$y' = 0x + y$$

These four coefficients of x and y can be expressed as a two by two matrix, the so-called **strain matrix**

$$\begin{matrix} x' \\ y' \end{matrix} = \begin{bmatrix} 1 & -\gamma \\ 0 & 1 \end{bmatrix} \begin{matrix} x \\ y \end{matrix} \tag{1.3}$$

We will see later (Session 2 and Appendix B) that this matrix is an extremely important basic mathematical expression from which all the properties of the strain may be derived.

The *absolute displacements* are not recorded by the strain matrix, but the fundamental internal deformation features are contained in it. For example, we could have lifted our shear box and transported it to another bench in our laboratory. In doing this we would have subjected all the points on our card surface to another displacement (technically termed a **body translation** perhaps together with a **body rotation**), but this movement would not have affected the results of our shear experiments inside the box in terms of changes of lengths and shear strains. A total displacement equation would add other terms in our equations to those of Equation 1.1. Technically these would add displacement vectors, but these additional vectors would not change the specific vector gradients which set up the internal distortions in our cards.

Geological significance of length changes: Boudinage and folding

We have seen that rather complex increases and decreases in length take place in different directions in a deformed material. Although these effects have until now mostly been discussed from a purely geometrical viewpoint, it should be clear that they have some special significance when we attempt to interpret the significance of structures seen in deformed rocks.

If the rock mass we investigate is more or less uniform, isotropic and homogeneous, then deformation will not lead to the initiation of any special mechanical instabilities, at least not in the early stages of deformation. We will see in our later sessions that such a deformation will be likely to produce overall modifications of shape and orientation of the constituent particles which can lead to the development of statistically preferred orientations and the production of planar and linear fabrics.

Many natural rocks, however, are not isotropic and homogeneous. They show variations in composition usually in the form of a planar anisotropy (bedding in sediments, banding in schists and gneisses, planar igneous dykes and sills etc.). These layers of differing composition generally possess differing **rheological properties,** that is to say they may be brittle, or ductile, and if ductile they can be of varying ductility. Often we do not know the exact type of rheology (e.g. whether Newtonian viscous, non-linear viscous, plastic etc.), but often we are aware that the layers had differences of ductility under the particular conditions at the time of the natural deformation, for they behaved in different ways and so acquired different geometrical features. When layered rocks with such different properties are stretched or shortened parallel to the layering anisotropy, the layers become mechanically unstable.

In geology we often use the term **competence** to describe the differences in rock properties which lead to mechanical instability. Those rocks which flow more easily than their neighbours are described as the least competent members, whereas those that are stiffer than their neighbours are termed the most competent rocks. Although the term competence is therefore rather vaguely defined in any exact rheological way, it gives a very useful comparative description of the components of a layered system, which the field geologist can use to give some idea of the competence contrast of the components of deformed layered rocks.

Figure 1.8B shows the results of a simple laboratory experiment in which a competent plastic layer was first surrounded by a layered plastic of lower competence, and then subjected to a progressive stretching parallel to the layers. The competent material became mechanically unstable during the deformation. Slight imperfections inside the competent layer gave high stress concentrations at certain points, and these higher stresses led to higher deformation rates at these locations. The competent layer, therefore, became preferentially stretched and thinned and eventually failure occurred on a fracture. Further extension led to a separation of the isolated fragments of the competent layer, and the surrounding more ductile material flowed into the so-called *neck zones* between the fragments. This process is called **boudinage** and the fragments, which are rectangular or sub-elliptical in section and form long flat rod-like masses in the third dimension, are termed **boudins**. Where the competence contrast is high, the boudins generally have block-like profiles, because their ends are controlled by cross fractures. As the surrounding material flows in towards the boudin neck, the form of the ends of the boudin blocks is often modified by the differential shear along the competent–incompetent rock interface. The boudins develop a characteristic *barrel-shaped cross section*, and, where the separation is large, the strong flow of incompetent material can completely modify the shape of the end of the boudin into a *fish-mouth form* (Figure 1.8). Where the competence contrast is low the competent layer generally undergoes strong, ductile flow around the initial stress concentrations and may preferentially stretch without rupture. The cross section of such boudins shows a more lenticular form known as **pinch and swell structure**.

The shortening of a competent layer embedded in a less competent matrix also leads to mechanically unstable situations. The layer is deflected sideways and develops a folded form. These folds, known as *buckle folds*, generally show a rather regular periodic wavelength (Figure 1.9). If the competent layer is isolated and separated from other competent layers by a considerable thickness of incompetent rock, then this wavelength is a function of competent layer thickness and a function of the competence contrast. An increase in the thickness of the competent layer and an increase in the competence contrast both lead to an increase in the wavelength of the initial buckle fold. If the competence contrast is high, the large initial wavelength–thickness ratio is translated into the later fold style as a long layer length relative to the distance between adjacent fold hinges. Such folds (Figure 1.9A layer a; B, C) are termed **ptygmatic structures** and are very characteristic of deformed aplite and pegmatite veins in gneissic terrain.

We will leave this important subject of the structural forms of layers in deformed rocks now, but we will return to the topic several times in later Sessions when we have built a stronger background in strain analysis.

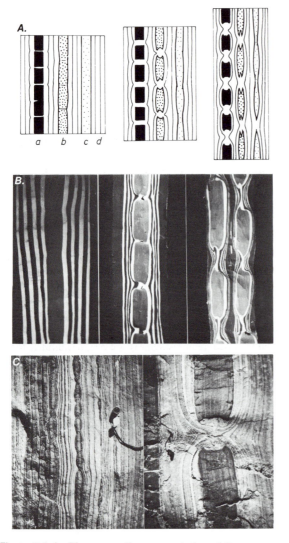

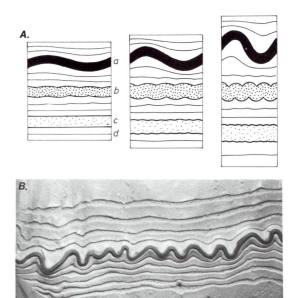

Figure 1.8. A: *Diagrammatic representation of the cross sectional forms of boudins; competence contrast: a > b > c > d. B: Three stages during a laboratory experiment of boudin structure in plastic model materials. C: Examples of boudin structures; competent rock calc silicate, incompetent rock marble. Khan Gorge, Namibia.*

Figure 1.9. A: *Types of buckle folds developed during the shortening of competent layers; competence contrast: a > b > c > d. B: Laboratory experiment producing ptygmatic structure in model materials of differing plasticity. C: Ptygmatic structure in a pegmatite vein contained in a metasedimentary matrix. Chindamora, Zimbabwe.*

Answer 1.6★

Consider an initial line OA of unit length, oriented at an angle α to the x-axis (Figure 1.10). Then, after displacement by a simple shear γ_{yx}, this line changes position to OA' so as to lie at an angle α' to the x-direction. Its length is now $(1 + e)$ units (e is the extension).

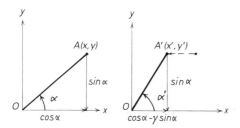

Figure 1.10. Changes in length of a line as a result of simple shear.

The coordinates of point A are (x, y) or $(\cos \alpha, \sin \alpha)$. A is displaced parallel to x by an amount $\gamma_{yx} \sin \alpha$ to new coordinates (x', y') given by $(\cos \alpha - \gamma_{yx} \sin \alpha, \sin \alpha)$. It follows from Pythagoras' theorem that

$$(1 + e)^2 = x'^2 + y'^2 = (\cos \alpha - \gamma_{yx} \sin \alpha)^2 + \sin^2 \alpha$$

or   (1.4)

$$e = (1 - 2\gamma_{yx} \cos \alpha \sin \alpha + \gamma_{yx}^2 \sin^2\alpha)^{1/2} - 1$$

Answer 1.7★

$$\tan \alpha' = \frac{y'}{x'} = \frac{\sin \alpha}{\cos \alpha - \gamma_{yx} \sin \alpha}$$

This can be put in two forms:

$$\tan \alpha' = \frac{\tan \alpha}{1 - \gamma_{yx} \tan \alpha} \qquad (1.5)$$

$$\tan \alpha = \frac{\tan \alpha'}{1 + \gamma_{yx} \tan \alpha'} \qquad (1.6)$$

Answer 1.8★

Substituting Equation 1.6 into Equation 1.4

$$e = (1 + 2\gamma_{yx} \sin \alpha' \cos \alpha' + \gamma_{yx}^2 \sin^2 \alpha')^{-1/2} - 1 \quad (1.7)$$

Answer 1.9★

Consider an initial line in direction p making an angle of α to the x-axis (Figure 1.11). Its slope is given by $m = \tan \alpha$. Line q, initially perpendicular to p, has a slope $-1/m$. From Equation 1.5 the slope of the initial line p after displacement to direction p' is

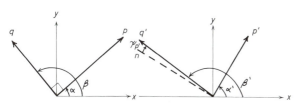

Figure 1.11. The angular shear strain $\psi_{p'}$ with reference to a direction p' after a simple shear displacement.

$$\frac{m}{1 - \gamma_{yx} m}$$

and the slope of q' is

$$\frac{-1/m}{1 + (\gamma_{yx}/m)} = \frac{-1}{\gamma_{yx} + m} = b$$

The slope of a line n perpendicular to p' is

$$\frac{\gamma_{yx} m - 1}{m} = a$$

which is generally not the same as the slope of q'. The difference in orientation defines the angular shear strain $\psi_{p'}$. Then

$$\gamma_{p'} = \tan \psi_{p'} = \frac{a - b}{1 + ab}$$

where a and b are the slopes of n, (the perpendicular to p') and of q' respectively, i.e.

$$\frac{(\gamma_{yx} m - 1)m^{-1} + (m + \gamma_{yx})^{-1}}{1 - [(\gamma_{yx} m - 1)/(m(m + \gamma_{yx}))]}.$$

which can be simplified to:

$$\gamma_{p'} = \frac{\gamma_{yx}(m^2 + \gamma_{yx} m - 1)}{m^2 + 1} \qquad (1.8)$$

where $m = \tan \alpha$. Two special solutions to Equation 1.8 are apparent:

(I) where $m = 0$ then $\gamma_{p'} = \gamma_{yx}$, a result we had discovered previously from our experiments considering the shear strain along direction CA.
(II) where $\gamma_{p'} = 0$ either $\gamma_{yx} = 0$ (a result implying no displacement)

$$\text{or} \quad m^2 + \gamma_{yx} m - 1 = 0$$

This second conclusion shows that in general two values of m are possible. There are, therefore, two directions in our deformed surface along which no shear strain occurs. The product of the roots of a quadratic equation $ax^2 + by + c = 0$ is c/a. Because the constant term of the quadratic equation is -1, the products of the two values of m must be minus one, implying that the *two directions were initially perpendicular*. It will be seen in the next Session that this result has a very great significance in terms of the principal features of deformation. The two orientations are the roots of the quadratic equation given by

$$m = \tan \alpha = \tfrac{1}{2}\left(-\gamma_{yx} \pm \sqrt{(\gamma_{yx}^2 + 4)}\right) \qquad (1.9)$$

Answer 1.10★

Using Equation 1.6 in 1.8

$$\gamma_{p'}' = \frac{\gamma_{yx}(m'^2 - \gamma_{yx} m' - 1)}{m'^2(1 - \gamma_{yx})^2 + 2\gamma_{yx} m' + 1} \qquad (1.10)$$

where $m' = \tan \alpha'$

By extending the previous discussion of the solutions of Equations 1.8 to 1.10 it will be apparent that *after deformation* the two lines showing no shear strain *are also perpendicular* and have orientations given by

$$m' = \tan \alpha' = \tfrac{1}{2}\left(\gamma_{yx} \pm \sqrt{(\gamma_{yx}^2 + 4)}\right) \qquad (1.11)$$

Answer 1.11★

See Figures 1.12 and 1.13 for the complete graphical solutions over a range of γ_{yx} from 0·0 to −2·0 to Equations 1.4 and 1.8 respectively. The longitudinal strains for lines lying between $\alpha = 0°$ and $\alpha = 90°$ (Figure 1.12) are all positive. The geological implications here are that any competent rock layers with such an initial orientation would show the progressive development of boudins. In contrast, the longitudinal strains for the lines between $\alpha = 90°$ and $\alpha = 180°$ show an initial history of contraction, and a later history of stretching. The implications of such changes will be discussed fully in Session 12: at this stage we should note the possibility of layers showing early formed folds which were subsequently boudinaged.

The changes in shear strain for different values of α are quite complex. This is because the rates of angular change of the reference direction and an initial perpendicular are, to some extent, independent. The shear strain variations are complex because they combine the effects of these two variably changing lines.

The development of mathematical expressions for describing length and angular changes leads to two types of equations. One set of equations describes the features of strain in terms of parameters existing *before the deformation took place*. These are known as *Lagrangian equations*. For example, Equations 1.4 and 1.8 are both Lagrangian forms involving the angle α before deformation. It is also possible to describe the same strain features in terms of *directions after deformation*; these are known as *Eulerian equations*. Equations 1.7 and 1.10 involve α' and are both Eulerian forms. The link between Lagrangian and Eulerian equations can be made by a single equation describing the relationships of α and α' before and after deformation respectively.

We will see that for many geological purposes the Eulerian equations are most useful, because in practice it is the deformed state which is our present day reference point. Lagrangian equations are useful, but they are mostly used for predicting the geometric effects of particular strain sequences.

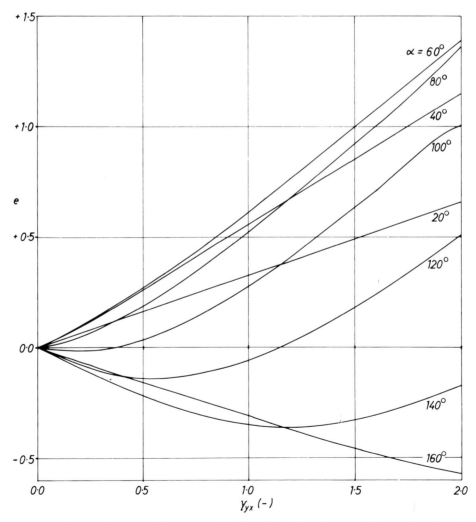

Figure 1.12. *Variations of longitudinal strain (extension e) with initial orientation α of a line undergoing simple shear γ_{yx}.*

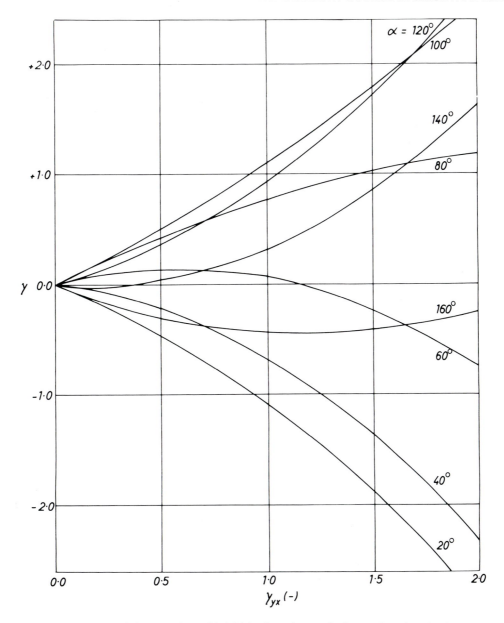

Figure 1.13. *Variations of shear strain γ with initial orientation α of a line undergoing simple shear γ_{yx}.*

KEYWORDS AND DEFINITIONS

Angular shear strain	A measure of the angular deflection between two initially perpendicular lines (Figure 1.4).
Body rotation	A type of displacement where all points in a body undergo an identical angular rotation about some fixed point.
Body translation	A type of displacement where the displacement vectors for all points are identical.
Boudinage	A structure produced during the extension of competent layers enclosed in an incompetent matrix. The competent layers show more or less regular regions of thinning (or breaking), and separate into sub-parallel pieces termed **boudins**. The term boudin was originally proposed by Lohest *et al.* (1909) because of the similarity of the forms of the sub-parallel competent layer pieces to a variety of French sausages (boudins) lying side by side in a butcher's shop window.

Brittle deformation — The failure of a stressed body, leading to the formation of fractures, when the elastic limit is exceeded.

Competence — A general term to describe the ease with which a material can deform. Competent materials are stiffer, flow less easily or break more readily than incompetent materials.

Displacement — The change in position of a point in a body from an initial position with coordinates (x, y) to a final position (x', y'). Displacement is the straight line vector joining (x, y) and (x', y'). Displacements for all the points in a body may be expressed in the form of **displacement equations**, and **coordinate transformation equations** (p. 7 and Appendix B).

Ductile flow — The property of certain materials whereby, when stressed, the material undergoes a permanent deformation without fracturing. The strains arising during ductile flow are generally larger than those of an elastic strain.

Elastic strain — A strain in a solid body set up as the result of application of a stress. Elastic strains are reversible and the body recovers its original shape when the stresses are removed. Elastic strains are generally small (longitudinal strains $e < 0.02$, and shear strains $\gamma < 0.02$).

Longitudinal strain — A change in length of a line element: a line of initial length l and final length l' has a longitudinal strain defined as **extension** e (or engineers' extension) where $e = (l' - l)/l$. The **finite longitudinal strain** e_f is a measure of the total strain from initial to final state. The **incremental longitudinal strain** e_i is the partial change of length of a line from its length l at the start of the increment to its length l'_i at the end of the increment $(e_i = (l'_i - l)/l)$.

Pinch and swell structure — A variety of boudinage in which the boudins are connected by narrow zones of competent material. This structure is characteristic of materials where the competence contrast is not strongly marked (Figure 1–8A, layer c).

Ptygmatic structure — Folds with the following geometric properties (see Figure 1.14):

1. Constant layer thickness t.
2. Distance between adjacent fold crests (or troughs) *measured along the layer* (l) large compared to layer thickness (>10).
3. Distance between adjacent fold crests measured along the layer (l) large compared to fold wavelength (w).

The form of the folded layer is characteristic of that produced by buckling a thin elastic sheet in air, known as an *elastica*. In deformed rocks ptygmatic structures are found where single, isolated layers of relatively highly competent material enclosed in a matrix of low competence are strongly shortened.

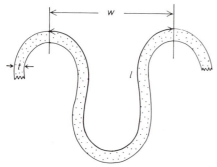

Figure 1.14. *Geometric features of ptygmatic structure.*

Rheology — The study of the relationships between stress imposed on a body and the resulting strains or strain rates.

Shear strain γ — The tangent of the angular shear strain: $\gamma = \tan \psi$.

Simple shear — A displacement which transforms an initial square into a parallelogram where the displacement vectors are all oriented parallel to one set of opposite sides of the square and parallelogram (Figure 1.5). The vector direction is known as the **shear direction** and the plane containing this and a normal to the plane of the parallelogram is the **shear plane**.

Strain   The change in shape or internal configuration of a body resulting from certain types of displacement. Mathematically it is a **second order tensor** quantity which requires four components for its definition in two dimensions and nine components in three dimensions (see later definitions in Session 2). In this book we will use strain for this tensor quantity, including features of **distortion** and **rotation**.

KEY REFERENCES

In each of the Sessions of this book a number of key references will be given to provide background reading which is especially pertinent to the topics under discussion. At the end of the book a more complete reference list will be found.

Important background reading on strain will be found in the following works:

Jaeger, J. C. (1956). "Elasticity, Fracture and Flow", 208 pp. Methuen, London.

This is a very useful small monograph setting out many of the basic features of displacement and deformation in a clear and concise way. For this and the following three Sessions the most useful background reading will be found in the accounts of displacement and two-dimensional strain (pp. 20–33).

Means, W. D. (1976). "Stress and Strain", 339 pp. Springer-Verlag, Heidelberg.

This book provides an excellent introduction for the mathematical approach to deformation. It has been written with the needs of geologists in mind so that the discussion and applications of the theory are especially valuable and informative. At this stage the most relevant sections are found under the headings of deformation, and especially pp. 130–139 and 168–173.

Ramsay, J. G. (1967). "Folding and Fracturing of Rocks", 568 pp. McGraw-Hill, New York.

This book, written for structural geologists, sets out many of the mathematical properties of strain. It develops the geological implications of the geometric features of deformation in terms of the structures of naturally deformed rocks. The following pages are recommended as being particularly relevant for the first four Sessions of this present work: pp. 50–69, 83–91, 94–96, 103–120.

Thompson, W. and Tait, P. G. (1879). "Principles of Mechanics and Dynamics", Part 1 (paperback version published 1962) 508 pp. Dover, London.

This reference is for those wishing to look at one of the classic historic works on mechanics. The sections most relevant for our studies will be found on p. 116, but the analysis rapidly proceeds to an investigation of three-dimensional features, and may be found to be a little hard going at this early stage.

Truesdell, C. and Toupin, R. A. (1960). The classical field theories. *In* "The Encyclopaedia of Physics", (S. Flugge, ed) 226–273. Springer-Verlag, Berlin, Heidelberg.

This work presents what is probably the most complete and mathematically most comprehensive account of strain to be found in the existing literature. Be warned, however, that this book does require a *very sound knowledge* of mathematical methods. A section on simple shear will be found on pp. 292–298.

SESSION 2

The Strain Ellipse
Concept
Distortion and Rotation

Experiments with the card deck model are used to illustrate the differences between displacements which set up homogeneous and heterogeneous strains, and how homogeneous strain can be analysed using the concept of the finite strain ellipse. The four components of strain in two dimensions are related to the two principal longitudinal strains and their orientations before and after displacement. Simple shear is shown to be a rotational deformation. Some further geological implications of simple shear are developed to illustrate the evolution of extension fissures in shear zones, and to show how the differences of forms of folds and of boudin structures arise from differences of rock layer orientation with respect to the directions of principal strains.

INTRODUCTION

The experiments of Session 1 showed that the strain that is developed by two-dimensional displacement has quite complex geometrical features. We saw that values of longitudinal strain and shear strain varied with the direction in which the measurements were taken. In this Session we will describe these variations and show that they form a systematic pattern.

How can we find an experimental solution to this problem with the shear box used in our previous experiments? One way would be to draw lines with many different orientations, say at 10° intervals, on the card deck surface. We could then study systematically how the longitudinal and shear strains vary with respect to the initial and final positions of these lines. However, we can use a simpler method to study these strain variations: we draw a circle on the card deck surface and see how the form of the circumference of this circle is modified when we shear the cards. Using a circular marker not only enables us to see how lines with *any* initial orientation (represented by all the possible radii joining circumference points to circle centre) change length and orientation, but helps us to formulate one of the most important concepts of strain analysis—the **finite strain ellipse**.

The concept of the strain ellipse is based on the *homogeneity of strain*. **Homogeneous strain** produced by simple shear displacement arises when the differential displacements between any 10 adjacent cards is the same as occurs across any other 10 adjacent cards. This differential displacement is termed the **displacement gradient**: homogeneous strains are set up where the displacement gradient is constant.

First group of experiments: The strain ellipse-homogeneous and heterogeneous strain

Question 2.1

Displace the cards using a pair of wedge shaped formers of constant displacement gradient. The large circle is deformed into an ellipse; if the *initial circle has a unit radius*, this ellipse is known as the **finite strain ellipse**. Measure the lengths of the major and minor semi-axis of this ellipse. These lengths relate to the maximum and minimum extensions of the strain (values $1 + e_1$ and $1 + e_2$, where e_1 and e_2 are known as the **principal finite extensions**—defined such that e_1 is always numerically greater than e_2). In our experiment e_1 is positive and e_2 is negative. The **ellipticity R** of the strain ellipse, recording certain aspects of the distortion component of the strain, is the proportion $(1 + e_1)/(1 + e_2)$; calculate its value. Determine the orientation of the strain ellipse by measuring the angle between the major axis and the direction of the edges of the cards (angle $\theta'\cdot$).

Question 2.2

Repeat these measurements on the ellipses derived by distortion of the smaller circles. Are the ellipticities and orientations constant? Why?

Question 2.3

With a hand lens observe the displacement at the edge of the large ellipse—note how they are discontinuous and

15

step-like. Is the form of the deformed circle really an ellipse?

Question 2.4

Derivation of the equation for the strain ellipse produced in simple shear.

By considering the displacement of any initial point (x, y) positioned on the circumference of a circle of unit radius and with centre at $(0, 0)$ given by the equation

$$x^2 + y^2 = 1$$

to some new position (x', y'), derive the equation of the strain ellipse after a simple shear displacement of γ_{yx}. Hint: displace the points on the circle using the coordinate transformation equations in the form:

$$x = x' - \gamma y'$$
$$y = y'$$

Question 2.5

Repeat the first experiment using a former with a curved displacement surface (Figure 2.1). What is the form of the large circle after it has been displaced: is it an ellipse?

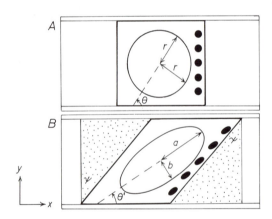

Figure 2.1. *View of the surface of the card deck model used for the experiments of Questions 2.1 to 2.10: A, before shearing, B, after shearing.*

What are the forms of the smaller circles? Are they more nearly elliptical than that of the large circle? How constant are the approximate ellipticities of the deformed small circles and how constant are their orientations?

Before proceeding to Question 2.6, check your answers and read the commentary.

Second group of experiments: Principal finite extensions and rotational component of the strain

The aim of this group of experiments is to investigate the so-called rotation effects of a displacement and strain system and to formulate more exactly how the form of the strain ellipse is a function of the displacement.

Question 2.6

Using the formers with values of $\gamma_{yx} = 0.8$, displace the cards to a state of homogeneous strain. On the large ellipse draw the major and minor diameters, the lines will be perpendicular.

With another set of formers with shear value $\gamma_{yx} = 0.4$ remove part of the shear. Are the lines of the major and minor axes of the previous ellipse still perpendicular?

Question 2.7

Completely remove the shear deformation. The strain ellipse will return to its originally circular form. Are the new positions of the lines of the original strain ellipse axes perpendicular?

Question 2.8

Measure the angle made by the long axis of the strain ellipse ($\gamma_{yx} = 0.8$) and the direction of the edges of the cards (angle θ') and the angle this line makes with the cards in the undisplaced state (angle θ). The angles are different; calculate the difference ($\omega = \theta' - \theta$) which is a measure of the **rotational component of the strain**. Although the strain is a rotational one, have the cards actually rotated?

Question 2.9

Repeat question 2.6 with formers of shear value $\gamma_{yx} = 1.6$. Do the directions of the axes of the ellipse coincide with those for a shear of $\gamma_{yx} = 0.8$? Calculate the rotational component of the strain for the shear of $\gamma_{yx} = 1.6$. Is it twice the rotation for the shear of $\gamma_{yx} = 0.8$?

Before proceeding to the third group of experiments, check answers and read the commentary on the significance of these results.

Third group of experiments: variations of principal finite extensions and rotation with displacement

This next group of experiments is aimed at evaluating more exactly the relationships between strain, rotation and displacement.

Question 2.10

With formers of variable γ_{yx} value evaluate the following data and tabulate them using the table overleaf.
Plot the data on to four graphs using scales $1.0\gamma = 5.0$ cm (abscissa), and ordinate scales ellipticity $R = 2.5$ cm per unit; θ', $10° = 2$ cm; θ, $10° = 2$ cm; and rotation ω, $10° = 2$ cm. Keep to these scales and plot data on to tracing paper. The results can then be compared with the exact solutions in the answer section.

Are any of the curves linear with increase in shear displacement γ_{yx}? The area of an ellipse is πab, derived from a circle of initial area πr^2. If we define the proportional change in areas by the Greek letter "delta" (Δ_A) then

$$1 + \Delta_A = (1 + e_1)(1 + e_2)$$

What do you deduce about area change in simple shear?

$\gamma_{yx} = \tan \psi$	$1 + e_1 = \dfrac{a}{r}$	$1 + e_2 = \dfrac{b}{r}$	$\dfrac{1 + e_1}{1 + e_2} = R$	$(1 + e_1)(1 + e_2) = (1 + \Delta)$	θ'	θ	$\omega = \theta' - \theta$
−0·2							
−0·4							
−0·6							
−0·8							
−1·0							

Extension structures developing in shear zones

Question 2.11

What is the orientation θ' of the maximum longitudinal strain of the *first* ellipse to develop by simple shear? Determine this by extrapolation of the θ' curve to the position $\gamma_{yx} = 0 \cdot 0$.

Figure 2.2 shows a shear zone with parallel sides cutting through displaced (but undeformed) wall rocks: A shows the initial development of the zone, and B and C illustrate later stages as the shears in the centre of the zone become stronger, and as the shear zone widens. In A, draw the system of extension cracks which would be expected to form perpendicular to the orientation of the maximum extension of the first formed strain ellipse. In B and C, show how these would be modified by later shear, and how they would extend into the less deformed wall areas (assuming that the propagation of an extension vein takes place at the crack tip in a direction perpendicular to the maximum incremental extension).

Derive an equation which enables you to calculate the total shear strain γ from the orientation made by the extension crack and the direction of the shear zone.

Now proceed to the Answers and Comments section, then continue with the starred questions, or move on to Session 3.

STARRED(★) QUESTIONS

Orientation θ' of the principal axes of the strain ellipse

Question 2.12★

In Session 1 we derived a general equation which expressed the value of extension e for all directions given by orientation α' in the deformed state (Equation 1.7). The orientation of the principal extensions after deformation can be derived from this general equation by finding its maximum and minimum values. Differentiate Equation 1.7 with respect to α' and find where $de_{\alpha'}/d\alpha'$ is zero (the standard method for finding the maximum and minimum values of any function). The simplest form for the final equation for θ' (values of α' giving maximum and minimum values of $e_{\alpha'}$) is found using double angle forms ($\sin 2\theta' = 2 \sin \theta' \cos \theta'$, $\cos 2\theta' = \cos^2\theta' - \sin^2\theta'$). Prove that there are

two solutions over the range of θ' 0°–180° (range of $2\theta'$ 0°–360°), and that these two directions are perpendicular. Show that the long axis of the strain ellipse can never pass through the shear direction (parallel to the x-axis).

Orientation θ of the principal finite extensions before deformation

Question 2.13★

In Session 1 we derived a general equation for the extension along any line with initial orientation α (Equation 1.4). Use the same mathematical technique as for Question 2.12★ to find an equation giving the orientation of the maximum and minimum extensions before displacement. Prove that the two directions are initially perpendicular.

Calculation of the rotation ω

Question 2.14★

The values of θ' and θ determined from questions 2.12★ and 2.13★ will be different, and we have seen that this difference defines the finite rotation ω. It should be clear that

$$\tan 2\omega = \tan (2\theta' - 2\theta)$$

expand this and find the value of $\tan 2\omega$. Then, using the identity

$$\tan 2\omega = \frac{2 \tan \omega}{1 - \tan^2\omega},$$

show that $\tan \omega = +\gamma/2$. Another mathematical solution to this equation will appear as $\tan \omega = -2/\gamma$, but this solution is not an appropriate geometrical solution to our problem. Discuss why it is inappropriate. Show that, in simple shear, the internal rotation can never exceed 90°.

Values of the principal finite extensions

Question 2.15★

By substituting the values of θ' in the original equation for $e_{\alpha'}$, find the values of the principal strains. The simplest final result is expressed in the quadratic extension (Greek "lambda" λ) and defined as $\lambda = (1 + e)^2$. Give your results in the form of the two principal quadratic extensions λ_1 and λ_2.

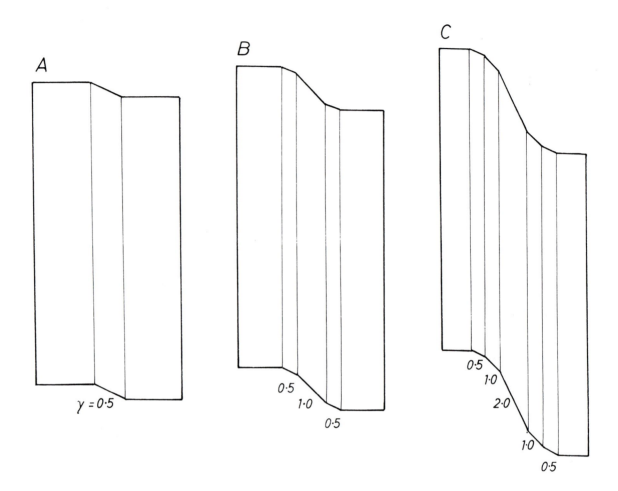

Figure 2.2. Progressive development of a shear zone. See Question 2.11.

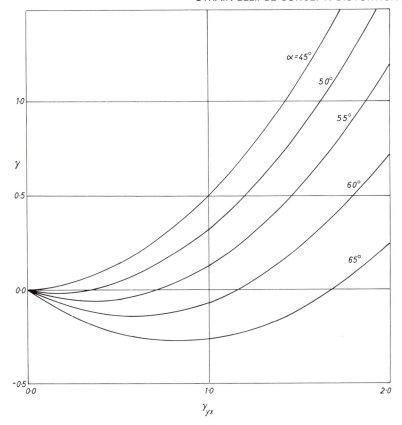

Figure 2.3. Values of shear strain γ along any line making an initial angle α with the direction of simple shear.

Question 2.16★

In Session 1 (Question 1.9★) we derived an equation expressing the value of the shear strain γ along any line making an initial angle α with the x-direction in terms of the amount of shear strain γ_{yx}. Figure 2.3 is a graphical representation of this function for various values of α between 45° and 65°.

The curve for $\alpha = 45°$ lies completely in the γ + ve sector, whereas the other curves initiate in the γ − ve sector, cross the $\gamma = 0$ abscissa and continue into the γ + ve sector. What is the significance of the locations where the various curves for different α values cross the abscissa axis? Discuss the evolution of the principal strain direction (θ') in terms of these graphs.

ANSWERS AND COMMENTS

The strain ellipse—homogeneous and heterogeneous strain

Answers 2.1, 2.2

When the cards are displaced using a pair of wedge-shaped formers with constant angle of shear ψ (corresponding to a shear strain in the card direction of γ_{yx}), the circle of unit radius is displaced into a perfect ellipse. The smaller circles are also transformed into ellipses with absolutely similar shapes to those taken up by the larger circle, and with their major and minor axes parallel throughout the surface. From the way that the card deck has been sheared it is obvious that the differential displacement taking place

between any 10 adjacent cards is the same as that taking place between any other 10 adjacent cards. In technical terms the displacement taking place over any small part of the surface is known as the **displacement gradient** and this gradient is clearly constant in our experiment. The distortion set up within any small (but not too small) area of the surface is identical to that in any other small area, as shown by the similarly shaped and similarly oriented ellipses in our experiment. A *constant displacement gradient leads to a state of homogeneous strain.*

Answer 2.3

If we view the ellipses on the surface of our model from a distance of a meter, the distortion appears very smooth and uniform, but these smooth outlines are an illusion and with a hand lens they are seen to be built up by a series of discontinuous steps. If we could have drawn circles of very small initial radius to cover only five of the cards (Figure 2.4) we would not have called the deformed shapes ellipses. The displacement over very small areas is clearly discontinuous and *in our card model* (and in real rocks) *the concept of homogeneous strain does not hold over fields of observation which are very small.* In naturally deformed

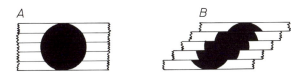

Figure 2.4. Shearing of a small circular marker (A) in the card deck model experiment. The resulting form (B) is not that of a simple strain ellipse.

Figure 2.5. *The concept of homogeneous strain and the strain ellipse is not always simply applicable to a rock and its component crystals. A illustrates the initial undeformed rock with crystals a–f and with a circular marker, and B shows the deformed shape of these crystals with new forms a'–f', and with the discontinuous locations of the marker (cf. Figure 2.4).*

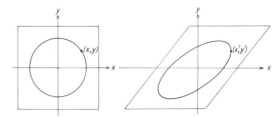

Figure 2.6. *Homogeneous simple shear displacement of a point (x, y) to new position (x', y') and the production of a strain ellipse.*

rocks the homogeneous strain theory can generally be applied over the surface of a single hand specimen, but because nearly all rocks are formed from an aggregate of crystal particles the theory cannot be applied when we can see the individual grains (Figure 2.5). In processes of natural rock deformation the overall shape change of the rock is accomplished by small discontinuous displacements between and within individual grains, and by changes of shape of single crystals by solution and growth. When the overall relationships of thousands of grains are considered we find a smoothing out of the effects of the discontinuities in the same way that in our cards we have found that the small discontinuous slips merge to give an overall picture of continuity. During natural deformation the crystal components generally keep the same neighbours, and there are usually no chaotic separations of adjacent elements. Although this is generally true in tectonic deformations, there are of course other natural processes in which this continuity (or "held togetherness") does not apply. In processes such as turbulent flow in sediments or magmas and in the selective solution, transfer and redeposition of crystal components, the displacements will not always obey the laws of ordered shape changes from which we develop the concepts of finite strain (i.e. the so-called laws of continuum mechanics).

Equation of the strain ellipse

Answer 2.4

Displacement equations can be expressed in two forms. The first tells us the positions of the new coordinates of a point (x', y') when we know the initial coordinates (x, y). In this form the equations have a reference which is the *initial* state, and they are known as *Lagrangian equations*:

$$x' = x - \gamma y$$
$$y' = y$$

This information can be easily reorganized so that x and y become the subject of the equations. Thus, knowing the final positions *after displacement*, we can compute the initial coordinates. Such equations which refer to the *final* state are known as *Eulerian equations*:

$$x = x' + \gamma y'$$
$$y = y'$$

To displace points on our initial circle we replace all the x and y coordinates to new positions (x', y') (see Figure

2.6) according to the Eulerian version of the displacement equations:

$$(x' - \gamma y')^2 + y'^2 = 1$$
$$x'^2 - 2\gamma x'y' + (1 + \gamma^2)y'^2 = 1 \qquad (2.1)$$

The general equation of an ellipse centred at the origin and oblique to the coordinate axes is given by

$$Ax^2 + 2Bxy + Cy^2 = 1$$

where A, B and C are constants. Equation 2.1 is clearly an ellipse of this type.

Answer 2.5

If we displace the cards with a former having a curving displacement edge the shift taking place between the 10 adjacent cards is generally *different* from that between any other 10 adjacent cards. The *displacement gradient is variable* and under these circumstances the large circle becomes deformed into a complex shape which is not that of an ellipse.

If, for example, the variable simple shear displacement induced into the card deck was of a sinusoidal nature so that the radius of the unit circle was displaced by a quarter wavelength of the sine function with coordinate transformation given by:

$$x' = x + \sin(2y/\pi)$$
$$y' = y$$

then the unit circle $x^2 + y^2 = 1$ is displaced into the form illustrated in Figure 2.7 given by the equation:

$$x'^2 - 2x' \sin(2y'/\pi) + \sin^2(2y'/\pi) + y'^2 = 1 \qquad (2.2)$$

Prove this equation yourself by displacing the unit circle by the equations describing the Eulerian form of the *non-linear displacement*. Any pair of non-linear displacement equations give rise to complex local displacement gradients, and, under these conditions, the internal dis-

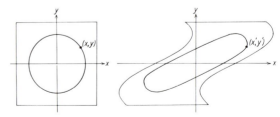

Figure 2.7. *Heterogeneous simple shear displacement leading to the deformation of an initial circle to a non-elliptical shape.*

tortions cannot be analysed by a single strain ellipse. The strain is then known as *heterogeneous finite strain*.

If we observe the forms of the circles of smaller diameter drawn on the model it will be seen that, although they are not exactly elliptical after deformation, they are more nearly elliptical than is the shape of the larger circle. This leads to an important concept: in a field of heterogeneous strain, the displacement gradients over a small element approximate to those necessary to induce a homogeneous strain in that element. If the displacement gradient varies smoothly, the field of *heterogeneous strain can always be subdivided into small elements which have almost homogeneous strain*. This is a most important conclusion for the structural geologist; it means that the distortion of the rocks in any large structure of heterogeneous strain, such as a fold, can always be analysed by applying the strain ellipse concept to individual outcrops or perhaps hand specimen-size pieces of rock. The work of Cloos (1947) on the strains seen in oolitic limestones in the South Mountain fold of Maryland provides a particularly clear example of the application of this principle (Figure 2.8). We will examine some of the methods of strain analysis of heterogeneous strain systems in the next session. For a complete examination of this rather complex problem we require the more complex differential equations relating heterogeneous finite strain to displacement; this topic will be examined in Volume 3 of this book series.

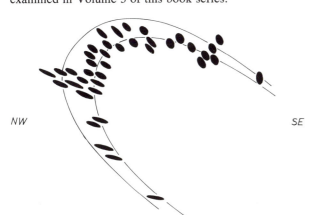

NW   SE

Figure 2.8. Strain ellipses in an anticlinal fold west of South Mountain, Maryland, USA. The ellipses show systematic variations in ellipticity and orientation related to position in the fold. Data from Cloos (1947).

It is worth emphasizing that practically all geological structures involve heterogeneous strain, and that if we wish to apply the strain ellipse concept we must be sure that *the area of our observation is sufficiently small for the conditions of homogeneity to be met, yet not so small that the behaviour of the individual crystal component of the rock interferes with the homogeneity*.

Principal strains and rotational component of the strain

Answers 2.6, 2.7

When displacement takes place we have seen that internal distortions are set up and that, if the displacement gradient is constant, the unit circle becomes distorted into the strain ellipse. The orientation and values of the principal strains

depend on the amount and type of displacement and can be expressed mathematically as functions of the displacement (see Answer 2.15★ below). In our experiments all the displacements are parallel to the card surfaces; as the amount of shear increases (γ_{yx} becomes larger) the proportions of the principal strains (ellipticity of the strain ellipse, R) increases. The angle θ' recording the orientation of the ellipse also changes, the ellipse's long axis coming to lie closer to the x-axis.

When the cards are unsheared from the state $\gamma_{yx} = 0.8$ to $\gamma_{yx} = 0.4$ the two perpendicular lines of the strain ellipse axes drawn on the cards loose their perpendicularity, but when the shear is completely removed and the strain ellipse has returned to its circular form the two lines regain their perpendicular relationships. In the *unstrained state the lines which become the principal axes of the strain ellipse are always perpendicular* and there is only one pair of initially perpendicular lines which remain perpendicular after deformation. This may be proved mathematically (see Answer 1.9★); you could also experiment by trial and error by drawing differently oriented sets of perpendicular lines on the cards to see if any remain perpendicular after deformation—you would not be successful! This special property of the principal axes of strain is very useful for determining the directions of principal finite strains in rocks. Many features in rocks are initially perpendicular: the angles between mud cracks and bedding, the angle between polygonal cooling fractures and dyke walls, fossils with initial bilateral symmetry, and certain worm borings are initially arranged perpendicular to the bedding planes. If we find these features remaining perpendicular in a rock we know to have been deformed, then their directions must coincide with the major and minor axes of the strain ellipse.

Answer 2.8

If the directions of the principal axes of strain do not change ($\theta = \theta'$) as a result of displacement, then that deformation is termed **irrotational finite strain**. In our card deck model, the angle θ always differs from the angle θ' and the difference defines the **rotational component** (ω) of our strain which is known as a **rotational finite strain**.

Before we discuss angular relationships we should note the sign convention for angles in our xy coordinate system. Angles measured in an anticlockwise sense from the positive direction of the x-axis are designated as positive, whereas those measured in a clockwise sense are designated negative. In our experiment, the angle θ' defining the long axis of the ellipse always makes a smaller positive angle with the x-axis than does the angle θ defining the initial orientation of this line. It therefore follows that the rotation ω (Greek "omega") ($\omega = \theta' - \theta$) is always negative.

Practically all deformations produced as a result of geological processes are rotational ones, but it is usually very difficult indeed to measure the rotational component in naturally deformed rocks. This is because, in order to measure ω, we have to know the initial orientation of the principal axes of the strain ellipse; when we have only deformed rocks to observe we generally have no way of determining this orientation. Although practically all geological deformations are rotational, it is worth emphasizing again that they are not all produced by simple shear: simple shear is just one of an infinite number of possible rotational strains.

Although simple shear is a rotational deformation, there has been *no actual rotation in space* of the card deck to produce this rotation. The development of a rotational strain does not necessarily imply that the body has to spin physically around some axis. Because of this lack of real rotation of the cards in space, some like to refer to the rotation that occurs in our experiment as an *internal rotation*, in contrast to an *external rotation* which would occur if we rotated the card deck box itself.

Because the concept of rotation can be easily misunderstood it is worth looking into more detail here. It is sometimes mathematically convenient to think of a finite strain as being made up of two components, an irrotational part which describes the *distortional aspect* of the displacement, and a *body rotation part* which rotates the mass about an axis until it comes to lie in the specific orientation that we see at the end of the displacement process (Appendix C, Section 5). It will be clear from the experiments carried out with our deck of cards that such a two-stage process has *not* actually occurred during simple shear. In fact, the concept would be false if we thought of displacement processes as being built up in this manner. Let us consider the two-stage model as applied to simple shear. We first have to distort the mass irrotationally about the two perpendicular directions defined by the principal axes of the strain ellipse and the angle θ (Figure 2.9B) and

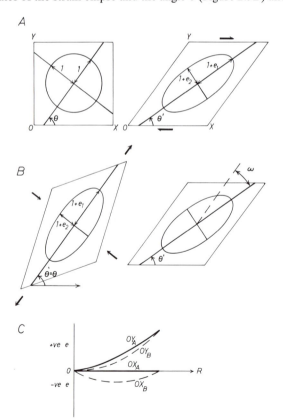

Figure 2.9. *Comparison of longitudinal strains arising* (A) *from simple shear and* (B) *from pure shear followed by a rotation where the resulting finite strains are identical.* C *illustrates the progressive changes in longitudinal strain along the line OX and OY as a result of these two deformations.*

then provide a solid body rotation through the angle ω. Are these changes in actual description of those which have gone on in the material? The end products are certainly identical, but the intervening changes of lengths, for example, are quite different. The differences can be realized if we study the length changes which take place along the direction OX and OY. In progressive simple shear (Figure 2.9A) lengths parallel to OX remain constant, whereas longitudinal strains along OY are always positive extensions given by $1 + e = (1 + \gamma_{yx}^2)^{1/2}$. If the deformation proceeded first by an irrotational strain the length changes along OX and OY would be quite different, and along OX the line would first be shortened and then lengthened, one effect exactly compensating for the other. The progressive length changes seen in our two processes and graphically recorded in Figure 2.9C are quite different. The mathematically convenient concepts for separating the distortional and rotational effects of displacement must be handled with care, and it is not possible to isolate one from the other in any progressive displacement sequence. Their separation is only valid if we refer to the geometric differences which describe the finite changes that have occurred as a result of the total process of displacement.

Answer 2.9

When the card deck model is deformed through a shear of value $\gamma_{yx} = 1 \cdot 6$, the directions of the axes of the finite strain ellipse are not coincident with the lines which were parallel to the axes of the strain ellipse developed by a shear of value $0 \cdot 8$. This effect is the rule in all rotational deformations because the deformation increments are not coaxial. We will analyse this effect in more detail later. There is no simple linear connection between the ellipticity of the two ellipses for shears of γ and 2γ, nor is there any simple doubling of the rotational component. At first sight you might find this rather surprising, for you might have expected that doubling the displacement would lead to twice the strain and twice the rotation. But, if you remember that the displacement gradient is expressed as a two by two matrix and that matrix products are more complex things than numerical products, then it becomes more apparent why there is a complicated relationship here.

Answer 2.10

The four curves which result from the laboratory model investigations are illustrated in Figure 2.10. None of the variations of ellipticity R, orientation θ and θ' or rotation ω are linear with increasing shear strain. The curve of ellipticity of the strain ellipse is concave upwards, implying that the distortion builds up more rapidly than the shear strain. This is a characteristic feature of most deformations taking place in geology (although it should be pointed out that it is not a universal feature of all steady state progressive deformation processes). In contrast to the increasing rate of ellipticity with progressive shear, the rates of change of the orientation parameters θ' and θ and of the rotation ω all decrease with shear. In fact θ' asymptotically approaches a value of $0°$, θ asymptotically approaches $90°$ and rotation ω also asymptotically approaches a maximum value of $90°$.

In simple shear the value of the area change Δ_A is zero for all values of γ_{yx}. It should also be apparent from the

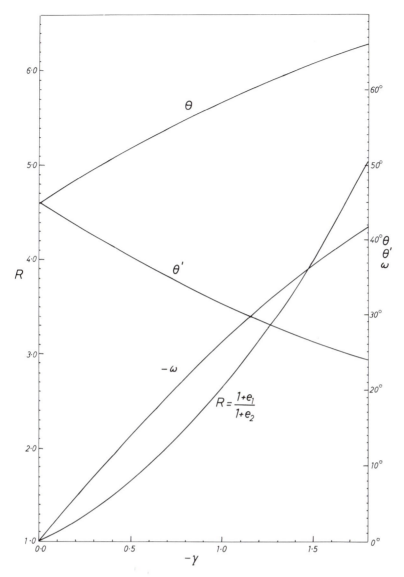

Figure 2.10. *Variations in initial θ and final θ' orientations of the finite strain ellipse axes, ellipticity R of the strain ellipse, and rotation ω as a result of progressive simple shear (γ_{yx}).*

geometry of the shear parallelogram made by the card surfaces (base length and perpendicular distance constant) that simple shear is a deformation without **area change** Δ_A. In simple shear no changes take place in lengths perpendicular to the card model surface. Such a deformation is termed a *plane strain*, and this, combined with the lack of surface area change, implies that in three dimensions the total deformation keeps a *constant volume* $\Delta_V = 0$. In most geological processes, two-dimensional planes do show area change and also show changes in the third dimension implying *non-plane strain* which may also imply **volume change or dilatation** Δ_V. We will discuss the geometry of these features in a later session of this book (Session 10).

Evolution of extension fissures in shear zones

Answer 2.11

From extrapolation of the θ' graph for strain ellipse orientation, the orientation of the first strain ellipse long axis is at 45° to the shear direction. The orientation of any incremental strain ellipse developed at any stage during the deformation history is also identically oriented because the geometric increments are the same. For example, if after a given total shear of say $\gamma_{yx} = 1.0$ we were to draw a new circle on the finite strain ellipse already formed from an initial circle, then the geometric changes taking place in this new circle during the next shear strain increment would be identical to those that took place during the very first deformation of the initial circle.

If an extension fissure system was to form during the initial shear displacement, then the fissures would be oriented perpendicular to the first maximum incremental extension, that is at 135° to the shear zone walls. They would form a parallel **en-echelon array** as shown in Figure 2.11A. Note carefully the relationships of the geometry of the en-echelon array to the shear zone displacement sense. Once formed, the fissures become carried along by subsequent displacements taking place in the shear zone, so that their initial formation angle of 135° becomes reduced by an amount depending upon the amount of subsequent shear strain. The first formed fissures therefore rotate, but it is most important to note that this **line rotation w** *is not*

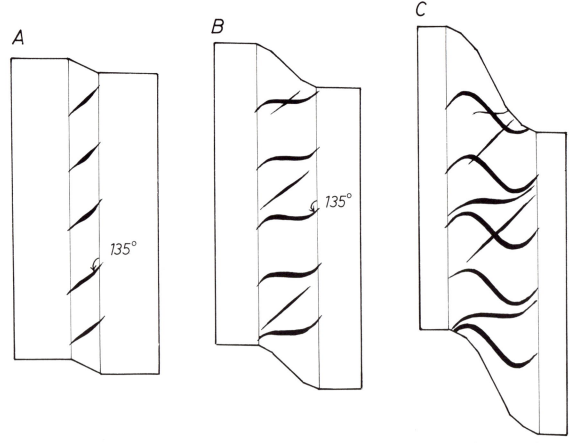

Figure 2.11. *Progressive development of extension fissures developing in a shear zone (Answer 2.11).*

the same as the rotation that we have defined previously as the **rotational component** ω of the strain ellipse. The geometry of line rotation is shown in Figure 2.12A. Any line starting with an initial orientation α is rotated to a new position α' such that

$$\cot \alpha' = \cot \alpha - \gamma \qquad (2.3)$$

Some curves for this function are illustrated in Figure 2.12B, including that of $\alpha = 135°$ which is of special interest in our study of fissure rotation. It should be clear that every line of different initial orientation undergoes a different **line rotation** ω for an identical later shear value, and that the **rotation** ω for the strain as a whole is a quite different function.

The initial fissures rotate according to the $\alpha = 135°$ curve of Figure 2.12B and clearly by measuring the change of orientation we could directly compute the amount of shear subsequent to fissure initiation. At the same time as rotating they often become opened and the fissures generally become filled with crystalline material deposited from pore fluids (fibrous crystals of calcite, quartz and chlorite are the most abundant filling species of *veins* in naturally deformed crack systems).

As the shear zone widens and the deformation front moves outwards, the fissure tips will propagate into the shear zone walls. The propagation direction will be controlled by the incremental strains and will therefore be oriented at 135° to the shear zone. The total fissure geometry now links the rotated central part of the vein with a 135° oriented tip, and will therefore exhibit a *sigmoidal shape*. Note carefully the shape of the sigmoidal

array and the shear zone displacement sense (Figure 2.11B). If the shear displacement at the zone centre becomes very large, it is possible that the central part of each sigmoidal vein will be rather badly oriented to allow the stretching that is necessary along the direction of maximum incremental extension. At this stage new *cross-cutting veins* may initiate at 135° to the shear zone walls, the "older" vein system becoming effectively dead as the new extension system develops (Figure 2.11C).

Figure 2.13 illustrates examples of en-echelon and sigmoidal vein arrays in different stages of development. Their geometry should be compared with the model we have developed from the progressive simple shear sequences; special attention should be paid to the possibilities of determining (a) shear zone displacement sense, (b) approximate value of the shear zone displacement (large or small), (c) how it would be possible to determine the value of the shear strain γ_{yx} at a point in the zone. The answers to these queries should be readily apparent from the model we have evolved.

Boudinage and folding and the strain ellipse

In Session 1 we looked into some of the geological consequences of changes of length in layered rocks containing layers of differing competence, and saw how lengthening of a competent layer could lead to boudinage, and shortening to buckle folding. Now that we have established the strain ellipse concept it will be useful to see what geometric implications follow from the orientations of the layers in respect to the axes of the strain ellipse.

A

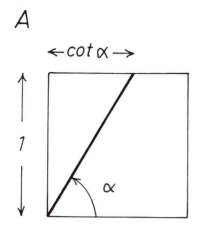
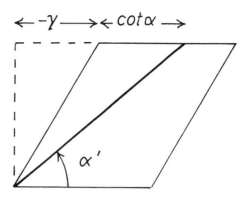

B

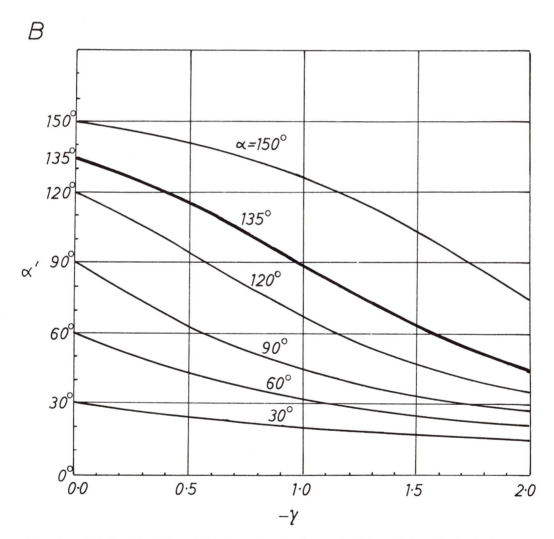

Figure 2.12. *Relationship of line rotation (w = α' − α) and amount of shear strain γ_yx in simple shear.*

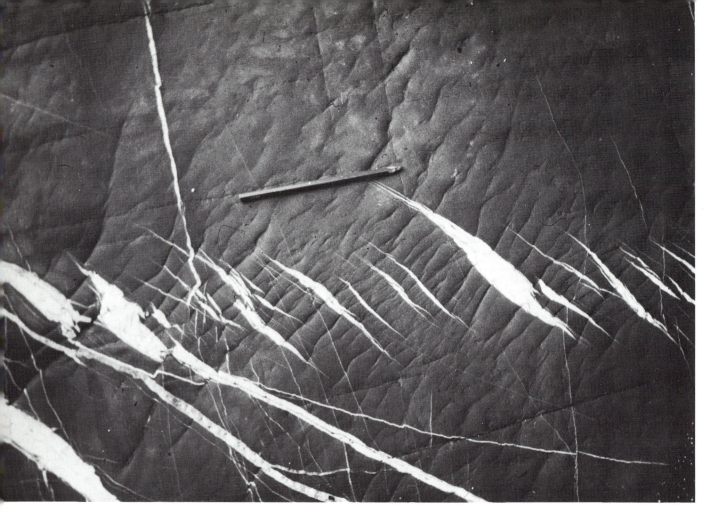

A

Figure 2.13. A: *En-echelon array of quartz filled veins developed in a shear zone in sandstone. Mullion, Devon, UK.* B: *Sigmoidally shaped en-echelon vein array in sandstone. Bude, Cornwall, UK.*

B

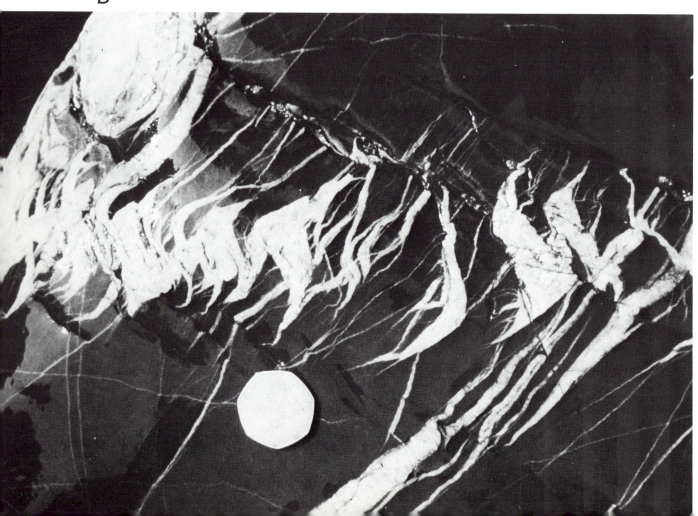

Perhaps we should first point out that the structures which form might be complex as a result of progressive length changes taking place during the strain history as we have seen in Session 1. We will discuss this problem later (Session 12). Here we will just look into the possibilities arising from total strain state.

The layer orientation within the strain ellipse can be varied so that it is symmetric or asymmetric with respect to the strain ellipse's principal axes (Figure 2.14). If the layering runs parallel or sub-parallel to the axis of principal extension it will be boudinaged, and these boudin structures are also symmetric: that is, the axes of the separated fragments of competent layer will be parallel to the layering of the less competent material and to the overall regional trend of the layering (Figure 2.14A').

If the layering is oblique to the principal strains, but still in a direction of stretching, then an interesting geometric effect is seen. The boudin fragments become asymmetrically aligned to the general trend of the layering and show an "en-echelon" distribution (Figures 2.14B', 2.16). They clearly have not been so strongly rotated (line element rotation) as the layering in the matrix. One finds in laboratory experiments that the difference in line element rotation between the boudin and its matrix is at a maximum when the cross sectional form of the boudins is approximately equi-dimensional, and is at a minimum when the boudins are long relative to their breadth.

If the layering is oriented parallel to the direction of maximum shortening in the ellipse, ptygmatic folds form, and these folds are symmetric and have their axial surfaces perpendicular to the average trend of the layers (Figure 2.14D'). However, where the layering is in an orientation where contraction takes place, but oblique to the maximum shortening, the folds are asymmetric. The first folds to form along the competent layer are symmetric, but with increasing shortening the orientations of the average layer direction and axial surface both move towards the maximum extension direction giving the fold a characteristic asymmetry (Figures 2.14C', 2.15).

As with Session 1, the reader has a choice of possibilities for continuing the study. After checking his conclusions of the results of Questions 2.1–2.11 he can either proceed directly to Session 3, after checking that all the key words and concepts set out in the summary list are fully understood, or answer the additional and more advanced questions of Session 2 set out in starred numbers below.

Orientation θ' of the principal axes of the strain ellipse

Answer 2.12★

$$(1 + e_{\alpha'})^2 = 1 + 2\gamma \cos \alpha' \sin \alpha' + \gamma^2 \sin^2 \alpha'$$

differentiating with respect to α'

$$2(1 + e_{\alpha'}) \frac{de_{\alpha'}}{d\alpha'} = 2\gamma(\cos^2 \alpha' - \sin^2 \alpha') + 2\gamma^2 \sin \alpha' \cos \alpha'$$

simplifying using double angle forms, equating $de_{\alpha'}/d\alpha'$ to zero and replacing α' by θ' we obtain the orientations θ' of the maximum and minimum extensions

$$0 = 2 \cos 2\theta' + \gamma \sin 2\theta'$$

$$\tan 2\theta' = -\frac{2}{\gamma} \qquad (2.4)$$

Figure 2.17 illustrates a tan $2\theta'$ graph. It will be seen that once we select a value for γ, this establishes values for $2\theta'$, and we automatically obtain two solutions over a 360° range for $2\theta'$. These two solutions are always given by $2\theta'$ and $2\theta' + 180°$ which implies that the two θ' directions are perpendicular. As γ becomes very large, θ' approaches zero. This means that the long axis of the ellipse comes closer to the x-axis, but can never pass through it.

Orientation θ of the principal strains before deformation

Answer 2.13★

This proof follows so closely on that of Question 2.12★ that it is unnecessary to give it in full. The initial orientations are given by

$$\tan 2\theta = \frac{2}{\gamma} \qquad (2.5)$$

and it is clear from the arguments developed above that these two directions must also be perpendicular.

It should be noted that in Session 1 (Question 1.9★) we established that there was *only one set of initially perpendicular lines which remain perpendicular after deformation*. This is true for all displacements which produce a homogeneous strain appropriately analysed using a strain ellipse.

Internal rotation

Answer 2.14★

$$\tan 2\omega = \tan(2\theta' - 2\theta) = \frac{\tan 2\theta' - \tan 2\theta}{1 - \tan 2\theta' \tan 2\theta}$$

Substituting appropriate values of tan $2\theta'$ (Equation 2.4) and tan 2θ (Equation 2.5)

$$\tan 2\omega = \frac{4\gamma}{\gamma^2 - 4}$$

tan ω is calculated from the identity

$$\tan 2\omega = \frac{2 \tan \omega}{1 - \tan^2 \omega}$$

which gives

$$0 = \tan^2 \omega - \left(\frac{\gamma^2 - 4}{2\gamma}\right) \tan \omega - 1$$

$$0 = \left(\tan \omega - \frac{\gamma}{2}\right)\left(\tan \omega + \frac{2}{\gamma}\right)$$

giving two solutions

$$\tan \omega = +\frac{\gamma}{2} \quad \text{or} \quad -\frac{2}{\gamma} \qquad (2.6)$$

The first of these roots is appropriate to our problem but the second is not. For example, when we consider the rotation at very low values of γ, the second result gives tan ω to be a very large negative number with a rotation approaching 90° or 270°, clearly a geometrically absurd result. Why does this solution appear? What does it mean, since mathematically it is a valid solution for tan ω? The

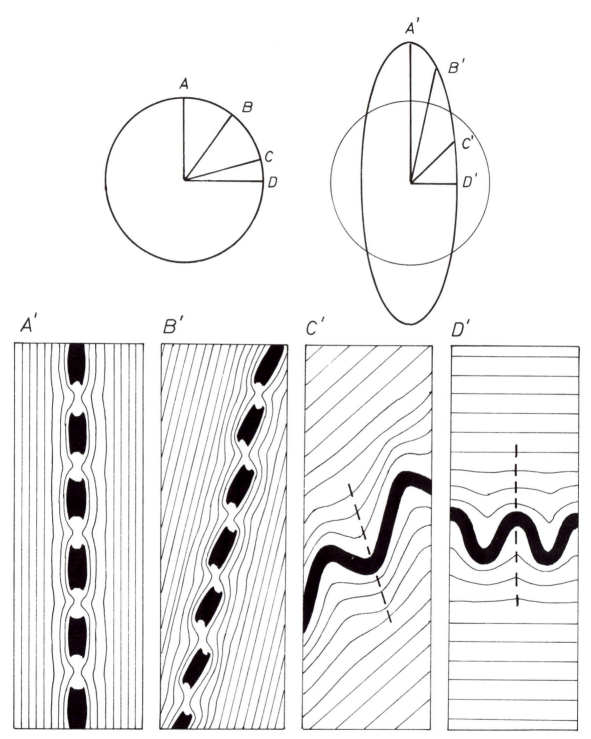

Figure 2.14. Geometric features of structures developed in a competent rock layer (black) embedded in an incompetent matrix: A', symmetric boudin structure; B', en-echelon boudin structure; C', asymmetric folds; D', symmetric folds.

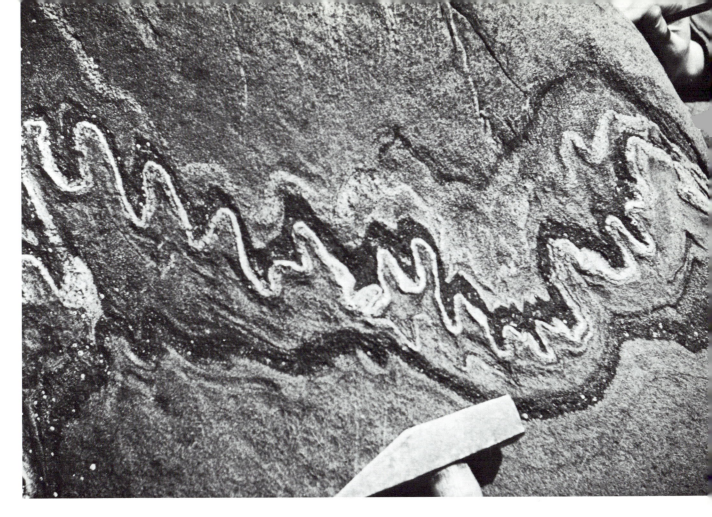

Figure 2.15. *Asymmetric ptygmatic folds developed in a competent pegmatite layer enclosed in Moine meta sediments. Mull, Scotland (cf. Figure 2.14C').*

Figure 2.16. *En-echelon boudins formed in competent calc-silicate layers surrounded by incompetent marble. Adamello massif, N. Italy (cf. Figure 2.14B').*

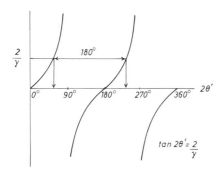

Figure 2.17. *Graphical solution of Equation 2.4 showing two solutions for $2\theta'$ oriented 180° apart.*

reason is apparent when we draw the physical orientations of θ and θ' (Figure 2.18). The axes which become the maximum (1) and minimum (2) axes of the strain ellipse with orientations θ and $\theta + 90°$ can be numerically taken away from the axes of the strain ellipse (1' and 2' with orientations θ' and $\theta' + 90°$) in several ways.

(a) $1' - 1 = \theta' - \theta = \omega$
(b) $2' - 2 = \theta' + 90° - (\theta + 90°) = \omega$
(c) $1' - 2 = \theta' - (\theta + 90°) = \omega - 90$
(d) $2' - 1 = \theta' + 90 - \theta = \omega + 90$

The first two solutions are geometric solutions to our problem, whereas the second two solutions, although mathematically correct, have no real significance for us.

It is clear from the solution $\tan \omega = + (\gamma/2)$ that the

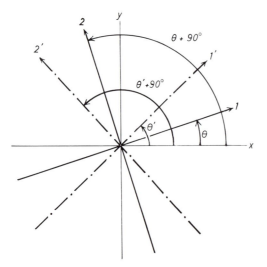

Figure 2.18. *Physical significance of the two solutions of Equation 2.6.*

anticlockwise rotation ω (implied by the positive sign) approaches a maximum value of 90° as γ becomes increasingly large.

Answer 2.15★

$$(1 + e_{\alpha'})^2 = 1 + 2\gamma \cos \alpha' \sin \alpha' + \gamma^2 \sin^2 \alpha'$$

Replacing α' by θ' (the directions of the axes of the strain ellipse) and using double angle forms

$$(1 + e)^2 = \frac{1}{2}(\gamma^2 \cos 2\theta' - 2\gamma \sin 2\theta' + 1)$$

substituting values of

$$\cos 2\theta' = \frac{1}{(1 + \tan^2 2\theta')^{1/2}}$$

and

$$\sin 2\theta' = \frac{\tan 2\theta'}{(1 + \tan^2 2\theta')^{1/2}}$$

and

$$\tan 2\theta' = -\frac{2}{\gamma}$$

and simplifying the result, we obtain

$$\lambda_1 \text{ or } \lambda_2 = \frac{1}{2}(\gamma^2 + 2 \pm \gamma(\gamma^2 + 4)^{1/2}) \qquad (2.7)$$

where λ is the quadratic extension $(1 + e)^2$.

Answer 2.16★

The curves cross the abscissa axis where the shear strain for the initial direction α has zero value. It follows from the discussion of Question 2.13★ that where this occurs the direction of the line now coincides with the axis or maximum elongation of the strain ellipse.

We have seen from our experiments (Question 2.11) that the principal elongation axis of the strain ellipse at the start of the deformation coincides with the $\alpha = 45°$ line. This could also be proved by evaluating $\tan 2\theta'$ using Equation 2.4 when γ is close to zero. With increasing shear, the line that had an initial orientation $\alpha = 50°$ becomes the direction of maximum extension and, with progressively increasing shear, the major axis of the strain ellipse moves successively through directions that were originally oriented $\alpha = 55°$, 60° and 65°. It is clear that the axes of successive strain ellipses never coincide with any fixed line direction in the body, but that they systematically sweep through material as it progressively deforms.

KEYWORDS AND DEFINITIONS

Area change or area dilatation Δ_A   The ratio of the area of the strain ellipse to the area of the initial circle from which it was derived. It can be expressed in terms of the principal finite extensions $1 + \Delta_A = (1 + e_1)(1 + e_2)$.

Displacement gradient   The gradient of displacement at different localities in a body. In simple shear, with shear strain γ_{yx} parallel to the x-coordinate axis, this can be expressed as $\partial \gamma_{yx}/\partial y$.

En-echelon array	An en-echelon array of fissures is one where the individual fissures show an echelon side step relative to each other (Figure 2.11A).
Finite strain ellipse	The ellipse derived from a circle of initial unit radius by a homogeneous strain. Its major and minor semi-axes are of lengths $1 + e_1$ and $1 + e_2$, where e_1 and e_2 are the **principal finite extensions** (sometimes termed the **principal finite strains** or **principal strains**). The **ellipticity R** is $(1 + e_1)/(1 + e_2)$. The **principal quadratic extensions** are λ_1 and λ_2 with $\lambda = (1 + e)^2$.
Homogeneous strain	A deformation such that the form of identically shaped initial markers (e.g. circles or squares) have their shapes modified into some new but constant form after deformation (e.g. ellipses or parallelograms). In contrast, a **heterogeneous strain** is one where the final shapes of the markers vary through the deformed body.
Line rotation (w)	The difference in angle between any line direction α before displacement and its orientation α' after displacement; $w = \alpha' - \alpha$.
Rotational component of strain (ω)	The difference in angle between the orientation of the axes of the finite strain ellipse (θ') and the orientation of these axes before deformation (θ); $\omega = \theta' - \theta$. Where $\omega \neq 0$ the strain is a **rotational finite strain**.
Volume change or volume dilatation Δ_V	The ratio of the volume of the strain ellipsoid to the volume of the initial sphere from which it was derived (see Session 10).

KEY REFERENCES

Many of the basic references are the same as those recommended at the end of Session 1. Further references relate mostly to the geological implications of strain, particularly with respect to boudinage and folding of layered rocks showing competence contrasts.

Boudinage

Cloos, E. (1947). Boudinage. *Am. Geophys. Union Trans.* **28**, 626–632.

A good review summary and description of field examples of boudins.

Griggs, D. and Handin, J. (1960). Rock deformation. *Geol. Soc. Am. Mem.* **79**, 382 pp.

This describes the results of boudin formation in laboratory experiments performed with various rock materials. Descriptions of the experiments will be found in pp. 355–358, and there are excellent photographs of these experimentally produced boudins in plates 6–10.

Ramberg, H. (1955). Natural and experimental boudinage and pinch and swell structure. *J. Geol.* **63**, 512–526.

This gives one of the best general modern accounts of boudin structure, describing field examples and laboratory model experiments. It also gives a mathematical analysis of elongation in ductile materials, and a discussion of the factors controlling boudin shape.

Wegmann, C. E. (1932). Note sur le boudinage. *Bull. Soc. Géol. France* **2**, 477–489.

This paper gives descriptions of the geometric features of boudins in naturally deformed rocks, discussing in detail the behaviour of the necks between boudins.

Ptygmatic folds

Kuenen, Ph. H. (1938). Observations and experiments on ptygmatic folding. *Bull. Comm. Geol. Finlande.* **123**, 11–27.

This paper gives many examples of ptygmatic structure and discusses the possible mechanisms of formation, namely, magma injection with buckling, or shortening of a pre-existing vein.

Ramberg, H. (1959). Evolution of ptygmatic folding. *Norsk. Geol. Tidsskr.* **39**, 99–151.

This is one of the most important contributions to understanding ptygmatic folds. It presents the first mechanical analysis of the structures in terms of shortening parallel to competent layers, and points out that ptygmatic folding and boudinage structure are complimentary phenomena which can develop simultaneously in a deforming rock.

Sederholm, J. J. (1913). Über ptygmatische Faltungen. *N. Jahrb. Min. Geol. Paläont.* **36**, 491–512.

This classic paper gives many examples of ptygmatic folds from the deformed Precambrian gneisses of Finland, and the folds are attributed to flow in a fluid rock medium.

Shear zones with extension veins

Beach, A. (1975). The geometry of en-echelon vein arrays. *Tectonophysics* **28**, 245–263.

En-echelon veins from a number of different geological environments are described and their geometric features related to different theoretical models. Relationships between the amount of dilatation and vein geometry are discussed and the significance of pressure solution as a source for the vein filling is stressed.

Hancock, P. L. (1972). The analysis of en-echelon veins. *Geol. Mag.* **109**, 269–276.

Various examples of en-echelon veins are analysed and it is shown that the angles between veins and the shear zone containing them can vary between 10° and 46°. These variations are attributed to differences in initiation mechanism by shear or by extension.

Roering, C. (1968). The geometrical significance of natural en-echelon crack arrays. *Tectonophysics* **5**, 107–123.

This gives a good discussion of the geometric forms of vein arrays with particular reference to quartz filled veins from the Witwatersrand and quartzites and the orientation of the stresses which might have produced the structures.

SESSION 3

An Introduction to Heterogeneous Strain

Experiments with the simple shear box made with curved formers result in patterns of heterogeneous deformation. The variably oriented and variably shaped strain ellipses are analysed using the concepts of direction fields and strain trajectories, and the compatibility principles which constrain the variable deformations are examined. The geometric and geological implications of the compatibility principles with reference to different types of ductile shear zones are discussed. Methods are developed for integrating strains to determine total displacements across shear zones. The effects of compatible volume changes in shear zones are related to the geometry of en-echelon vein arrays.

INTRODUCTION

We have seen from our laboratory experiments with simple shear that a non-linear coordinate transformation sets up rather complex deformations within the surface which cannot be analysed by a single strain ellipse over the whole card deck surface. We have also seen that the smaller we make the unit of our analysis the closer in that element do the concepts of homogeneous strain and the strain ellipse apply. We will prove mathematically later that if we reduce the region of investigation to that of a point in a heterogeneously strained body (mathematically a region which has no areal existence), then the displacement gradients existing at that point may be resolved into the properties of a homogeneous strain at that point. From the viewpoint of analysis of strain in geological structures this means that if we choose a small enough domain to investigate, then the strain may be analysed by the strain ellipse concept that we developed in the last session. How small is "small"? The answer to this question depends upon the scale of the overall heterogeneity. In Appendix B we prove that if any intersecting sets of initially straight, parallel, equally spaced lines remain straight and parallel and equally spaced after deformation, then the deformation is homogeneous. In geology nature generally provides us with only a single set of initially sub-parallel markers such as bedding surfaces or lithological banding. If such surfaces have a curved form, then it is clear that we are in a situation of heterogeneous strain (see Figure 2.8). If the surfaces are parallel we may or we may not have a situation of strain homogeneity. Inhomogeneous strain could arise by varying the displacement gradients across the surfaces, imposing heterogeneities of strain state without changing the orientation of the surfaces in much the same way that when the simple shear displacement gradient was varied in our experiments this led to heterogeneous strain without changing the direction of the cards. However, if the bedding planes are parallel there is a possibility that the strain will be homogeneous.

We should state now that, in detail, the analysis of heterogeneous strain is somewhat complicated. However, because so many geological deformation phenomena are the result of displacements which have led to heterogeneous strain, we must look into some of the geometric features arising in such structures at an early stage in our exploration of techniques so that we know how to practically analyse the commonest tectonic structures.

The aim of this session is to look at some more or less easily resolved problems of heterogeneous strain and to make come to life a very important notion, which we term **strain compatibility**. The concept of strain compatibility places important constraints on strain variation in heterogeneously strained bodies. The basic concept is very simple and depends upon the general coherence of matter in a continuum such as the deformed rock masses investigated by the structural geologist. A rock mass consists of an aggregate of connected particles before deformation, and as a result of a complex but *orderly displacement* there are rules which relate how the displacement gradients and therefore the strains from point to point are interconnected. If you were to draw on a sheet of paper a group of differently oriented and differently shaped strain ellipses it would be almost certain that they would not be compatible. That is to say, the displacements that each ellipse implies would not be mathematically linked—the body would have gaps or overlaps between elements or differences in rotations which make it incoherent: the elements would not join together. In Figure 3.1 we illustrate in simple visual terms the main features of strain compatibility. Four originally undeformed elements A B C D which fit together have been strained to produce four other elements A′ B′ C′ and D′. The distortion of each elemental piece is indicated by the centrally located strain ellipse. The strains from A′ to B′, B′ to C′ and C′ to D′ are compatible. Their half ellipses share a common plane with the next block, signifying the property of common features which enable the blocks to fit compatibly together. However, there is no way of fitting the strains of A′ and D′: the half ellipses in block A′ and D′ show geometric differences in shape, size and orientation of their common half plane. The whole body shows strain incompatibility, and in order to exist with the strains as indicated in this figure there would have to be a hole developed between block A′ and D′. We should now extend this concept to a body with much finer gradational variations resulting from a two-dimensional smoothly non-linear displacement field. All the strain ellipses from point to point must show gradual transitions in all directions into adjacent ellipses and these variations must obey well defined mathematical laws, the so called **rules of strain compatibility**. In mathematical terms the variations in the various features of strain (principal strain values e_1 and e_2, orientations θ' and rotation ω) must be smooth functions of initial positions (x, y) or final positions (x', y') capable of being expressed in equations like the Lagrangian set (in terms of initial position (x, y)) given by

$$\frac{\partial e_1}{\partial x} = f_1(x, y) \qquad \frac{\partial e_1}{\partial y} = f_5(x, y)$$

$$\frac{\partial e_2}{\partial x} = f_2(x, y) \qquad \frac{\partial e_2}{\partial y} = f_6(x, y)$$

$$\frac{\partial \theta'}{\partial x} = f_3(x, y) \qquad \frac{\partial \theta'}{\partial y} = f_7(x, y)$$

$$\frac{\partial \omega}{\partial x} = f_4(x, y) \qquad \frac{\partial \omega}{\partial y} = f_8(x, y)$$

where functions $f_1(x, y)$ through to $f_8(x, y)$ are smoothly varying continuous single-valued functions. The compatibility rules are expressed mathematically as the **compatibility equations** and imply that, for the body to cohere

as a single mass, these eight equations are not independent. At this early stage in our study an investigation of the compatibility equations is somewhat premature. The important point here is to realize what compatibility implies, especially in terms of the geometric property of elements which enables them to fit together. We will return to the mathematical analysis of heterogeneous strain and the compatibility equations in the third book of this series on advanced methods of analysis.

QUESTIONS AND EXPERIMENTS

Construction of finite strain trajectories

Question 3.1

Figure 3.2 shows the results of three laboratory experiments. A card deck was subjected to a variable simple shear displacement producing states of heterogeneous strain giving rise to structures known in geology as **shear zones** (A and B) and a **similar fold** (C). Before the card deck was sheared some 437 small circles were drawn on the surface. You could set up a similar experiment yourself, but, be warned, it does take a long time to prepare the circular markers! As a result of the shear we see that the circles are deformed into approximate ellipses; the strain ellipses vary in orientation (principal axes θ'), in strain ratio (ellipticity R) and in internal rotation (ω).

Cover the figures with transparent paper and draw the long and short axes of each of the strain ellipses. Such a diagram records a **direction field of finite strain**. Now construct smooth curves so that they integrate these directions: at any point they should be parallel to the lines of the direction field. We will have two sets of such curves, one integrating the ellipse long axis directions, and an orthogonal set integrating the ellipse short axis directions. These sets of curves are known as the **finite strain trajectories**. Note that *they do not join the centres of the adjacent ellipses recorded at individual localities in our*

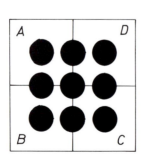

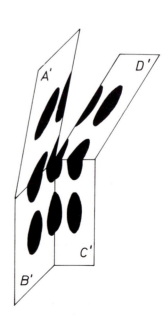

Figure 3.1. An illustration of compatible and incompatible strains.

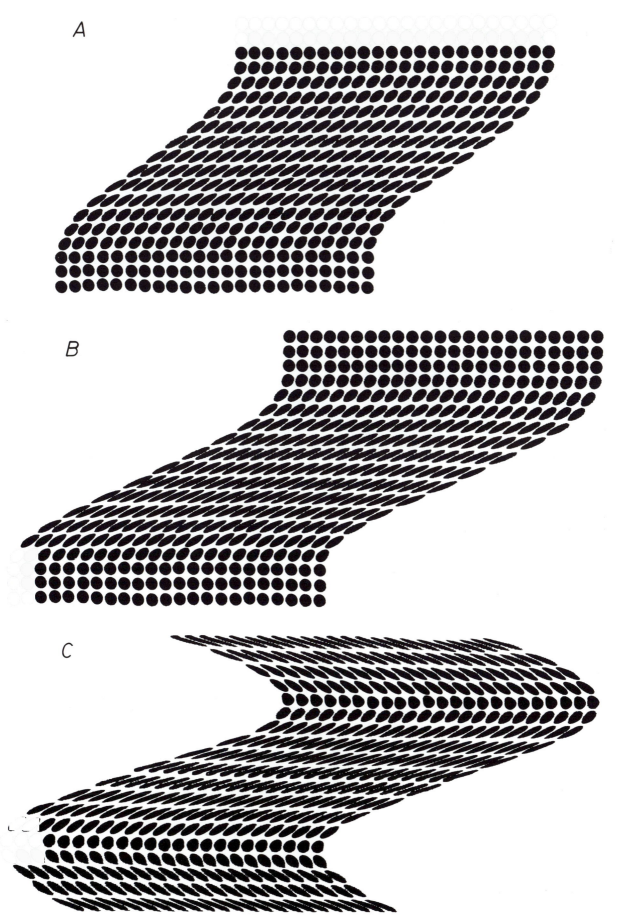

Figure 3.2. *Experimental models of deformed circular markers. A and B are shear zones, C is a similar fold. See Questions 3.1 and 3.2.*

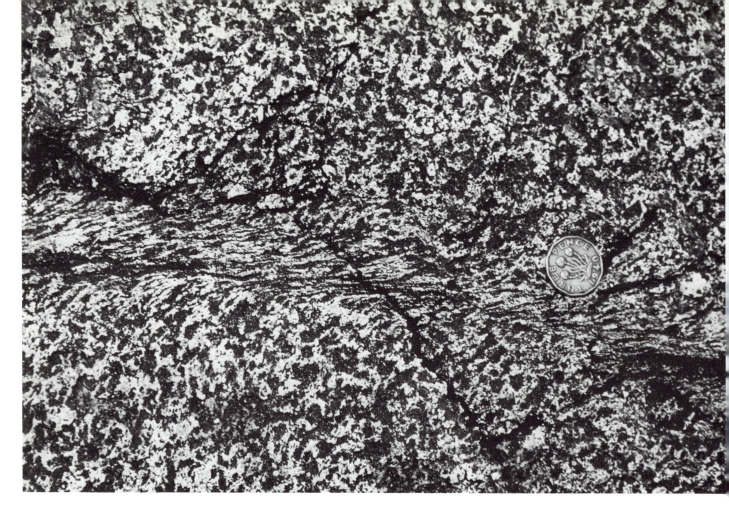

Figure 3.3. *Metagabbro deformed by a shear zone, Castell Odair, N. Uist.*

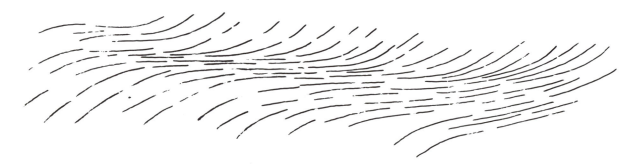

Figure 3.4. *Fabric trajectories of Figure 3.3. See Questions 3.3 and 3.4.*

experimental model. Finite strain trajectories offer an excellent and practical method of recording certain features of strain variation in a heterogeneously deformed body. Describe the general geometrical features of these trajectories. Note especially the convergence or divergence of adjacent curves of any one set.

Question 3.2

Record at each point the ellipticity (R) of the strain ellipse, and make a contoured map of variations of R through the structures. The strain trajectories constructed in Question 3.1 show convergences and divergences between adjacent trajectories. Can you see any relationships between strain trajectory geometry and changes in values of R? Make a graphical plot of ellipticity R (abscissa) and ellipse orientation θ'. The data points should fall on a single line. Why?

Question 3.3

Figure 3.3 shows a shear zone passing through a meta-gabbro consisting predominantly of amphibole (dark), feldspar and quartz (light). In the shear zone the shapes of the statistically equi-dimensional mineral components and mineral aggregates seen in the walls of the shear zone show significant changes: the mineral aggregates show a preferred orientation, and the intensity and preferred orientation of the mineral fabric varies through the shear zone. What do you think could be the significance of these features? Do you see any resemblances between these data and the geometric features of simple shear zones we discovered from Questions 3.1 and 3.2?

Now check your results and conclusions in the Answers and Comments section (p. 43), then proceed to Question 3.4.

Calculation of displacement in a shear zone

Question 3.4

Figure 3.4 shows lines connecting the crystal fabric orientations of Figure 3.3 known as **fabric trajectories**. Draw lines connecting points where the fabrics have the same direction. These are known as **isogon lines** (lines joining points where the surfaces have equal inclination) (see Figure 3.5; here they are termed **fabric trajectory isogons**. Are these isogons parallel? If the fabric orientation reflects the changes in orientation of the long axes of the finite strain ellipses, and the fabric trajectories coincide with the $1 + e_1$ finite strain trajectories, draw a shear strain

v. distance profile across the shear zone to indicate how the simple shear strain γ varies across the shear zone. For this construction you will need to use the graphs obtained experimentally in Session 2 relating strain ellipse orientation θ' to shear strain γ (Figure 2.10) and Equation 2.4 for high shear values. Discuss any problems arising from this calculation of γ from θ'.

When we have a shear zone shear strain v. distance profile we can integrate this graphically and reconstruct the total displacement across the walls of the shear zone. The technique is to draw at different points across the zone the orientations of lines that were initially perpendicular to the zone (after a shear of γ they will be oriented at an angle $\tan^{-1}\gamma$ to the shear zone walls). We can then graphically integrate these directions to arrive at a total shear displacement (see Figure 3.6). Can you devise a mathematical method for making this integration directly from the shear v. distance profile?

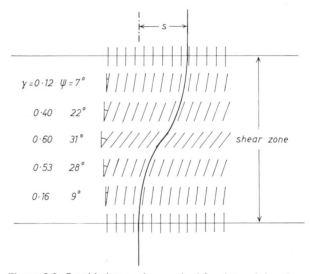

Figure 3.6. *Graphic integration method for determining the total displacement s across a simple shear zone.*

If you have time it is a very good exercise to make shear v. distance profiles across the experimentally formed shear zones illustrated in Figure 3.2, integrate these for the total strain and see if this coincides with the actual total displacement recorded in the experiment.

Now check the answers to this question before proceeding to Question 3.5.

Question 3.5

The geological map shown in the upper half of Figure 3.7 illustrates the exposed surface outcrops of deformed basic lavas and dyke rocks overlain unconformably by uniformly dipping sandstones. The lavas contain spherical to subspherical calcite filled vesicles. They are cut by a shear zone. In this shear zone the vesicles are elongated horizontally and shortened horizontally. Hand specimens have been collected at six localities (A to F) and the appearance of the deformed elliptical shapes of the vesicles on polished horizontal surfaces is illustrated in the lower half of Figure 3.7.

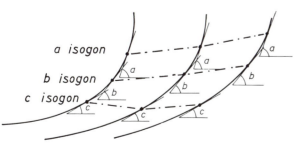

Figure 3.5. *Method of isogon construction. Isogons join points of equal inclination.*

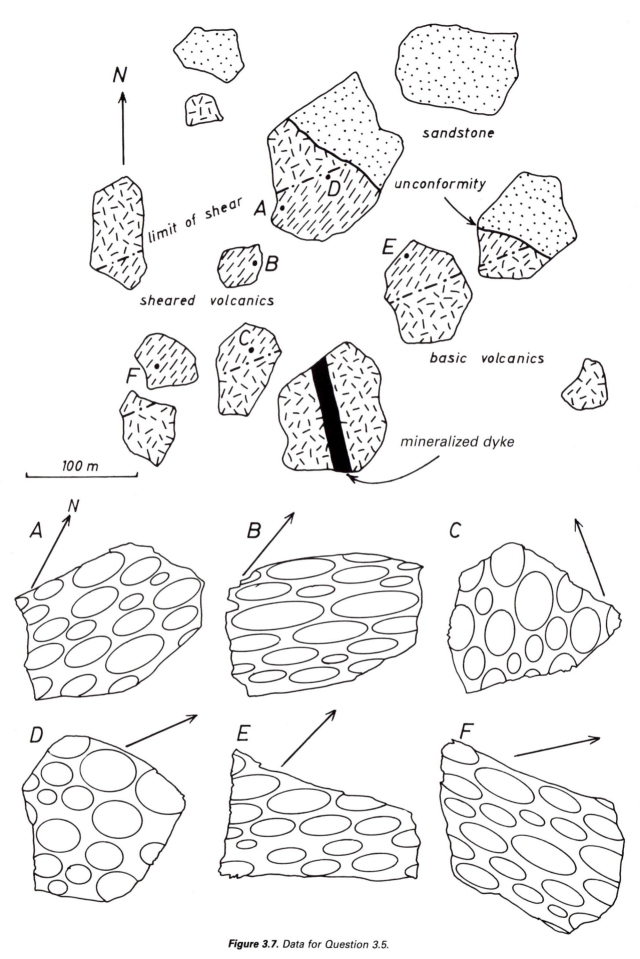

Figure 3.7. Data for Question 3.5.

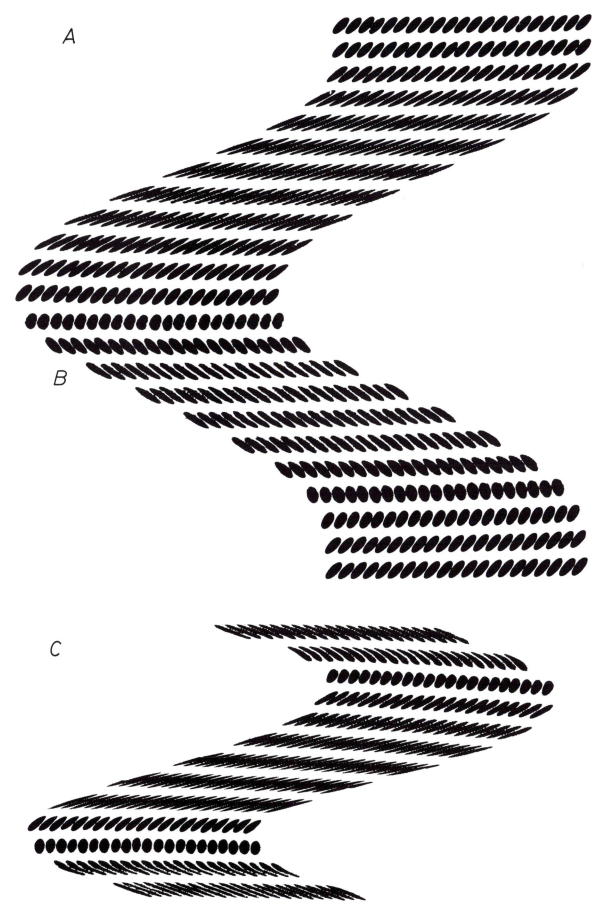

Figure 3.8. *Shear zone experiments—simple shear superposed on a homogeneous strain.* A: *a right-handed shear zone.* B: *a left-handed shear zone.* C: *a similar fold. See Question 3.6★.*

Determine the direction and shear sense of the shear zone displacements. Measure the long axis orientations and ellipticities of the deformed vesicles and see if these are compatible with heterogeneous simple shear in the zone. Where on the north west side of the shear zone might we expect to find the continuation of the vertically oriented mineralized dyke? Select a locality for a practical borehole site to check your predictions.

This question concludes the basic part of this session. Check your results with the Answers and Comments section (p. 43–52), then either go on to the starred questions below, or proceed to Session 4.

STARRED (★) QUESTIONS

Shear zones with wall deformations

Question 3.6★

Figure 3.8 illustrates some further experiments performed in the shear box somewhat similar to those already discussed (Figure 3.2 and Questions 3.1 and 3.2). However, these new experiments have one very important difference: namely, the card deck surfaces were initially imprinted with elliptical and not circular markers. This leads to a very different deformation pattern when the cards are subjected to simple shear. These initial ellipses had a constant ellipticity of $R = 2.77$, and a constant initial axial orientation of $\theta' = 45°$. The shear box experiments are therefore applicable to *simple shear superimposed on a pre-existing homogeneous strain*. In experiment A we developed a shear zone with a negative shear sense, then in experiment B we developed an identical shear zone, but with a positive shear sense. Finally (experiment C) we produced a similar fold by shearing in both positive and negative senses.

In experiments A and B measure the long axis directions (θ') and the ellipticities (R) of the finite strain ellipse. Include data from the unsheared (but deformed) wall material. Plot a graph of ellipticity (abscissa R) and orientation (ordinate θ'). Note that the angle θ' ranges from about $-15°$ to $-150°$ in our sign convention. Do these points coincide with those we produced in answer to Question 3.2? Why do they not coincide? In experiment B do the strain ellipses pass through an $R = 1$ (circular) state as the imposed shear strain works against the original strain? How could you determine the shear zone sense from the ellipse shapes and strain trajectories?

Draw strain trajectories through the similar fold of experiment C. What are the principal differences between this trajectory pattern and that of the folds we made by simple shear alone illustrated in Figure 3.2? Does the shape of the boundaries of a folded layer define completely the internal strains within the fold?

Shear zones with area (or volumetric) change

Question 3.7★

We saw in the Answers and Comments section in the discussion headed "Further comments on strain compatibility" that area changes arising by contraction (or expan-

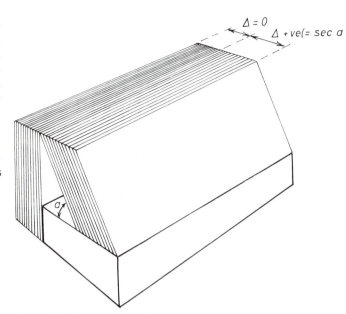

Figure 3.9. *Arrangement for a shear box experiment with surface area increase.*

sion) of the rock mass perpendicular to the shear zone walls could also produce strain states which were heterogeneous but compatible, provided that these area changes were only functions of distance from the shear zone wall. Figure 3.10 illustrates an experiment performed in the shear box aimed at illustrating this effect. A part of the card deck was allowed to tilt, keeping the surface of the cards horizontal (see Figure 3.9 for the geometric arrangement). This tilt induces a positive area increase of the initial circles drawn on the card edges relative to that seen in the untilted cards (Figure 3.10A), and the circles become elliptical, with the long axes oriented perpendicular to the card traces. The cards were then sheared sideways in the normal way of forming a shear zone so that we could study the geometric effects of superposing shear on a positive shear zone dilatation (Figure 3.10B).

Measure the ellipse shapes and orientations and graphically record the data (abscissa R, ordinate θ') on the graph previously made for Question 3.6. Do the points coincide with those for simple shear alone, or for shear superposed on a homogeneous wall strain?

What are the principal differences between the R–θ' plots for shear superposed on an area change and those for simple shear alone? In Question 2.11 we showed that the extension crack arrays likely to form in simple shear zones had a characteristic geometric form: the cracks initially form en-echelon arrays, the extension fissures oriented at 135° to the shear zone walls. What differences in en-echelon crack geometry might be expected where a shear zone contained a positive dilatation?

Shear zones with deformed walls and dilatation

Question 3.8★

Figure 3.11 illustrates a similar experiment to that shown in Figures 3.9 and 3.10, but here the card deck surface had elliptical markers like that used in the experiments of

A   B

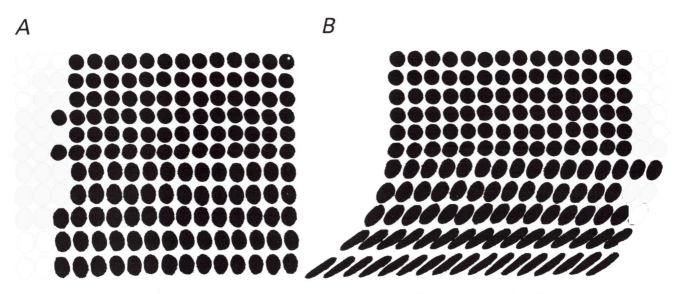

Figure 3.10. *Shear zone experiment—simple shear superposed on a dilatation. See Question 3.7★.*

A   B

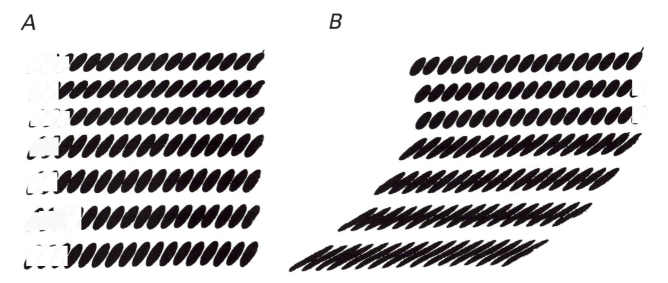

Figure 3.11. *Shear zone experiment—simple shear superposed on a dilatation and a homogeneous strain. See Question 3.8★.*

Figure 3.8. In this experiment we therefore have a shear zone containing three different deformation components: homogeneous wall rock strain, an area increase and a simple shear. Again measure the ellipticity and orientations of the strain ellipses and plot them on the graphs previously prepared. Are there any special geometrical differences?

What can be concluded in general terms about strain variations in parallel sided shear zones of the type we can produce in our shear box? How might we distinguish different types of strain combinations in naturally deformed rocks?

ANSWERS AND COMMENTS

Answer 3.1

Figure 3.12 illustrates curves known as **finite strain trajectories**. These orthogonal curves are parallel to the long and short axes of the strain ellipses at the points through which they pass. The trajectory patterns of the shear zones of Figure 3.12A and B show characteristic sigmoidal forms, the sigmoidal form being strongest developed in the shear zone showing the highest central strain. The trajectories terminate at the undeformed walls at angles making 45° and 135° with the shear zone boundary (these being the orientations of the infinitesimal principal strains in the deformed elements at the zone boundary). The trajectories in the simple shear similar fold (C) show a change over from an e_1 trajectory into an e_2 trajectory along the fold axial surfaces. The positions where this happens are known as **isotropic points**. At an isotropic point the strains in all directions are equal, that is the strain ellipse has a circular form. At the isotropic points in this fold the strains in all directions are zero, but that is not a necessary criterion for the existence of an isotropic point. In tectonic structures such isotropic points are not common, but do occur. We will see later how such points (known as **neutral points**) can occur in certain types of folds, and can combine together to form **neutral lines** or **neutral surfaces**. In experiment C the axial surfaces mark neutral lines of zero strain in our model surface (and neutral surfaces in three dimensions).

An interesting feature of strain trajectories arises from the geometric characteristics of the strain axes discussed in Session 2, that the directions of the strain ellipse axes were initially perpendicular. This geometrical property means that the strain trajectories, when undeformed, also form orthogonal sets of curves, which differ only in their spacing and in their orientations. The spacing changes relate to the strain values and the orientation differences are given by the differences in internal rotation of the strains from point to point.

If we can express the values of strain axis orientations as a function of position in our body as $\theta' = f(x, y)$, the strain trajectories represent the family of curves found by solving the differential equation

$$\frac{dy}{dx} = f(x, y)$$

because tan θ' is clearly the slope of the trajectory family. We will discuss these equations in Book 3.

Answer 3.2

A special geometric feature of strain trajectories is that they converge or diverge. In fact, if they are parallel it means either that the strain is homogeneous, or that they coincide with special types of area change such as is illustrated in Figure 3.17B. Apart from these special cases the e_1 trajectories always converge as we pass from a region of low strain into a region of high strain, whereas the e_2 trajectories diverge as we pass from low to high strains.

The variation of ellipticity with ellipse orientation plots on a single curve (Figure 3.13). The reason is that ellipticity is a function of shear strain γ and orientation of the ellipses is also a function of γ (see experimental results of Figure 2.10, Equations 2.4 and 2.7). It therefore follows that, in simple shear, ellipticity and orientation must themselves be related. This functional relationship provides a very good test to see if a shear zone contains only simple shear components. We will return to this later (Question 3.5).

Answer 3.3

The preferred orientation of the minerals in this shear zone show sigmoidal curves that resemble very closely the geometric features of the e_1 strain trajectories of Figure 3.12A and B. At the edge of the shear zone the mineral alignments make an angle of about 40 to 45° to the zone boundary, and here the fabric is very weakly developed (suggesting a low finite strain). As the fabric strengthens towards the centre of the shear zone (into a zone of higher deformation) the fabric lines come closer together as do e_1 strain trajectories passing from low to high strain regions.

Answer 3.4

The fabric trajectory isogons are shown in Figure 3.14A. As a first stage of defining the orientation of the shear zone, isogons were drawn with isogonal angle values chosen at random. These isogons defined a series of lines from which the average direction of the shear zone was obtained. Then isogonal lines were drawn at angles making $\theta' = 10°$, 20°, 30°, 40° and 45° to the best fit line for the shear zone direction. The isogons are not absolutely parallel: they show slight curvatures, and in the shear zone centre higher angles were found at the left-hand end of the zone than seen at the right-hand end. Two profiles were chosen perpendicular to the zone and the shear v. distance graphs constructed using the isogon intersections and the relationship $\gamma = 2/\tan 2\theta'$ (Figure 3.14B). The main problems that arise here concern the difficulty of measuring θ' at the centre of the zone, for at low values of θ' very slight differences give large variations in γ-value. Graphical integrations of these data are shown in Figure 3.14C. Profile PQ gave total displacements of 7·0 units, RS gave 7·6 units in a shear zone about 4 units wide.

Another method of integrating the shear strain data can be developed when we appreciate that mathematically, to determine the total shear we perform the integration

$$s = \int_0^x \gamma \, dx$$

where x is the distance across the shear zone. This simple integral is evaluated by finding the area under the shear

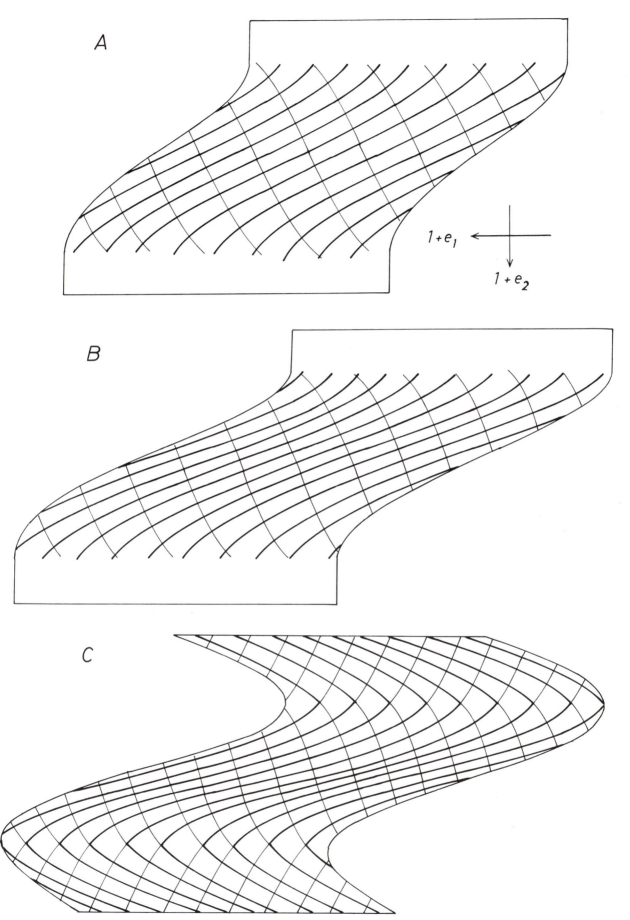

Figure 3.12. *Finite strain trajectories derived from the experimental models of Figure 3.2.*

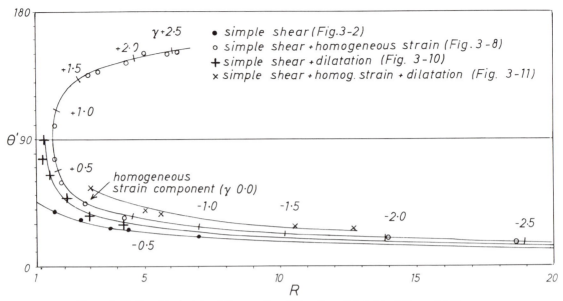

Figure 3.13. Graphical plots of the results of Questions 3.1, 3.6★, 3.7★ and 3.8★.

strain v. distance curve. For example, profile PQ gives a total displacement of 6·6 units and RS gives 7·9 units, in reasonable accord with (and probably more accurate than) the results obtained by graphical integration. The difficulty noted above of giving an exact value for γ at the zone centre does not lead to great computation errors, because the peak of the graph is comparatively narrow, influencing only slightly the area under the curve. However, if the central peak is wide and of high strain then grave computational errors are likely to occur in determining the total displacement across the zone.

In solving this problem we made two principal assumptions, first that *the deformation in the zone was that of simple shear*, and second *that the fabric trajectory coincided with the e_1 finite strain trajectory*. The assumption of simple shear is clearly a critical one if our integration technique is to be valid, and the reader would be right to ask if we can be sure that no other types of displacement take place in the zone. There are in fact other types of displacement plans which produce similar (but not identical) structural geometry to those of simple shear zones. A good way of checking the correctness of the assumption is to find more than one way of evaluating the shear strain at each locality. If different methods give the same result, then it is proven that the assumption is a correct one (Question 3.2). We will use this method in Question 3.5, and we will discuss further the complexities of other types of possible strain plans in shear zones in Questions 3.6★, 3.7★ and 3.8★. The second assumption is also a critical one. The geometrical comparison looked convincing, but clearly a better check would be to have some markers from which we could measure strain and see if the maximum elongations of the strain ellipses and the fabric directions coincided. Figure 3.15 illustrates an outcrop of deformed xenolithic granite in the Pennine zone of the Swiss Alps where we can check this. The photograph shows the margin of a shear zone, the granitic material on the right-hand side being almost undeformed, whereas that on the left-hand side is very strongly deformed. The progressive develop-

ment of the fabric known as schistosity (produced by the alignment and streaking out of the mineral components) from right to left is very striking, and it clearly goes together with the deformation of the xenoliths. The xenolith in the upper right is almost undeformed, that in the lower centre has a "tear-drop" form because it is located where the displacement gradients are rapidly changing at the shear zone boundary (cf. some of the ellipses near the fold hinges in Figure 3.2C). Xenoliths on the left all show strongly elongate forms with their long axes lying in the plane of schistosity. In this example there is no doubt that the fabric produced during the deformation does coincide with orientation of the strain ellipse, and that its intensity is related to the ellipticity of the strain ellipse.

Methods of measurement of shear strain in shear zones

In simple shear zones we have seen that it is possible to evaluate the shear strain from locality to locality using various methods, namely:

1. The ellipticity of the strain ellipse.
2. The orientation of the strain ellipse, a direction which may coincide with the orientation of planar fabrics produced during the shear displacement.

Other methods for calculation of shear strain can be obtained from:

3. The deflection of pre-existing markers such as bedding planes, lithological layering in gneisses, and dykes using the relationships of Equation 2.3.
4. The change in orientation of extensional veins formed early during the shearing event, and which were later deflected by further shearing.

These techniques of integrating shear zone displacements have produced outstanding results in the measurement of regional displacements, particularly in Precambrian gneissic terrains. Calculations of γ-values across very

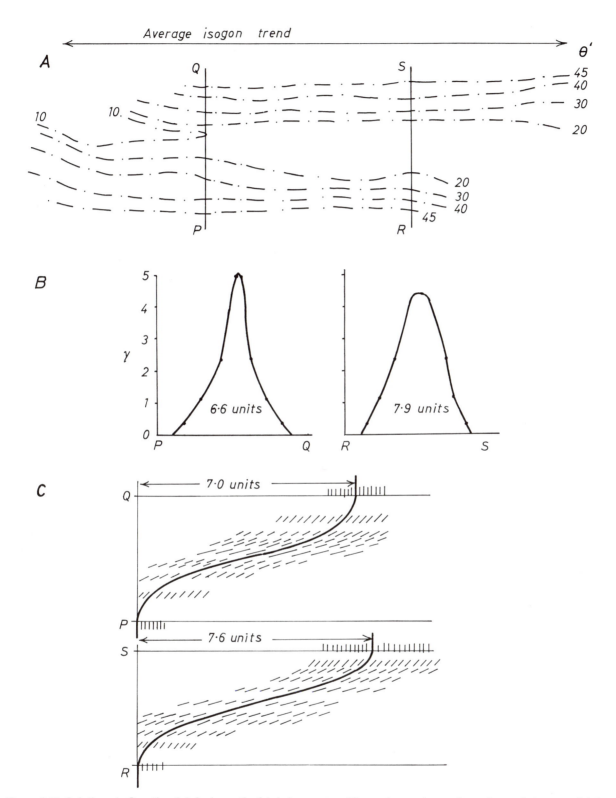

Figure 3.14. Solutions to Question 3.4. A shows the fabric isogon trend lines: the numbers refer to the angle between fabric direction and the shear zone walls. B shows the shear strain profiles across sections PQ and RS. C illustrates the graphical method of integrating the shear strain profile.

Figure 3.15. *Deformed xenoliths at the edge of a shear zone in a deformed granite. Laghetti (Fusio), Lepontine Alps, Switzerland.*

large shear zones in West Greenland using dyke displacements led Escher *et al.* (1975) to conclude that the Nagssugtoqidian orogenic front shows an average horizontal shortening of some 66% over many tens of kilometres. Shear zones are also one of the most important deformation styles in the Precambrian terrain of West Scotland, and Beach (1974) was able to demonstrate that the movements along the Laxfordian orogenic front by shear zone displacements could be resolved into a total horizontal displacement component of 18 km, and a vertical displacement component of 16 km.

Answer 3.5

Figure 3.16B gives the data of ellipse orientations and ellipse shapes. They lie on a curve which is compatible with that for simple shear. The location of the dyke continuation is shown in Figure 3.16A. Because the dyke is vertical and because there is always some uncertainty arising from the shear integration construction it would not be advisable to site a vertical drill hole above the predicted position of the dyke. It would be more sensible to locate a 45° inclined drill hole somewhere at the location indicated in Figure 3.16A.

Further comments on strain compatibility

The strain states arising during heterogeneous simple shear

displacement are clearly compatible because they arise from a displacement plan which keeps the body coherent. In terms of the ability of the strain ellipses to fit together, it is clear that all the ellipses have in common uniformly directed lines of no longitudinal strain parallel to the shear zone walls. This means that they can be compatibly linked together (Figure 3.17A).

What other compatible strains could exist in a shear zone where the strain profile across the zone at different localities was constant? We could clearly impose a homogeneous strain across the shear zone and its walls (or we could impose a shear zone on a state of homogeneous strain). The strain geometry could not then be analysed by simple shear alone (see Question 3.6★ for an examination of this problem). Another possibility is that we could change the surface area of the shear zone in the manner shown in Figure 3.17B provided that the area change axis was perpendicular to the shear zone walls. In practice such a change would probably go together with a three-dimensional volume loss or volume gain. Provided that such variations take place in an orderly way such that the dilatation value Δ is a function only of distance from the shear zone wall, then the strains so produced are themselves compatible. Area change (and three-dimensional volumetric change) of this type does occur in natural systems. Because it is internally compatible, it can be combined in any way with the compatible strains of heterogeneous simple shear to produce a strain geometry

A

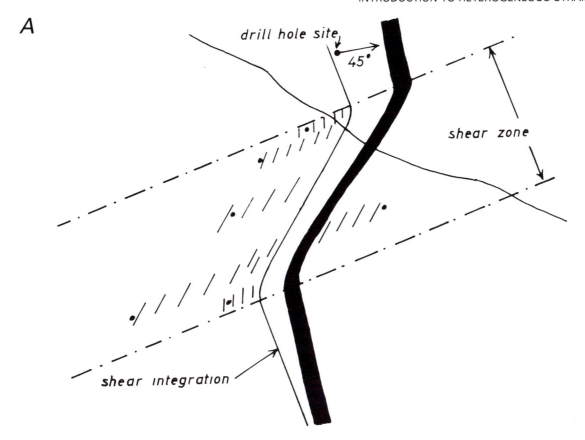

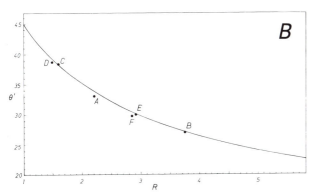

Figure 3.16. *Solution to Question 3.5. The θ'/R data for the various specimens A to F is indicated on the graph (Fig. B). These data fall close to the relationships for a simple shear deformation.*

which is also internally compatible. The modifications of simple shear strain by such dilatation is discussed further in Question 3.7★.

In conclusion we can see that *shear zones showing constant strain profiles can be produced by any combinations of the three following factors:*

1. Heterogeneous simple shear.
2. Heterogeneous dilatation normal to the shear zone walls.
3. Homogeneous strain across the shear zone and its walls.

No other strain systems are possible. For example, consider the effects of flattening normal to the shear zone with

extension parallel to the shear zone (Figure 3.17C). Such an effect would lead to a "toothpaste" extension between the shear zone walls. To make such variable flattening compatible we have to induce a differential shear of increasing value along the shear zone—an effect which would clearly add additional displacements and strains to the system, and lead to the development of strong variations in the strain profiles drawn at different cross sections of the zone. It should be clear from Figure 3.17C that heterogeneous irrotational strain (except for the area loss types discussed above), where the axes of the strain ellipses are all parallel is *not* a strain compatible combination.

The concept of "factorizing" different strain components in a shear zone (factors 1, 2 and 3 above) is only possible where the factors show independent strain compatibility. In some of the geological literature wall parallel extension/wall perpendicular contraction factors of the type described in the last paragraph have been considered as possible shear zone components. This is incorrect, because variation in such factors are not themselves internally strain compatible features.

Shear zones with wall deformation

Answer 3.6★

The data from the variable ellipticities and ellipse orientations are illustrated in Figure 3.13. They fall on a curve, but it is a different curve from that of simple shear superposed on undeformed wall material. The different positions of the data points clearly reflect the inheritance of a component of strain which differs from that which could be produced from a simple shear displacement alone. The data points move in two directions away from the point

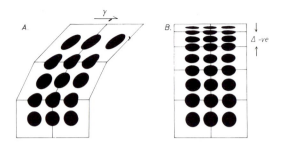

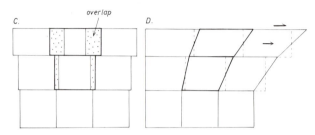

Figure 3.17. *Compatibility and incompatibility in shear zones. A illustrates the compatible strains of heterogeneous simple shear and B is a compatible strain variation resulting from a dilatation Δ_A. C is an example of incompatible shear zone parallel pure shear (irrotational flattening) which necessitates overlap of the elements. D is a compatible form of C illustrating how additional shear zone components are required for a compatible solution.*

representing the initial wall strain. In experiment A the strain ratio increases with progressive (negative sense) shear, whereas in experiment B the strain ratio decreases to a minimum value, then starts to increase. The minimum value is not zero but has an ellipticity R of 1.57. The simple shear can never completely efface the original strain because the original wall strain did not have the characteristics of a simple shear strain. If, for example the wall strain was of a type where the R and θ' values could be located on the simple shear curve, then such a strain could be completely removed by a later simple shear displacement.

If one knows the wall rock strain it is possible to determine the geometrical effects of superposing a simple shear, and it is technically feasible from any of the measured ellipses of Figure 3.8 to compute the superposed shear value γ. The calculations need a computer to evaluate the precise locations of the points on the R/θ' curve together with the corresponding γ-value (see Figure 3.13). If these calculations can be made it is clearly possible to reconstruct the shear zone shear profile in the way we discussed earlier for simple shear alone, and hence determine the total shear displacement across the whole zone.

The sense of the simple shear displacement can be deduced from the way the principal strain directions in the wall are modified as they are traced into the shear zone. Although the actual strains and ellipse orientations are rather complex, where the superposed shear has a negative shear sense, the ellipse long axes move in a negative sense with increasing shear, and vice versa when the superposed shear has a positive sense. Figure 3.18 shows an example of a left-handed shear superposed on previously deformed gneisses which illustrates this principle.

Figure 3.19 shows the orientations of the finite strain trajectories drawn from Figure 3.8C. They show some similarities with those of Figure 3.12C; however, the trajectories continue unbroken across the axial plane of the fold with no isotropic points. The boundaries of the folded layer of Figures 3.2C and 3.8C are identical, yet the strains inside the fold differ. It is therefore clear that from the geometric form of a folded layer we are not able to exactly specify the internal strain pattern. This point is especially important when we discuss later how we can classify fold geometry.

Shear zones with dilatation (area or volume change)

Answer 3.7★

The ellipticity and ellipse orientation data derived from the experiments of Figure 3.10 are plotted in Figure 3.13. Because of the dilatation effect the strain data do not lie on the simple shear R/θ' curve. They lie on one of a family of curves lying above or below the simple shear curve (if the dilatation Δ_A is positive the data lie above the simple shear curve, if negative they lie below the simple shear curve). These curves are shown in Figure 3.20, and illustrate the variations in orientation and form of ellipses produced by variable simple shear and variable dilatation. We have already seen that because variable simple shear and variable area (volume loss) contraction normal to the shear zone walls are both in themselves compatible strain plans (Figure 3.17A and B), then any combination is also compatible. It therefore follows that, if we can measure R and θ' of the strain ellipses in a parallel sided shear zone with undeformed walls, it is possible to "factorize" the strain into a simple shear component γ, and an area change component Δ_A. Any point plotting on the graphs of Figure 3.20 enables the two components to be directly read off from the two sets of intersecting curves.

En-echelon extension fissures

The presence of an area change component in the shear zone influences the orientation of the long axis of the strain ellipse. This is true for incremental or small deformations as it is for large strains. This means that the geometric features of cracks and fissures initiated at early stages of deformation will be controlled by the area (volumetric) changes which accompany the shear. The maximum extension directions of a system with a positive dilatation always make a higher angle with the shear plane than those arising by simple shear alone, whereas the maximum extension directions in a system with a negative dilatation make a smaller angle than those of simple shear. Figure 3.21 illustrates the consequences of this geometric relationship. Figure 3.21A shows an initiating shear zone with a positive dilatation, corresponding to the θ'/R curve with $\Delta_A = +0.1$. If extension fissures develop after a shear strain of $\gamma = 0.1$, then the orientation of these will be controlled by the direction of maximum stretch at the time of failure. With these parameters the maximum stretch oriented at 66° must give rise to en-echelon extension fissures at an angle of 24° to the zone. In contrast, if there is a negative dilatation of $\Delta_A = -0.1$, then extension fissures forming at a shear strain of $\gamma = 0.1$ form at 90° minus 21° = 69° to the shear

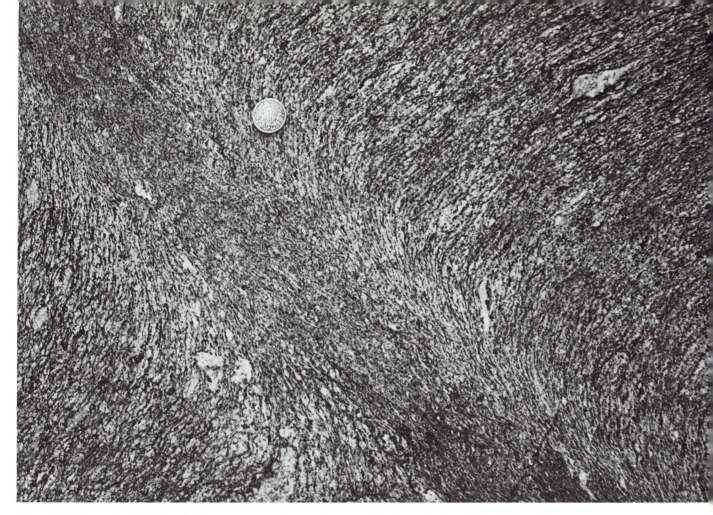

Figure 3.18. Example of a left-handed shear zone superimposed on a previously strained gneissic rock. Cristallina, Lepontine Alps, Ticino, Switzerland.

Figure 3.19. Finite strain trajectories for the similar fold of Figure 3.8C.

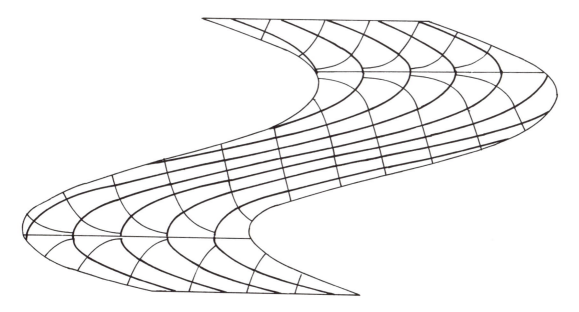

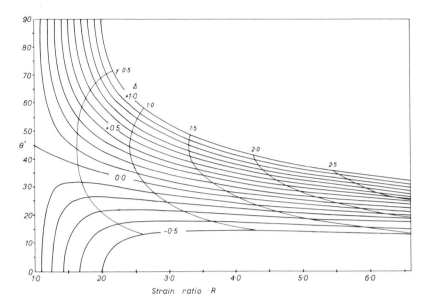

Figure 3.20. Curve showing variations in ellipse orientation θ' and ellipticity R for differing simple shears (γ) and dilatations (Δ).

zone walls (Figure 3.21B). The actual angle of initiation of these fissures is controlled by the failure criteria, and in particular the relative proportions of influence of deformation by shear, and deformation by dilatation. However, it is geometrically correct (except for failure only at high finite strains) that en-echelon fissure systems will make angles of less than 45° to the shear zone in positive dilatational situations, and angles of more than 45° in zones with negative dilatation. Subsequent deformation in the shear zone will modify these initial angles and lead to the production of distorted and sigmoidal veins in the same way as discussed in Session 2.

Examples of naturally deformed **shear zone vein arrays** which illustrate the deductions of Figure 3.21A and B are illustrated in Figure 3.22A and B respectively. The en-echelon vein zone of Figure 3.22B shows a darkening of the country rocks within the zone as a result of the activity of **pressure solution** processes which have led to an overall volume reduction in the shear zone. Many shear

zones with en-echelon vein systems show evidence of such pressure solution where the more soluble components of the rock (calcite or quartz for example) are taken out of the body of the rock, to be deposited nearby as vein filling. This solution is sometimes localized along particular surfaces oriented sub-perpendicular to the maximum shortening direction. These local dark **pressure solution seams or stripes** occur where solution processes become concentrated and the dissolved material can be more efficiently removed along the planar channel-way. The dark colour arises from the accumulation of the relatively insoluble components in the rock (clay, phyllosilicates, carbon, iron ores, etc.). In calcareous rocks the pressure solution seams often show a characteristic saw-tooth cross-section form known as **stylolites**. Stylolites are well known in certain sedimentary rocks which have been subjected to gravitational loading during diagenetic compaction and **diagenetic stylolites** are usually parallel or sub-parallel to the bedding surfaces of the sediment. In contrast, **tectonic**

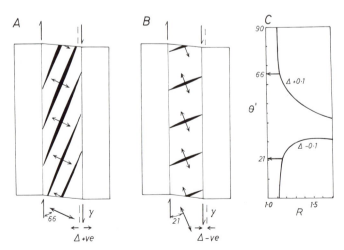

Figure 3.21. Orientations of shear zone vein arrays as a result of: A, simple shear together with positive dilatation Δ; and B, simple shear together with negative dilatation Δ. C shows the relationship of the strain ellipse orientations at failure (failure chosen at γ = 0·1) and how these relate to the curves of Figure 3.20.

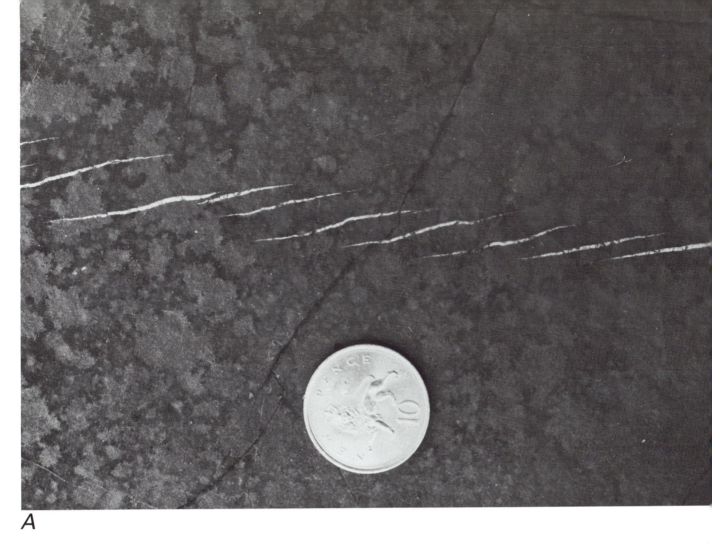

A

Figure 3.22. Shear zone extension vein arrays—with a positive dilatation (cf. Figure 3.21 A, Lyme Regis, Dorset, UK). and with a negative dilatation (cf. Figure 3.21 B, Pralognan, Vanoise, W. Alps).

B

Figure 3.23. *Tectonic stylolites developed with extension fissures in part of an en-echelon shear zone vein array.*

stylolites form under tectonic stress situations and are probably initiated perpendicular to the maximum compressive tectonic stress. As a result, they usually cross-cut the bedding surface, and intersect and displace tectonic features of the rock. Figure 3.23 shows examples of tectonic stylolites in a limestone forming normal to, and displacing, calcite-filled extension veins. The tectonic stylolites and veins are probably forming more or less synchronously: the stylolite is the source of the calcite which becomes deposited in the vein. This is a very good example of the process of rock deformation by chemical activity rather than by physical mechanisms. The stylolite enables the rock to contract by preferential solution of minerals, whereas the vein enables the rock to elongate by preferential deposition of the same material in extension veins.

Shear zones with deformed walls and dilatation

Answer 3.8★

The R/θ' data from the experiment of Figure 3.11 are plotted in Figure 3.13. They fall on one of the curves of Figure 3.20. Inside the shear zone, *the ellipse geometry of simple shear together with a dilatation is therefore indistinguishable from that of simple shear with dilatation superposed on a homogeneous strain.* The only geometric difference is the state of the walls, whether undeformed (for the former case) or deformed (for the latter case). The reason for the coincidence of the data curves for this experiment and those of Figure 3.20 is that, although the strain in the walls is not necessarily corresponding to a simple shear alone, the wall strain ellipse R and θ' values can always be produced by the combination of a particular wall normal dilatation together with a simple shear. If we know the state of wall rock strain, it is possible to compute the particular set of curves for differing dilatation and simple shear components and hence factorize the total strain into its three components. However, each particular wall strain has its own set of R/θ' curves for various values of γ and Δ, so it is not possible here to give a unique solution as a single set of graphs. All the curves will coincide with those of Figure 3.20, but the actual Δ- and γ-values will not be the same as those illustrated in this figure.

KEYWORDS AND DEFINITIONS

Direction field of finite strain A graphical representation of the principal strain axes at points in a deformed body.

Fabric trajectories Lines drawn on a deformed rock which integrate the directions of the fabric (mineral alignments, mineral elongations, etc.) through the surface.

Fabric trajectory isogons	Isogons drawn on curving fabric trajectories.
Finite strain trajectories	Lines which integrate the principal strain axes of a direction field into two orthogonal sets of smooth curves.
Isogon lines	Lines joining points where surfaces have identical orientations.
Isotropic point	A location in a strained body where extensions are equal in all directions (i.e. the strain ellipse has a circular form).
Neutral line	A line connecting points of zero extension.
Neutral point	An isotropic point where extensions are zero.
Neutral surface	A surface connecting points of zero extension.
Pressure solution	A deformation process whereby material under stress goes into solution at a localized point in a material. This material is transported by flow or diffusion and is usually deposited at some other locality in the rock system, a process termed *solution-transfer*.
Shear zone	A zone with sub-parallel walls in which high deformations are localized. Such zones can show strain discontinuities across a plane (a *fault*), or the strain variation may be continuous across the zone such that, on the scale of a rock outcrop, no geometrical discontinuities can be seen (a *ductile shear zone*).
Shear zone vein array	A geometrically grouped series of mineral filled fissures arranged along a shear zone.
Similar fold	A folded layer with identically curving layer boundaries.
Strain compatibility	The geometric constraints which relate variations in finite strain (variations in e_1, e_2, θ' and ω) at all points in a heterogeneously strained material so that the body is coherent (deformation without discontinuities or holes).
Stylolites	Pressure solution localized along surfaces which show a characteristic saw-tooth profile, and an interdigitating cone-like form in three dimensions. Generally found in limestones and quartzites.

KEY REFERENCES

Beach, A. (1974). The measurement and significance of displacement on Laxfordian shear zones, North-West Scotland. *Proc. Geol. Ass.* **85**, 13–21.

This paper sets out the results of measurement techniques in shear zones in a Precambrian orogenic zone, and shows how a regional strain pattern can be resolved into horizontal and vertical displacement components.

Escher, A., Escher, J. C. and Waterson, J. (1975). The reorientation of the Kangâmiut dike swarm, West Greenland. *Can. J. Earth Sci.* **12**, 158–173.

This beautifully illustrated paper describes the development of shear zone formation at the Nagssugtoqidian orogenic front over an area of some 15 000 square kilometres. It applies some of the methods of strain integration described in Session 3 to the solution of regional displacements at this orogenic front.

Ramsay, J. G. (1980). Shear zone geometry: a review. *J. Struct. Geol.* **2**, 83–89.

This review describes the various types of brittle, brittle-ductile and ductile shear zones found in naturally deformed rocks and gives good photographs of many different types of shear zone. It summarizes the geometric features seen in simple shear zones, and extends the investigations into situations of shearing with volume loss and volume gain and to problems involving other types of strain combinations. Equations are presented for the calculation of displacements in all types of shear zones, including those, investigated in Session 3 under Questions 3.6★ and 3.7★.

Ramsay, J. G. and Graham, R. H. (1970). Strain variation in shear belts. *Can. J. Earth Sci.* **7**, 786–813.

This paper was the first to discuss some of the geological implications of strain compatibility in heterogeneously strained rocks. The first part gives a mathematical analysis and this might be found to be rather tough going at this stage in our Session. The conclusions on the possible strain states in shear zones (p. 798) and the sections which follow on methods of computing strain and displacement are directly relevant to Session 3.

Schwerdtner, W. M. (1977). Geometric interpretation of regional strain analysis. *Tectonophysics* **39**, 515–531.

This paper discusses some of the problems of heterogeneous strain and provides some interesting visual models on the problems of fitting together, in a way which satisfies strain compatibility, the elemental units in a heterogeneously deformed rock mass.

SESSION 4

Displacement Vector Fields and Strain

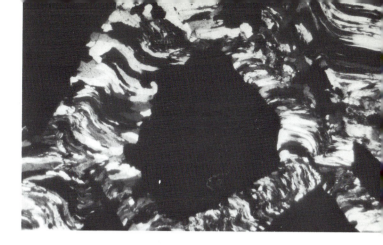

The geometric properties of different types of displacement and resulting displacement vector fields are examined. The differences between absolute-, relative- and local-displacement vectors are defined and their relationship to problems arising in geology discussed. The four components of the general linear transformation equations are related to the strain matrix and the four components of homogeneous strain. Superposition of strains is shown to be a non-commutative process. The concept of two-dimensional strain ellipse fields is established, and the characteristic structural features of competent layers in such deformation fields are described.

INTRODUCTION

In the previous three sessions we have concentrated primarily on the deformation and internal rotation characteristics of a special type of displacement, simple shear. The aim of this session is to widen our view by studying more general types of displacement and the various strains that arise from them.

In two dimensions we have seen how it is possible to represent the movements of points in a body by means of two equations, the coordinate transformations. There are many different types of coordinate transformations which lead to a variety of patterns of **displacement vectors** joining the original and final positions of certain points in the body. We will now investigate the main types of vector patterns, or, as they are known technically, the different types of **displacement vector fields**.

QUESTIONS

Question 4.1

Figure 4.1 illustrates a number of different types of displacement. The thin-line, orthogonal grid gives the initial location of the points in the body, and the thick-line grid represents these points after displacement. In each of the six examples, A to F, draw the vectors connecting the original and final positions of the 25 original grid intersections. These are known as **displacement vector fields**. Describe these vector fields. Which of the fields have parallel vectors throughout? Which of the fields lead to internal straining in the body, and which lead to no internal distortion? Of those fields which lead to deformation differentiate the ones leading to homogeneous strain from those which develop a heterogeneous strain. Write the

coordinate transformation equations relating the final positions (x', y') to initial positions (x, y) for each of the displacement vector fields.

Before proceeding to Question 4.2, check the results of Question 4.1 in the Answers and Comments section.

Question 4.2

Because the concept of homogeneous strain is of such great importance in analytical problems in structural geology we should be quite clear what the geometrical effects of a general linear coordinate transformation exactly are.

Construct an orthogonal grid of lines with initial positions given by $x = -1.0, -0.8, -0.6, -0.4, -0.2, 0.0, 0.2, 0.4, 0.6, 0.8$ and 1.0 and $y = -1.0, -0.8, -0.6, -0.4, -0.2, 0.0, 0.2, 0.4, 0.6, 0.8$ and 1.0. Transform these grid lines according to the coordinate transformation equations given by

$$x' = x + 0.6y$$
$$y' = 0.4x + 1.5y$$

On the original orthogonal grid construct a circle with centre $(0, 0)$ and unit radius. Determine the positions in the deformed grid of points which lay on the initial circle. What is the form of the displaced circle? The form is that of a perfect ellipse, the **finite strain ellipse**.

A. Measure the two principal strains e_1 and e_2.
B. Calculate the ellipticity of the strain ellipse $R = (1 + e_1)/(1 + e_2)$.
C. Measure the orientations θ' of the axes of the ellipse (NB the positive and negative signs of the angles. Angles measured anticlockwise from the +ve x-axis are positive, those clockwise are negative).

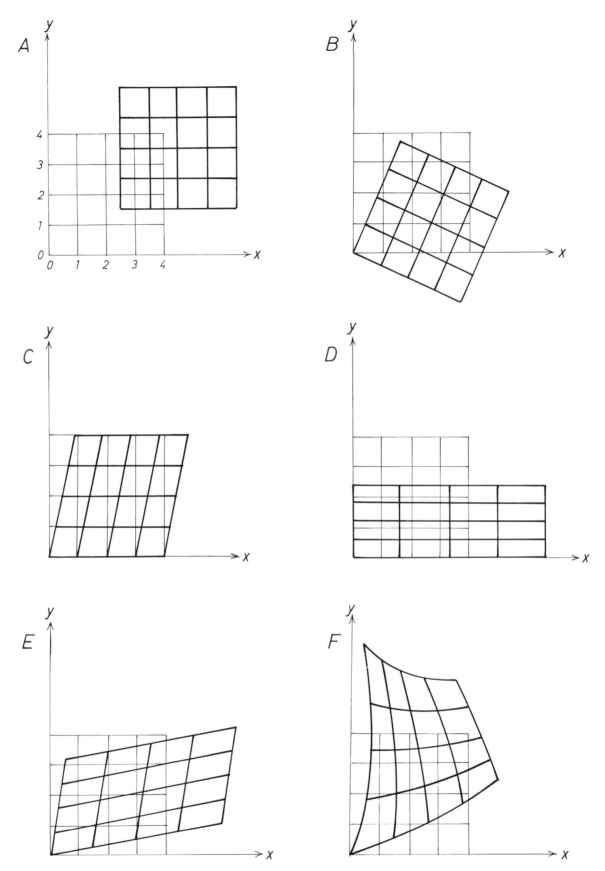

Figure 4.1. Six different types of displacement. Determine the different displacement vector fields. See Question 4.1.

D. By reconstructing the positions of these lines before displacement, measure the initial orientations θ of the lines which become the strain ellipse axes. Check that the lines are initially perpendicular.

E. Determine the internal rotational component of the deformation $\omega = \theta' - \theta$ (NB sign).

F. Determine the area change (Δ_A) of the deformation by comparing the area of an initial square and the area of the subsequent parallelogram.

G. Using the general linear coordinate transformation equations

$$x' = ax + by$$
$$y' = cx + dy$$

in their Eulerian form

$$x = \frac{dx' - by'}{ad - bc}$$

$$y = \frac{-cx' + ay'}{ad - bc}$$

transform points initially (x, y) on the unit circle to new positions (x', y') on the strain ellipse. Determine the general equation for this strain ellipse, and the particular equation for our graphical investigation. Transform points on the straight line and parabola given respectively by

$$y = 2x + 1$$
$$y = x^2$$

using the general coordinate transformation, and find particular numerical solutions for our graphical constructions.

Question 4.3

Using the Eulerian form of the coordinate transformation equations, determine the general form of a deformed ellipse centred at the origin given by

$$Ax^2 - 2Bxy + Cy^2 = 1$$

If this ellipse was the finite strain ellipse of a previous homogeneous deformation, and the transformation represents a displacement producing a second finite strain what can be deduced about the nature of the strain produced by superimposed homogeneous deformations?

Question 4.4

Transform the special ellipse given by

$$(a^2 + c^2)x^2 + 2(ab + cd)xy + (b^2 + d^2)y^2 = 1$$

using the coordinate transformation equations. Simplify the resulting expression, and determine the form of the deformed ellipse. Discuss the geological significance of the results.

Question 4.5

Deform the circle with unit radius and centre $(0, 0)$ by successive transformations:

A. simple shear (value γ) parallel to the x-axis, followed by

B. pure shear (stretch value k parallel to x, shortening value $1/k$ parallel to y) using the transformations

A. $x' = x + \gamma y$ B. $x' = kx$

   $y' = y$ $y' = \dfrac{y}{k}$

Note that both of these transformations really have four components. The deformation matrices are given by

A. $\begin{bmatrix} 1 & \gamma \\ 0 & 1 \end{bmatrix}$ B. $\begin{bmatrix} k & 0 \\ 0 & \dfrac{1}{k} \end{bmatrix}$

Reverse the order of deformations (i.e. B first followed by A second). Is the finite strain ellipse of this second result the same as the first? What effect does the order of coordinate transformations have on the final deformation product. Determine the total deformation matrices for A followed by B, and for B followed by A.

Now go to the Answers and Comments section (pp. 58–66). Then proceed to the starred questions below, or to Session 5.

STARRED (★) QUESTIONS

Question 4.6★

Two periodic functions given by

$$y = \sin x$$
$$y = \sin x + k$$

(where k is a constant term) represent the form of the lower and upper boundary of a folded rock layer (Figure 4.2). This fold is known as a **similar fold** in that the spacing between the layers measured parallel to the fold axial planes (parallel to the y-axis) is constant.

Deform the layer by a homogeneous pure shear using the coordinates transformation

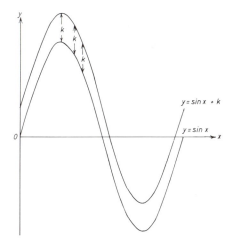

Figure 4.2. *Geometry of a similar fold: the distance k between the layers is constant. See Question 4.6★.*

$$x' = \frac{x}{a}$$

$$y' = ay$$

and compute the new positions of the folded layers. How is the geometry of the folded layer modified? Is the fold still a *similar fold*?

Determine how the fold geometry is modified by a homogeneous simple shear (value γ) parallel to the x-axis using the coordinate transformation equations

$$x' = x + \gamma y$$

$$y' = y$$

Express the equations of the boundary surfaces in their simplest form and discuss their significance. Prove that the folds remain similar—this is not too easy, but remember the definition of a similar fold is that the distance measured parallel to the fold axial planes remains constant.

Question 4.7★

What is the significance of the strain matrices given by the following?

A. $\begin{bmatrix} 1 & 0 \\ 0 & 1 \end{bmatrix}$

B. $\begin{bmatrix} -1 & 0 \\ 0 & 1 \end{bmatrix}$

C. $\begin{bmatrix} -1 & 0 \\ 0 & -1 \end{bmatrix}$

Can these represent situations likely to arise in structural geology?

Question 4.8★

The angle α between any two straight lines with slopes of m_1 and m_2 is given by

$$\tan \alpha = \frac{m_1 - m_2}{1 + m_1 m_2}$$

Consider the deformation represented by the matrix

$\begin{bmatrix} \cos \beta & -\sin \beta \\ \sin \beta & \cos \beta \end{bmatrix}$

What are the effects of this transformation on the straight lines through the origin given by $y = m_1 x$ and $y = m_2 x$? Give the equations of the deformed lines. What is the angle between the deformed lines? What can therefore be deduced about the nature of this transformation?

ANSWERS AND COMMENTS

Answer 4.1

Figure 4.3 shows the displacement vector fields for the six types of displacement shown in Figure 4.1.

A. **Body translation:** The displacement vectors are all parallel and of constant length throughout the body.

Such a displacement plan moves the body in space without internal deformation or rotation. The coordinate transformation equations are

$$x' = x + A$$

$$y' = y + B$$

where A and B are constants defining the components u (parallel to x) and v (parallel to y) of the uniform displacement vectors.

B. **Body rotation:** The displacement vectors vary in direction and length with initial coordinate position. The effect here is of a rotation around the origin $(0, 0)$ without internal deformation, through an angle $-\omega$ (the negative sign arises because the rotation has a clockwise sense). The coordinate transformation for this rotation through an angle $-\omega$ is

$$x' = \cos \omega x + \sin \omega y$$

$$y' = -\sin \omega x + \cos \omega y$$

The proof of this is set out in Appendix C, section 2.

C. **Simple shear:** This displacement is the one that we have investigated in the previous three Sessions. The displacement vectors are all parallel, but of unequal length. The transformation shown here leads to a homogeneous strain because the shear gradient is constant across the body. The strain is rotational. If lines originally parallel to y-axis are deflected through an angle ψ, the coordinate transformation equations are

$$x' = x + y \tan \psi$$

$$y' = y$$

D. **Pure shear:** This displacement has led to a homogeneous shortening parallel to the y-axis, and a homogeneous stretching parallel to the x-axis. The deformation set up is rather simple, but the displacement vector field is quite complex, the vectors varying in length and orientation through the body. The strain produced is homogeneous and irrotational (the strain ellipse long axis is parallel to x; the short axis is parallel to y. These have the same orientation at the original directions of these lines). In the example illustrated here since there has been no change of area the strain is known as a pure shear. The coordinate transformation equations are

$$x' = kx$$

$$y' = \frac{y}{k}$$

where k is a constant term (in the case of Figure 4.1D, k is 1·65).

E. **General homogeneous rotational strain:** This displacement vector field is quite complex, the vectors varying in length and in direction with initial position. If, however, one selects any parallelogram element and removes the vector component related to the lower left corner from each of the other three corners, then the vector differences for the three corners will be found to be identical to that for the parallelogram

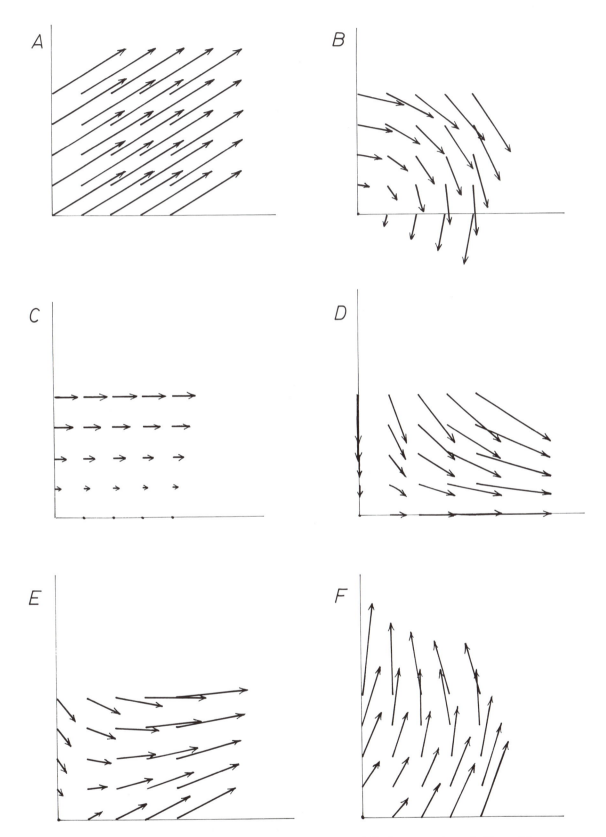

Figure 4.3. Displacement vector fields of the six types of displacement shown in Figure 4.1.

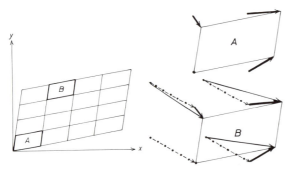

Figure 4.4. *Analysis of the vector components of regions* A *and* B *in the homogeneous deformation of Figures 4.1 E and 4.3 E. The total displacement vectors of the corners of parallelogram* A *are indicated in thick continuous lines. Those of parallelogram* B *are given by thin continuous lines. Each of the total displacement vectors of parallelogram* B *can be resolved into a constant body translation vector (dash–dot line) and a variable differential vector identical to that of the corners of parallelogram* A.

whose lower left corner lies at the origin (Figure 4.4). This implies that the **displacement vector gradients** are constant through the body. Although the total vector field looks complex, the shape changes induced by this displacement field are simple. Each originally square-shaped element is transformed into a parallelogram. Each of the parallelograms arising from the deformation of an identically shaped original element is identically shaped and identically oriented. The deformation set up is therefore homogeneous. Generally the directions of the principal axes of the strain differ from those of the originally perpendicular lines which become the strain ellipse axes. The strain is generally rotational. The coordinate transformations are given by

$$x' = ax + by$$

$$y' = cx + dy$$

where *a*, *b*, *c* and *d* are constants. The geometric significance of these four coefficients is shown in Figure 4.5. A thorough understanding of the properties of this transformation is basic to the understanding of deformation geometry. One especially important feature of this displacement vector field is that any *initial set of parallel straight lines* changes its orientation and spacing distance, but *remains straight*

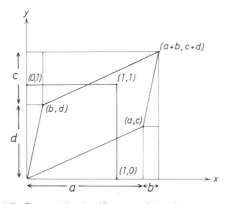

Figure 4.5. *Geometric significance of the four components* a, b, c *and* d, *of the coordinate transformation equations.*

and parallel. Such a transformation is therefore not appropriate for describing the overall straining in folded stratified rocks, although it is appropriate for the analysis of small elements in a fold.

F. **General heterogeneous strain:** The displacement vector field is complicated, and mathematically it is more complex than is the general homogeneous strain transformation. If we were to analyse the displacement vector gradient in the way that we analysed it for general homogeneous strain we would find that this gradient varies from point to point in the body. The originally square-shaped elements are transformed into shapes which are not parallelograms, and each element has a different shape after displacement. Generally, any *initial set of straight parallel lines* before deformation *becomes curved*, and clearly this is the type of displacement which occurs in geological situations of folding. There are two ways of studying the geometrical properties of the distortions set up in such a situation. The first concept is one that we have already briefly encountered in Sessions 2 and 3; it is to focus attention on small elements in the body. The smaller the field we investigate, the closer the local distortions approach those of the homogeneous strain model for that element. A heterogeneously strained body can be envisaged as an agglomeration of a large number of elemental units with approximately homogeneous strain. The strain values, orientations of the principal axes and rotations vary from element to element. The second way of tackling the problem is to examine mathematically how strains at a point relate to the general coordinate transformation equations, a method we will investigate in Volume 3.

The coordinate transformation equations for situations such as that illustrated in Figures 4.1F and 4.3F, where there are no discontinuities, must clearly be smooth single-valued non-linear functions of position *x* and *y*. In general

$$x' = f_1(x, y)$$

$$y' = f_2(x, y)$$

For example, non-linear functions given by

$$x' = x$$

$$y' = k \sin x + y$$

would set up periodic sinusoidal folds in lines parallel to the *x*-axis. These folds have amplitudes *k* and wavelength 2π, and are of the type experimentally produced in our shear box in the previous session and illustrated in Figure 3.2C.

The "movement direction" concept

In the literature of structural geology one often encounters a concept of a general movement direction to express the overall direction of transport of a rock mass during orogenic activity. Those geologists following the work of Bruno Sander often employ the term "*a*-direction" for this movement direction, a terminology derived from coordinates applicable to simple shear displacement such as we produced in our experiments with card deck models. The direction *a* is defined as the slip direction, *ab* is the slip plane (with *b* perpendicular to *a*) and *c* is normal to *ab*. Clearly such a concept is only applicable to simple shear displacements. Because deformations arising from tectonic processes are not generally those of simple shear, it seems that the movement, or *a*-direction, concept is unlikely to be of use to geologists wishing to clarify their description of tectonic processes.

Types of reference schemes to measure displacement

It is always important in science to be aware of the nature and limitations of the measurements we make. This is especially critical when considering displacement. Geologists sometimes make claims about displacements that are impossible to substantiate. For example, you may read that an actively moving African continental plate advanced towards a stable European plate causing the Tethean sediments to be pushed northwards as the Alpine fold belt. The terms active, stable and direction of transport are generally impossible to determine, because all depend upon a knowledge of the absolute positions of parts of the crust before the movements took place.

The determination of the **absolute displacement vectors** depends upon a knowledge of the initial and final position of a point relative to a fixed coordinate frame with unmoved origin (Figure 4.6). We have no such fixed anchors from which we may chart geological displacements and thus we can never measure **absolute displacements**. In our study of the earth's crust we have "moving anchors"; it is only possible to say how one point has been displaced relative to another. There are two main types of "moving anchors" we may use, both are of practical application to geological problems. The first depends on relating the displacement of one point relative to some other point acting as a moving origin for a new coordinate frame. We can define a **relative displacement vector** of point P' (components u and v, Figure 4.6) such that if the origin has moved through an absolute displacement with vector components u_0 and v_0, and the point P' through an absolute displacement with coordinates u_{abs} and v_{abs}, then

$$u + u_0 = u_{abs}$$

$$v + v_0 = v_{abs}$$

Thus we relate all displacements relative to that of some conveniently chosen point in the body in the same way that

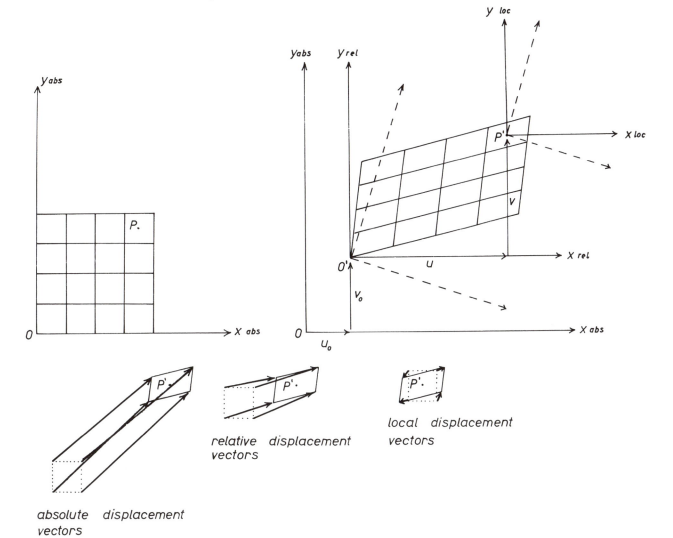

absolute displacement vectors

relative displacement vectors

local displacement vectors

Figure 4.6. *Different types of coordinate systems referring to different types of displacement vectors measurements: absolute-, relative- and local-displacement vectors.*

positions of longitude on the earth's surface are fixed relative to the arbitrarily selected zero meridian passing through Greenwich, UK. Using this scheme, it is entirely a matter of relativity whether Africa has moved towards the European plate (origin chosen in Europe) or Europe has moved towards Africa (fixed origin in Africa). Closely connected with this concept is that of the displacement difference vector between any two points. This is defined by the differences in vector components (either absolute or relative) of any two points and provides a measure of the well-known geological ideas of **crustal shortening** (difference in u components) and **relative uplift** (difference in v components). The second "moving anchor" makes use of the displacements which have taken place around any single point relative to that point. If an observer were stationed within the rock mass and carried along with the deformed body, he would observe quite different displacements of the material surrounding him from those observed by an outsider at some fixed position (Figure 4.6B). In order to evaluate these **local displacement vectors** it is necessary to consider the vector field around the point P' by erecting a new coordinate frame with its origin at P' (x_{loc} and y_{loc}) or by relating these vectors to the location of point P' given in terms of coordinates defined by the axes x_{rel} and y_{rel}. These local displacement vectors around the point P' are related to the **vector gradients**. If u and v are changing through the body (and they generally are unless the example is body translation of the type shown in Figure 7.4) then both u and v can vary both in the x_{loc} and y_{loc} directions. This concept of local displacements is extremely important to the geologist and it is closely connected with the concept of strain at a point. For example, a fossil embedded in a rock mass will often have its shape changed as a result of the local displacement. By measuring these changes it is possible to evaluate the strain at the point, but not the absolute displacement it has undergone.

A further complication arising from the concepts of absolute and relative positions of the displaced rock mass concerns rotational effects. We have seen that since it is not possible to know the initial position of any point it is impossible to determine the origin of the initial coordinate frame. It is generally impossible to fix the initial *directions* of the x_{abs} and y_{abs} axes, except in situations where we have palaeomagnetic data. It is generally very difficult or impossible to determine the correct orientation of the deformed mass of O' relative to its initial position, and therefore the directions of the x_{rel} and y_{rel} (and x_{loc} and y_{loc}) axes are chosen in any direction that is convenient (Figure 4.6, dashed lines). Once chosen, however, they enable the orientation of the deformed material at any point P' to be related to that at O'.

Three important conclusions can be summarized:

1. In geology, it is impossible to measure the *absolute displacement* of any point. This means that it is neither possible to differentiate between "active" and "passive" elements in the crust, nor to evaluate absolute body translation or absolute body rotation.
2. It is possible to determine the *relative displacements* of any two points. It is therefore possible to compute values for crystal shortening and differential uplift

between two points, and to evaluate the differences in body translation and body rotation between the two points.
3. It is possible to measure the *changes (gradients) in displacement vectors* in a local region around a point, and so to define the state of strain around that point.

Answer 4.2

The graphical solution to the problem is illustrated in Figure 4.7. The circle is transformed into a strain ellipse.

A. The lengths of the semi-axes of the strain ellipse are $1 + e_1 = 1\cdot8$ and $1 + e_2 = 0\cdot7$ giving values of the principal extensions $e_1 = +0\cdot8$ and $e_2 = -0\cdot3$.

B. The ellipticity of the strain ellipse $R = (1 + e_1)/(1 + e_2) = 2\cdot6$.

C. The orientations of the axes of the strain ellipse after deformation are $\theta' = 56°$ and $-34°$.

D. The orientations of the axes of the strain ellipse before deformation are $\theta = 61°$ and $-29°$. These lines are perpendicular.

E. The rotational component of the deformation $\omega = \theta' - \theta = -5°$. The minus sign shows that the rotational sense is clockwise.

F. The area of the parallelogram shown in Figure 4.8 of a general displacement is

$$(a + b)(c + d) - 2cb - 2\left(\frac{bd}{2} + \frac{ac}{2}\right) = ad - bc$$

The original area of the rectangle joining $(0,0)$, $(0,1)$, $(1,1)$ and $(1,0)$ was $1\cdot0$, and the proportional change in area Δ_A is

$$\Delta_A = \frac{(ad - bc) - 1}{1}$$

or

$$1 + \Delta_A = ad - bc$$

Note that this is the *determinant of the strain matrix*. In our example

$$1 + \Delta_A = 1\cdot26$$

and the area increases by 26%.

G. Points (x, y) on the initial unit circle given by

$$x^2 + y^2 = 1$$

are transformed to

$$\left(\frac{dx' - by'}{ad - bc}\right)^2 + \left(\frac{-cx' + ay'}{ad - bc}\right)^2 = 1$$

which is an ellipse centred at $(0, 0)$ given by

$$(c^2 + d^2)x'^2 - 2(ac + bd)x'y' + (a^2 + b^2)y'^2 = (ad - bc)^2$$

The ellipse of Figure 4.7 is

$$1\cdot52\,x'^2 + 1\cdot63\,x'y' + 0\cdot86\,y'^2 = 1$$

The straight line $y = 2x + 1$ is transformed to

$$\frac{-cx' + ay'}{ad - bc} = 2\left(\frac{dx' - by'}{ad - bc}\right) + 1$$

That is another straight line given by

$$y' = \frac{(2d + c)x'}{a + 2b} + \frac{ad - bc}{a + 2b}$$

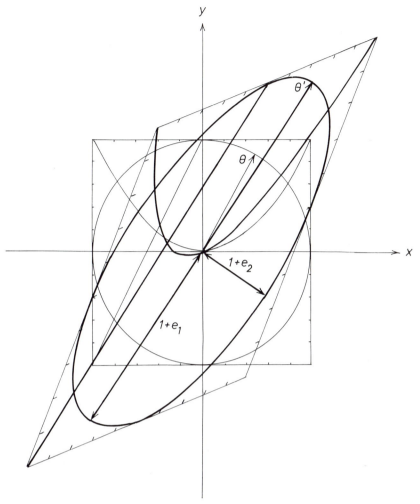

Figure 4.7. *Graphical solution to Question 4.2.*

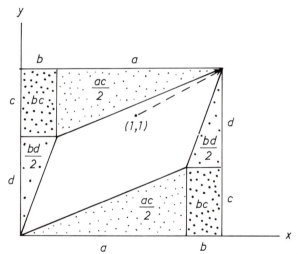

Figure 4.8. *Relationships of the components a, b, c and d of the coordinate transformation equations (and strain matrix) and the area change Δ_A during deformation.*

or

$$y' = 1 \cdot 55x' + 0 \cdot 57$$

The parabola given by $y = x^2$ transforms to

$$\frac{-cx' + ay'}{ad - bc} = \frac{(dx' - by')^2}{(ad - bc)^2}$$

or

$$\frac{b^2}{ad - bc}y'^2 - \frac{2bdx'y'}{ad - bc} - ay' + \frac{d^2}{ad - bc}x'^2 + cx' = 0$$

a quadratic equation in x' and in y'. This means that for specific chosen values of x' there will be two real solutions for y'. In the translation of Figure 4.7 the simplest form of the solution is given by

$$y' = 1 \cdot 43x' + 1 \pm (-0 \cdot 04x'^2 + 2 \cdot 40x' + 1)^{\frac{1}{2}}/0 \cdot 58$$

The general coordinate transformation and strain

The four linear coefficients of the coordinate transformation equations define a matrix known as the **strain matrix**.

$$\begin{bmatrix} a & b \\ c & d \end{bmatrix}$$

and this matrix sets up distortions and rotations in the body. Because there are always four components to this matrix there are always four characteristic features of the strain state: two are the principal strains e_1 and e_2, the third is the orientation of the strain ellipse axes (θ') and the fourth is the orientation of these lines before deformation (θ) related to the internal rotation component of the strain

$(\omega = \theta' - \theta)$. These four features can always be expressed mathematically as functions of the four components of the deformation matrix. They are:

$$(1 + e_1)^2 = \tfrac{1}{2}(a^2 + b^2 + c^2 + d^2 +$$
$$((a^2 + b^2 + c^2 + d^2)^2 - 4(ad - bc)^2)^{1/2}) \quad [B.19]$$

$$(1 + e_2)^2 = \tfrac{1}{2}(a^2 + b^2 + c^2 + d^2 -$$
$$((a^2 + b^2 + c^2 + d^2)^2 - 4(ad - bc)^2)^{1/2}) \quad [B.19]$$

$$\tan 2\theta' = \frac{2(ac + bd)}{a^2 + b^2 - c^2 - d^2} \quad [B.14]$$

$$\tan 2\theta = \frac{2(ab + cd)}{a^2 - b^2 + c^2 - d^2} \quad [B.15]$$

$$\tan \omega = \frac{c - b}{a + d} \quad [B.16]$$

The proofs of these important basic equations are set out in Appendix B and the equation numbers refer to the relevant section in this Appendix. Using these equations we can check numerically the results of the graphical solutions to Questions 4.2 with $a = 1 \cdot 0$, $b = 0 \cdot 6$, $c = 0 \cdot 4$, $d = 1 \cdot 5$, we find

$$1 + e_1 = 1 \cdot 81$$
$$1 + e_2 = 0 \cdot 70$$
$$\theta = -29 \cdot 43° \text{ and } +60 \cdot 57°$$
$$\theta' = -34 \cdot 00° \text{ and } +56 \cdot 00°$$
$$\omega = -4 \cdot 57°$$

It should be clear that we can determine the components of the deformation matrix if we know the four characteristic features of the strain. These equations are set out in Appendix C. There is much mathematical material in Appendices B and C, probably too much to take in in one Session. However, we must stress the importance of this material. It forms the basis of many of the following Sessions and is of key importance to those wishing to develop new techniques and new approaches.

Answer 4.3

The ellipse $Ax^2 - 2Bxy + Cy^2 = 1$ is transformed to

$$A\left(\frac{dx' - by'}{ad - bc}\right)^2 - 2B\frac{(dx' - by')(-cx' + ay')}{(ad - bc)^2}$$
$$+ C\left(\frac{-cx' + ay'}{ad - bc}\right)^2 = 1$$

or

$$(Ad^2 + 2Bcd + Cc^2)x'^2$$
$$-2(Abd + B(cb + ad) + Cac)x'y'$$
$$+ (Ab^2 + 2Bab + Ca^2)y'^2 = (ad - bc)^2$$

which is in the form of an ellipse centred at the origin

i.e. $Px'^2 - 2Qx'y' + Ry'^2 = 1$

(P, Q and R are constants).

It follows that two superposed homogeneous deformations combine to give a single homogeneous finite strain which may be analysed using the strain ellipse concept.

Answer 4.4

Transforming the ellipse given in this question and simplifying the coefficients we find that the coefficients of x'^2, $x'y'$ and y'^2 become 1, 0 and 1 respectively and the ellipse transforms to

$$x'^2 + y'^2 = 1$$

the equation of a circle of unit radius. The given ellipse which transforms to this circle is known as the **reciprocal strain ellipse**. It has axial directions parallel to the lines in the undeformed state which eventually become the axes of the finite strain ellipse, and the values of these axes are the reciprocal values of those of the principal strains.

It is therefore possible for two finite strains to be superposed such that, if the first deformation is the reciprocal strain ellipse of the second deformation, the resulting strain becomes zero. Although such a possibility might seem rather unlikely (and it probably is a rather rare occurrence) certain situations do occur whereby a state of no strain can arise in this way. We will see later, in our studies of strain variations in folded strata that so-called **neutral strains** occur in certain positions in and around folded layers. Many of these **neutral strain surfaces** and **neutral strain points** arise by the build up of strain during the early part of the fold development which is subsequently removed during the later stages of folding.

A second implication of the reciprocal strain ellipse concept occurs when an aggregate of initially variably oriented elliptical objects such as conglomerate pebbles or oolites are subjected to deformation. It is not uncommon for certain of these objects to be initially shaped and oriented like the reciprocal strain ellipse and, in their deformed condition, to take on a circular form.

Answer 4.5

The unit circle

$$x^2 + y^2 = 1$$

is transformed by simple shear to

$$(x' - \gamma y')^2 + y'^2 = 1$$

i.e. $x'^2 - 2\gamma x'y' + y'^2(1 + \gamma) = 1$

This strain ellipse is transformed by pure shear into

$$\left(\frac{x''}{k}\right)^2 - 2\gamma\frac{x''}{k}ky'' + k^2y''^2(1 + \gamma) = 1$$

or

$$\frac{x''^2}{k^2} - 2\gamma x''y'' + k^2y''^2(1 + \gamma) = 1$$

If we reverse the order of the deformations the circle is first transformed into the ellipse

$$\left(\frac{x'}{k}\right)^2 + (ky')^2 = 1$$

and this is transformed by simple shear to

$$\frac{(x'' - \gamma y'')^2}{k^2} + (ky'')^2 = 1$$

or

$$\frac{x''^2}{k^2} - \frac{2\gamma x'y''}{k^2} + y''^2\left(k^2 + \frac{\gamma}{k^2}\right) = 1$$

These strain ellipses are different. Mathematically we

say that *strains are not commutative* (in comparison to the commutability of number multiplications where, for example, x multiplied by y is identical to y multiplied by x). Those used to matrix multiplication will be aware that matrix products must always be handled with care, and that the order of multiplication is of extreme importance.

The deformation matrix for simple shear followed by pure shear is derived by relating the final position of a point (x'', y'') to the original position (x, y) using two coordinate transformations

$$x' = x + \gamma y \quad x'' = kx'$$
$$y' = y \qquad y'' = \frac{y'}{k}$$

giving

$$x'' = kx + k\gamma y$$
$$y'' = \frac{y}{k}$$

In matrix multiplication notation we can write

$$\begin{bmatrix} k & 0 \\ 0 & \dfrac{1}{k} \end{bmatrix} \times \begin{bmatrix} 1 & \gamma \\ 0 & 1 \end{bmatrix} = \begin{bmatrix} k & k\gamma \\ 0 & \dfrac{1}{k} \end{bmatrix}$$

pure shear × simple shear
(second) (first)

If we reverse the order then

$$\begin{bmatrix} 1 & \gamma \\ 0 & 1 \end{bmatrix} \times \begin{bmatrix} k & 0 \\ 0 & \dfrac{1}{k} \end{bmatrix} = \begin{bmatrix} k & \dfrac{\gamma}{k} \\ 0 & \dfrac{1}{k} \end{bmatrix}$$

simple shear × pure shear
(second) (first)

Since the deformation matrices are different, the strains are not identical (Figure 4.9). The important general rules

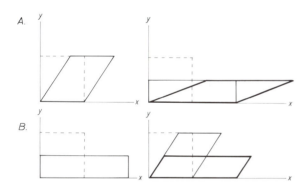

Figure 4.9. A: *Simple shear followed by pure shear.* B: *Pure shear followed by simple shear. The final parallelogram shapes are different.*

for matrix products from which the effects of superposing any two coordinate transformations or strains are set out in Appendix C.

Strain ellipse fields and their geological significance

Strain ellipse shapes can be simply and conveniently classified by making a graphical plot of the major semi-axis of length $1 + e_1$ as abscissa, and the minor semi-axis $1 + e_2$ as ordinate (Figure 4.10). By definition $e_1 \geqslant e_2$, so it follows that only the part of the positive quadrant lying between the $1 + e_1$ axis and a line at 45° to it through $(0, 0)$ can be

occupied by points representing strain ellipses. The point $(1, 1)$ represents a unit circle, and all other circles (with positive or negative dilatation) lie along the 45° line through $(0, 0)$. Ellipses lie in three fields in the diagram. **Field 1** contains those ellipses where both e_1 and e_2 are positive, so that these strain ellipses lie completely outside the unit circle from which they were derived. They clearly have a marked area increase or positive dilatation, Δ_A. **Field 2** ellipses have e_1 positive and e_2 negative, so they intersect the unit circle in two lines symmetrically related to the principal strain axes. These lines represent directions of **no finite longitudinal strain**, and their orientations (ϕ) before and after (ϕ') deformation are characterized by the values of e_1 and e_2 (see Appendix D, Equations D.9 and D.10). Field 2 ellipses can show either positive or negative dilatation (see Figure 4.10). **Field 3** ellipses have both e_1 and e_2 negative so that they lie completely inside the unit circle. They show large negative dilatations.

All three fields of strain ellipses can be produced during geological deformations. Field 1 ellipses generally occur where there has been a strong contraction sub-normal to the ellipse, and develop with a flattening or "pancake-like" three-dimensional strain. In contrast Field 3 ellipses generally develop where there has been a large stretching sub-normal to the ellipse and they are indicative of constrictive or "cigar-like" three-dimensional strains. Field 2 ellipses can be found in practically all types of three-dimensional strain. We will discuss these relationships in more detail in Session 11.

In layered rock sequences, where the layers have differing competence, stretching and contraction lead to the development of boudinage and buckle folding respectively. Competent layer surfaces will show one of the three types of strain ellipse and various combinations of interrelated boudins and folds will therefore develop. Each strain ellipse field has its own characteristic combination of boudins and folds:

Field 1: All directions in the surface are stretched, therefore all directions could potentially develop boudin necks. A Field 1 ellipse can form during a single deformation of flattening type, or it can develop as a result of the superposition of two or more phases of non-coaxial deformations. Both possibilities lead to the formation of rather complex boudinage, with crossing boudin neck zones. The result is the formation of a structure known as **chocolate tablet boudinage**. The origin of this name will be clear from the geometry shown in Figure 4.11A and B. It should be emphasized that rarely are the boudin necks as regularly developed or as mutually perpendicular as seen in a block of good Swiss chocolate! If the strain arises during a single deformation, the boudin necks will be generally rather irregular, although they will occur more frequently and show a wider separation where they are sub-perpendicular to the maximum extension direction (B). Chocolate tablet boudinage arising from the superposition of different deformations or different phases within one overall deformation often shows periodic or sequential development, and the boudins often intersect at angles other than 90° (Figure 4.12A). We will discuss later (Session 13) how the special geometric features of such boudinage enable us to unravel a complex deformation sequence.

Field 2: These ellipses show a dominant shortening in

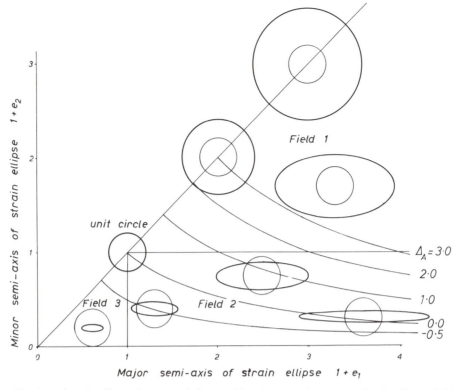

Figure 4.10. Classification of strain ellipse shapes made by graphing $1 + e_1$ against $1 + e_2$. Strain ellipses fall into one of three main Fields. Lines of equal area change Δ_A are shown.

one sector and a dominant stretching in another. Competent layers involved in such a strain field will therefore show sub-perpendicular developments of boudins and buckle folds (Figures 4.11C and 4.12B).

Field 3: All directions in the surface show contraction; therefore all potentially have perpendicularly oriented fold axes. The geometry of the structures which develop depends on whether the deformation is a single constriction or the result of the superposition of contractions of different phases. If the deformation is one phase, then the folds are often highly irregular in shape like the form of a table cloth crumpled from all sides (Figures 4.11E and 4.13B). The fold axes and axial planes will generally have many differing orientations, and the intersections and mergings of the folds will be unsystematic. In contrast, where the finite strain is built up by the superposition of two or more separate contractions, the folds generally show a more systematic interference geometry with characteristic interference patterns (Figures 4.11D and 4.13A). The fold axis orientations will be variable, but this variation is ordered. Similarly, one will be able to identify sets of differently oriented fold axial surfaces: those of the earliest folds having a folded form, those of the latest folds being more regular in orientation. Further discussion of these geometric features is best left until we have examined fold geometry in more detail in later sessions.

Answer 4.6★

The boundary of the folded layer with original form $y = \sin x$ is transformed by pure shear into

$$\frac{y'}{a} = \sin ax' \quad \text{or} \quad y' = a \sin ax'$$

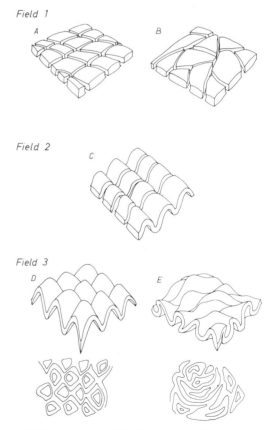

Figure 4.11. Structures arising in competent layers as a result of deformations in differing strain ellipse fields. A and B show chocolate tablet structure, C shows the folds and boudins developed in Field 2, and D and E show dome–basin folds developed in two-phase and one-phase deformation.

A

Figure 4.12. A: *Chocolate tablet structure developed in a competent layer of dolomite enclosed in a less competent limestone, Modane, W. Alps.* B: *Folds and sub-perpendicular extension boudinage in a competent limestone layer. Ilfracombe, N. Devon, UK.*

B

A

Figure 4.13. A: *Dome and basin folds arising from a Field 3 type strain ellipse developed during two phases of deformation. Loch Monar, Inverness, Scotland; cf. Figure 4.11D. B: Irregular dome and basin folds in deformed pegmatite dykes. Chindamora Batholith, Zimbabwe; cf. Figure 4.11E.*

B

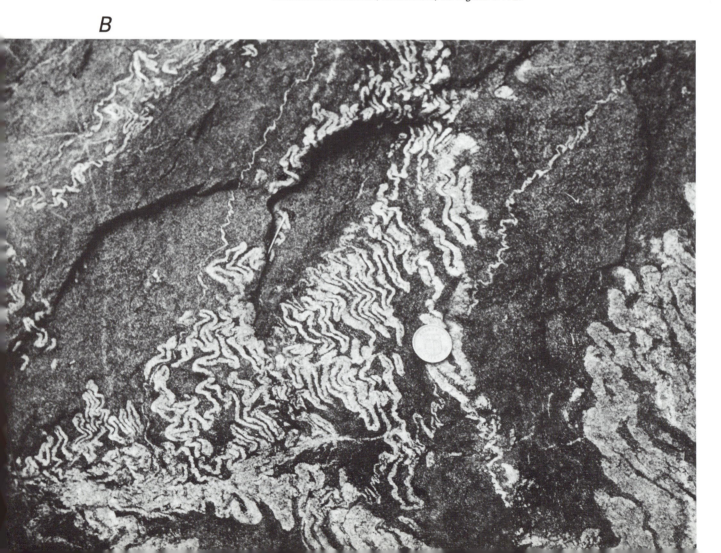

that with an original form

$$y = \sin x + k \quad \text{to}$$

$$y' = a \sin ax' + ak$$

The form of these two boundary layers is identical, and they are separated only by a constant term ak. The folded layer therefore remains similar, although its wavelength has been decreased and its amplitude increased.

The two boundary layers are displaced by a homogeneous simple shear into the forms given by

$$y' = \sin (x' - \gamma y')$$

$$y' = \sin (x' - \gamma y') + k$$

whereas $y = \sin x$ has a single value of y for a chosen value of x, the form of these equations is not so limited. For example, expanding the second we have

$$y' = \sin x' \cos \gamma y' - \cos x' \sin \gamma y' + k$$

For a specific value of x we could write this

$$y' + P \cos \gamma y' + Q \cos \gamma y' = k$$

where P and Q are constants.

This function can have either one, two or three solutions depending on the choice of x' and the subsequent values of P and Q. This is clearly seen when we look at the shape of the sheared original fold (Figure 4.14, Solutions A_1, B_1B_2, $C_1C_2C_3$).

To prove that the fold remains similar after simple shear we must prove that the distance between the boundaries *when measured parallel to the axial surface* remains constant. One convenient way of doing this is to rotate the coordinate frame so that it coincides with the axial surface orientation and then see what the new spacing of the curves parallel to the new ordinate direction is.

If the axial surfaces of the folds have been sheared through an angle ϕ, then it follows from the geometry of Figure 4.13 that

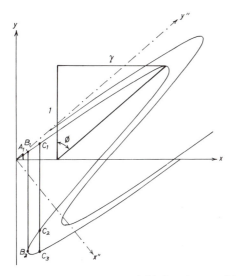

Figure 4.14. *Sheared similar fold. See Answer 4.6★.*

$$\sin \phi = \frac{\gamma}{(1 + \gamma^2)^{1/2}}$$

$$\cos \phi = \frac{\gamma}{(1 + \gamma^2)^{1/2}}$$

We now rotate the folds through an angle of $+\phi$ using the body rotation coordinate transformation given by

$$x'' = x' \cos \phi - y' \sin \phi$$

$$y'' = x \sin \phi + y \cos \phi$$

or

$$x' = x'' \cos \phi + y'' \sin \phi = \frac{x'' + \gamma y''}{(1 + \gamma^2)^{1/2}}$$

$$y' = -x'' \sin \phi + y'' \cos \phi = \frac{-\gamma x'' + y''}{(1 + \gamma^2)^{1/2}}$$

Replacing the x' and y' terms in the two equations representing the sheared folds and simplifying, we obtain

$$y'' = \gamma x'' + (1 + \gamma^2)^{1/2} \sin(x''(1 + \gamma^2)^{1/2}) + k(1 + \gamma^2)^{1/2}$$

and

$$y'' = \gamma x'' + (1 + \gamma^2)^{1/2} \sin(x''(1 + \gamma^2)^{1/2})$$

These two curves are identical, except for a constant term, and it therefore follows that the folded layer retains a similar fold form.

Answer 4.7★

A. The matrix is known as the *unit matrix*. It relates only to a possible body translation without rotation or internal deformation.

B. This matrix changes the position of points only according to their value of x. The negative sign change produces a reflection about the y-axis. It has no geological significance for no displacement processes could lead to a mirror image effect.

C. This matrix changes the sign of both x and y coordinates. The body is therefore transposed into the opposite quadrant through a body rotation of 180°. It is a possible geological transformation.

Answer 4.8★

The straight lines $y = m_1x$ and $y = m_2x$ are transposed into

$$y' = \left(\frac{m_1 \cos \beta + \sin \beta}{\cos \beta - m_1 \sin \beta}\right) x'$$

$$y' = \left(\frac{m_2 \cos \beta + \sin \beta}{\cos \beta - m_2 \sin \beta}\right) x'$$

Replacing the two new line slopes into the equation for $\tan \alpha'$, we find (after much simplification)

$$\tan \alpha' = \frac{m_1 - m_2}{1 + m_1m_2}$$

This implies that the angles between any two lines are unchanged by this transformation, the matrix must therefore represent a body rotation without internal distortion.

KEYWORDS AND DEFINITIONS

Body rotation	The body moves as a rigid mass by rotation about some fixed point.
Body translation	All points in the body have identical displacement vectors. The body moves without rotation or internal distortion.
Chocolate tablet structure	A type of boudinage where the competent mass is stretched in all directions and produces a complex of crossing boudin necks (Figure 4.11A and B).
Displacement vector or absolute displacement vector	The line joining the initial point $P(x, y)$ with its final position $P'(x', y')$. The vector has two components defined as $u_{abs} = x' - x$ and $v_{abs} = y' - y$ parallel to the x- and y-coordinate axes respectively.
Displacement vector field	The mathematical description of all the variously oriented and various valued vectors in a displaced body (Figure 4.3). If the coordinate transformation equations are given by:

$$x' = f_1(x, y)$$
$$y' = f_2(x, y)$$

the vector components defining the displacement vector field are given by:

$$u = f_1(x, y) - x$$
$$v = f_2(x, y) - y$$

Displacement vector gradient	A mathematical expression which describes how the vectors (or vector components) change their values from point to point. Both u and v defining the displacement vector can change through the body in directions x and y, and so the displacement vector gradient can be expressed in matrix form by

$$\begin{bmatrix} \dfrac{\partial u}{\partial x} & \dfrac{\partial v}{\partial x} \\ \dfrac{\partial u}{\partial y} & \dfrac{\partial v}{\partial y} \end{bmatrix}$$

If all four terms of this matrix are constant and independent of x and y the displacement vector gradient is constant and the displacement vector field leads to a state of homogeneous strain, homogeneous body rotation or body translation. If the terms of the matrix are not constant, heterogeneous strains result.

Local displacement vector	The displacement vectors of points with reference to a local origin situated at the point P' (Figure 4.6).
Movement- or a-direction	A concept only applicable to the displacement vector fields of simple shear and body translation. The displacement vectors are parallel, and this direction defines the unique movement direction of the displacement.
Pure shear	An irrotational strain where the area dilatation Δ_A is zero.
Reciprocal strain ellipse	The ellipse before deformation which transforms into a circle with unit radius after deformation.
Relative displacement vector	The displacement of a point P' relative to that of some conveniently chosen reference point O' in the body (Figure 4.6). If the absolute displacement of the reference point O' has vector components u_0 and v_0, the components of the relative displacement vector u and v are given by:

$$u = u_{abs} - u_0$$
$$v = v_{abs} - v_0$$

Similar fold A fold where the distance between the layer boundaries, measured in a direction parallel to the axial surface, remains constant (Figure 4.2: k).

Strain ellipse fields The three types of strain ellipses defined according to the values of the principal longitudinal strains e_1 and e_2: **Field 1,** $e_1 > e_2 > 0$; **Field 2,** $e_1 > 0 > e_2$; **Field 3,** $0 > e_1 > e_2$ (Figure 4.10).

Strain matrix A two by two matrix which can be used to define the state of strain at a point in a body. If the coordinate transformation equations are linear and given by:

$$x' = ax + by$$

$$y' = cx + dy$$

the strain matrix is:

$$\begin{bmatrix} a & b \\ c & d \end{bmatrix}$$

It can be related to the displacement vector gradient matrix (see definition above) by:

$$\begin{bmatrix} a & b \\ c & d \end{bmatrix} = \begin{bmatrix} 1 + \dfrac{\partial u}{\partial x} & \dfrac{\partial u}{\partial y} \\ \dfrac{\partial v}{\partial x} & 1 + \dfrac{\partial v}{\partial y} \end{bmatrix}$$

KEY REFERENCES

Hobbs, B. E., Means, W. D. and Williams, P. F. (1976). "An Outline of Structural Geology." Wiley International, New York.

A well illustrated, simple but rigorous account of displacement and strain can be found in pp. 18–31.

Ramsay, J. G. (1969). The measurement of strain and displacement in orogenic belts. *In* "Time and Place in Orogeny" (P. E. Kent *et al.*, eds) 43–79. Special Publication 3. Geological Society, London.

A discussion of the effects of displacement in terms of body translation, body rotation and strain is presented and the different types of displacement vectors are related to the geologically useful concepts of crustal shortening and differential uplift.

Ramsay, J. G. (1976). Displacement and strain. *Phil. Trans. R. Soc. Lond.* A **283**, 3–25.

This gives an account of strain and a general review of strain measurement techniques relevant to geological problems. It would make a good recapitulation of the material covered so far in this present volume and would provide an introduction to the subsequent sections on strain measurement.

SESSION 5

Practical Strain Measurement:

1. Initially Circular and Elliptical Markers

This session describes methods of strain determination using objects that had initially circular, sub-circular, sub-elliptical or elliptical shapes. The basic technique for initially elliptical objects employs measurements of the elliptical shape R_f and orientation ϕ' of the deformed objects and the construction of an R_f/ϕ' data graph. Variations of ϕ' determine the fluctuation which is also shown to be a useful parameter to check results. It is shown how rapid methods of strain determination can be made using arithmetic, geometric and harmonic data means. The effects of a preferred orientation of the initial particles is discussed.

INTRODUCTION

This is the first of a series of Sessions devoted to strain measurement, in two, and then later in three dimensions. In analysing homogeneous two-dimensional strain we have already shown that an initially circular object is transformed into a perfect ellipse, the strain ellipse. The first approach to strain measurement is to seek initially circular objects (or spherical objects in three dimensions) from which we should be able to measure directly the form and orientation of the strain ellipse. Like the coordinate transformation equations, the strain state has four components: principal strains e_1 and e_2, orientation of the strain ellipse axes θ' and rotation component ω. In practice not all of these components can generally be determined. Although we can often measure directly the strain ellipse derived from an initial circle, we generally do not know the initial size of that circle. This means that we are not always in a position to measure e_1 and e_2, but only the ellipticity or aspect ratio as defined by the ratio $R = (1 + e_1)/(1 + e_2)$. The determination of the ellipse orientation is generally straightforward, but to determine the rotation we need to know the initial orientations of the strain ellipse axes. In most practical forms of strain analysis in two dimensions we therefore have to be content with the determination of only two of the four strain parameters. We will see later that, in special circumstances, it is sometimes possible to determine e_1 and e_2, but we have to acknowledge that these circumstances are encountered rarely.

The second part of this session discusses how we measure strain when the initial objects have an elliptical (or approximately elliptical) initial form. This is a very common problem arising in strain determination because markers of truly circular form are rather rare in nature. We have already seen in Question 4.3 that when we deform an ellipse we obtain another ellipse, but clearly this new ellipse combines components of the initial shape with the strain

ellipse. The main problem which will concern us is how we separate these two components.

Strain determination using initially circular objects

Question 5.1

Figure 5.1 illustrates an outcrop with elliptically shaped, deformed objects. The elliptical shapes arise from the deformation of initially circular cross sections of cylindrical worm borings (the trace fossil *Scolithus*) found in quartzites of Cambrian age from northwest Scotland. In undeformed quartzites the axes of the cylindrical tubes are always perpendicular to the bedding plane, and therefore the tube sections on the bedding surface must have been circles, but circles of unknown radius. After deformation the cylindrical tubes become transformed into cylinders with elliptical cross sections, and generally the axes of these cylinders are no longer perpendicular to the bedding planes (Figure 5.2). The elliptical shapes on the bedding surface are derived from initially circular cross sections and therefore do have the same form and orientation as the strain ellipse. Because we do not know the size of the original tube radius, we cannot compute the principal strain values e_1 and e_2 directly, but we can establish the ellipticity R and orientation θ' of the strain ellipse.

For each ellipse, measure the length of the short axis $k(1 + e_2)$ and long axis $k(1 + e_1)$ and plot these graphically as abscissa and ordinate respectively (k is a constant of unknown value depending on the initial radius of the circular cross section). The data points should lie on a straight line passing through $(0, 0)$ and the slope of this line gives the ellipticity R. Any deviations from this line should record only imperfections in measurement. Determine the arithmetic mean of the abscissa and ordinate

Figure 5.1. *Bedding plane surface of Cambrian Quartzite showing the cross-sectional forms of deformed worm tubes* Scolithus. *Loch Eribol, Scotland. See Question 5.1.*

Figure 5.2. *Section of Figure 5.1 perpendicular to the bedding planes (which run from upper left to lower right) showing the variable inclinations of the* Scolithus *tube axes to the bedding surfaces resulting from heterogeneous strain. Originally this angle was 90°.*

data—the quotient should give a close approximation of the average ellipticity. Determine the mean of the orientation values of the ellipse's long axes: this will give the best value of θ'. Those who understand regression methods for determining "best fits" to data might statistically process the data to see if the improved statistical average justifies the extra effort involved.

Is the form of the elliptical section of the worm tubes normal to the axis of the deformed tube similar to that of a strain ellipse?

Strain determination using initially elliptical objects

Question 5.2

When an initial ellipse with ellipticity R_i is homogeneously deformed, the resultant form is also elliptical. The shape of this final ellipse R_f is a function of four factors: the initial shape (the *form* and *orientation* of the initial ellipse) *and the form and orientation* of the strain ellipse. Figure 5.3A illustrates a surface containing a number of constantly shaped ($R_i = 2 \cdot 0$) initial ellipses, oriented at different angles ϕ to an initial marker direction x ($\phi = 90°$, 75°, 60°, 45°, 30°, 15°, 0°). The graph shows a plot of constant R_i against ϕ for a range of values of ϕ. Figure 5.3B shows what happens to the form and orientation of these initial ellipses when we impose a homogeneous strain with strain ellipse ratio $R_s = 1 \cdot 5$. The shapes of the initially constant ellipses are changed depending on their orientations with respect to those of the axes of the strain ellipse, and the orientations of the axes of the new ellipses (ϕ') also change (except with those with initial orientations $\phi = 90°$, 0°), so that the long axes of the combined ellipses come to lie

closer to the direction of the long axis of the strain ellipse. The initial ellipse which had its long axis parallel to the maximum elongation of the strain ellipse ($\phi = 0°$) takes on a new form which is *more elliptical than that of the strain ellipse*, whereas the ellipse which had its long axis parallel to the minimum elongation of the strain ($\phi = 90°$) becomes *less elliptical than the strain ellipse*. The ellipticities R_f of all the other elliptical markers lie between these two extreme values, but most have a higher ellipticity than that of the strain ellipse. This implies that if we were to make an average of the ellipticities by finding the arithmetic mean $(R_{f_1} + R_{f_2} + R_{f_3} \ldots + R_{f_n})/n$, this mean value would be considerably higher than the true ellipticity of the strain ellipse. It will be clear that we can never determine the tectonic strain by making a simple arithmetic average of the data.

Figure 5.3C shows the same ellipses as 5.3A, but subjected to a strong deformation, with a strain ellipse shape greater than the ellipticity of the initial elliptical form. Many of the overall features of the deformed ellipses are similar to those of Figure 5.3B. However, the range of orientation of the particle ellipse long axes is much more restricted. The range of orientation of long axes is known as the **fluctuation F**. The initial fluctuation before deformation was 180°. Where the tectonic strain ellipse has a lower ellipticity than the original elliptical objects, the fluctuation after deformation is also 180°. Where the ellipticity of the strain ellipse equals the ellipticity of the initial objects, an interesting geometrical effect occurs. The initial marker ellipses that lie with their long axes coinciding with the greatest tectonic shortening take on a circular form, because they have the property of lying in the direction of the reciprocal strain ellipse of the tectonic system (see

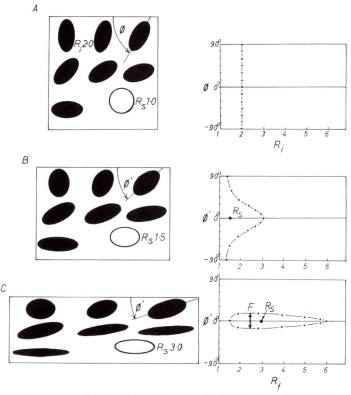

Figure 5.3. Effect of deformation on a series of elliptical objects with initial orientation ϕ and ellipticity R_i. After deformation (ellipticity R_s of strain ellipse) the original marker ellipses change shape (ellipticity R_f) and orientation (ϕ'). F is the fluctuation.

Answers and Comments for Question 4.4). All other ellipses come to lie with long axis orientation in the range $45° > \phi' > -45°$, and the fluctuation becomes 90°. As deformation becomes stronger than the initial elliptical form of the objects (as in Figure 5.3C) the fluctuation decreases to less than 90° and the R_f/ϕ' plot becomes closed.

R_f/ϕ' plots are the key to the analysis of the geometry of deformed elliptical markers, and provide an excellent way of separating the components of the tectonic strain from the initial shapes of the markers. In general we will find that elliptical markers usually have a range of ellipticities as well as a range of orientations. The R_f/ϕ' plot for each group of ellipses with a fixed initial shape gives a single curve, and as we vary the initial shape so we obtain data that lie on other related R_f/ϕ' curves. The total R_f/ϕ' data, therefore, form a data field. Examples of R_f/ϕ' curves for different initial values R_i of the elliptical objects and different tectonic strains R_s are illustrated in Figure 5.4. These curves were constructed using a computer. The method was as follows: for a given initial shape ratio and ellipse orientation, the coordinate transformation matrix was determined which could have produced this shape from an initial circle (Appendix C.4, Equation C.17). The strain matrix was then computed (Appendix C.4, Equation C.18), and the two matrices multiplied together (Appendix

C.6, Equation C.22). The ratio of the principal lengths and directions of the ellipse axes were then calculated (Appendix B.8, Equation B.14, Appendix B.11, Equation B.20). By successively varying the orientation of the initial ellipses a number of R_f/ϕ' points were calculated and linked to form a continuous curve.

There are various ways of using the R_f/ϕ' field and curves to separate initial shape from tectonic strain. Generally, one tries to find the curve which forms the best fitting envelope to the R_f/ϕ' data points and which relates to initial ellipses with high R_i values. The curves are always symmetric about the direction of the long axis of the strain ellipse, and on each curve the points derived from an initially randomly oriented group of ellipses tend to concentrate towards high R_f values (see the locations of the line of 50% data in Figure 5.4).

Two principal situations can arise:

A. *Maximum R_i is greater than the strain R_s*: The data points show a 180° fluctuation in ϕ' and will be bounded on the right-hand side of the graph by a curve like that shown in Figure 5.5A. The data will concentrate around the maximum R_f value. The direction of this maximum concentration of data points gives the orientation of the strain ellipse long axis. The distribution of data points on either side of this direction should be symmetric. If it is not

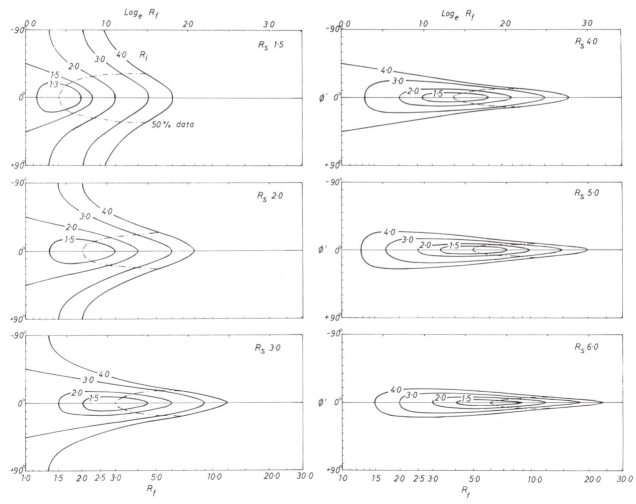

Figure 5.4. Standard R_f/ϕ' reference curves for different values of initial ellipticity R_i and strain ellipse R_s.

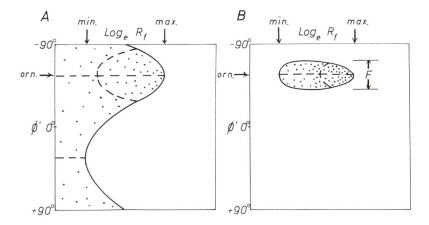

Figure 5.5. *Principal features of R_f/ϕ' plots used for computing the strain R_s. In A, where $R_i > R_s$ the data envelope is symmetric about the orientation of the long axis of the strain ellipse (orn.) and shows maximum and minimum R_f values. In B, where $R_s > R_i$, the data envelope is closed and the data points show a limited range of orientations defining the fluctuation F.*

symmetric, then the initial distribution of ellipse orientations was not random. The maximum and minimum values of the R_f/ϕ' envelope are determined. These values have special mathematical relationships. From the geometry shown in Figures 5.3B and 5.5A the maximum and minimum ellipticities of R_f relate to the product and quotient respectively of the two component ellipticities.

$$R_f\,\text{maximum} = R_s R_i\,\text{maximum}$$

$$R_f\,\text{minimum} = R_i\,\text{maximum}/R_s$$

Cross multiplying or cross dividing these values we find

$$(R_f\,\text{maximum}\;R_f\,\text{minimum})^{1/2} = R_i\,\text{maximum}$$

$$(R_f\,\text{maximum}/R_f\,\text{minimum})^{1/2} = R_s$$

An alternative technique is to overlay the data plot with the R_f/ϕ curves of Figure 5.4 to find which curve gives the best fit envelope. This curve then has characteristic R_i and R_s values.

B. *Maximum R_i is less than the strain R_s:* This situation is recognized by a fluctuation of data points of less than 90° and by the very restricted nature of the R_f/ϕ' field. The maximum orientation frequency coincides with the orientation of the long axis of the strain ellipse, and the distribution of data points should be symmetric about this direction. The maximum and minimum R_f values of this field again have special characteristics (Figures 5.3C and 5.5B).

$$R_f\,\text{maximum} = R_s R_i\,\text{maximum}$$

$$R_f\,\text{minimum} = R_s/R_i\,\text{maximum}$$

Cross multiplying or cross dividing these functions we obtain:

$$(R_f\,\text{maximum}\;R_f\,\text{minimum})^{1/2} = R_s$$

$$(R_f\,\text{maximum}/R_f\,\text{minimum})^{1/2} = R_i\,\text{maximum}$$

As with the analysis (A) above, we can use a best fit

R_f/ϕ' envelope curve to discover the original shape and tectonic strain components. We can also use the value of the **Fluctuation F** to assist the calculation. The fluctuation F is a function only of R_i and R_s given by

$$F = \tan^{-1}\frac{R_s(R_i^2 - 1)}{((R_i^2 R_s^2 - 1)(R_s^2 - R_i^2))^{1/2}} \qquad (5.1)$$

The mathematical proof can be found in Ramsay (1967, p. 207). This function is plotted in Figure 5.6, and from these curves possible pairs of values of R_s and R_i can be found to check out the other methods described above.

The main problem in this analysis is to find the best fit envelope curve of the data. Natural data often show quite a considerable spread on account of initial variations in the shapes of the original elliptical markers. Another likely problem concerns R_f/ϕ' plots arising from situations where the original ellipses were not randomly oriented. Such situations occur where there was a statistically significant orientation in the initial fabric (for example, where the particles showed an orientation due to current activity). Preferred orientations may be statistically related to bedding surfaces, or the particles may show an "imbrication" of these long axes across the bedding planes. These initial fabrics lead to a final R_f/ϕ' distribution that shows asymmetric data point locations. The maximum and minimum R_f values are then not always so simple to analyse in terms of the component R_s and R_i as we have discussed above. The fluctuation may also be considerably smaller than that expected for the particular R_s and R_i values of the system. There are methods to analyse these more complex data and we will return to these problems in Book 3.

Figure 5.7 shows a thin section of a deformed ironstone oolite from a formation of Ordovician age from North Wales, and our practical work will consist of making a strain analysis of this section. From a first inspection of this rock, how do we know that the ooids had an initial shape that was not circular (i.e. not spherical in three dimensions)? Can you, by inspection, decide if the tectonic strain ellipse has a higher or lower ellipticity than did the initial, sub-elliptical shapes of the ooid sections?

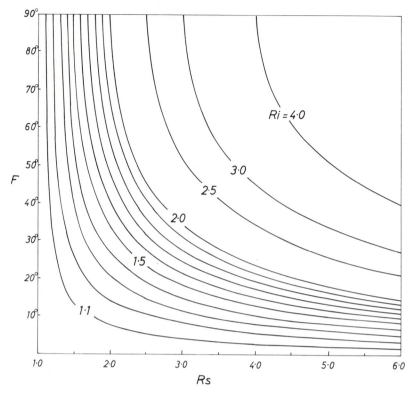

Figure 5.6. *Graphs of fluctuation F for differing values of initial object ellipticity R_i and ellipticity of the strain ellipse R_s.*

A marker reference line has been drawn on Figure 5.7. For each ooid determine the orientation of the long axis from this reference line (angle ϕ' measured such that ϕ' + ve is anticlockwise, ϕ' − ve measured clockwise from the reference direction). For each ooid measure the longest and shortest shape axes and determine the ratio or ellipticity R_f. Plot each ellipse as a single point on a R_f/ϕ' graph. Use the same scales as those shown in Figure 5.7.

You will find that this is a time-consuming exercise, but ultimately you will be in a much better position to see what is actually involved in an exact strain analysis of naturally deformed rock material. Fifty data points will probably give a reasonably good result. There are some 280 oolites in the section shown, and a complete analysis of all these is given in the Answers and Comments section to enable you to see how the conclusions from a few data compare with conclusions derived from a large amount of data.

When you have plotted the data points, discuss the significance of their distribution in terms of initial shape and tectonic strain, and determine the orientation and ellipticity of the strain ellipse using as many techniques as possible. Compare the results obtained by different techniques.

Now check your results with the Answers and Comments section below. Further questions are posed below, or you can proceed directly to Session 6.

STARRED (★) QUESTIONS

Rapid methods for determination of the tectonic strain

Question 5.3★

Because of the considerable time required to produce a detailed R_f/ϕ' analysis we should now consider the possibility of analysing the data more rapidly to obtain an answer which, though not perfect, may be good enough for practical strain analysis on a regional scale. Whether one aims at high precision or deliberately employs more data of lower accuracy from more localities depends upon the problem one is trying to investigate. For example, if the problem is to compare very exactly the optical or shape orientations of mineral grains with the total strain of the rock as a whole, one must work to the highest standards of analytical precision. Often, however, in regional problems one needs to find the overall strain pattern, and it is often quite good enough to have each datum point to within 10% (or even 20%) of the true value. It should be emphasized that for many regional studies it is often more rewarding to spend time in the field collecting a lot of data of a relatively low degree of accuracy at many localities, rather than to concentrate on obtaining a few strain data with an extremely high degree of accuracy.

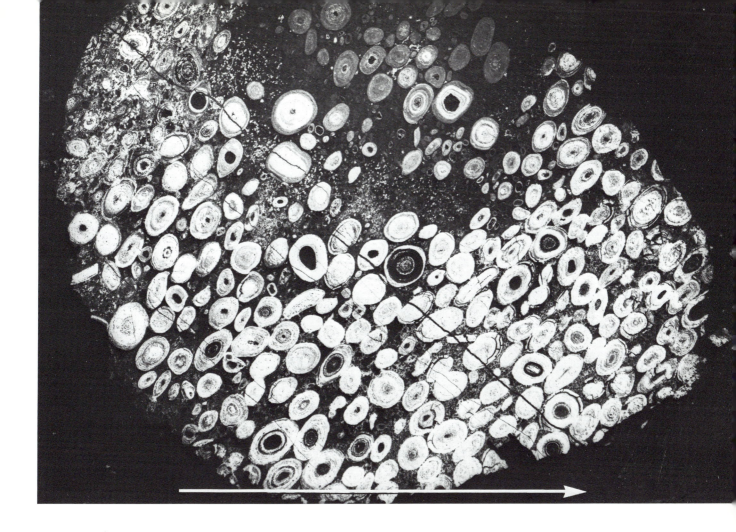

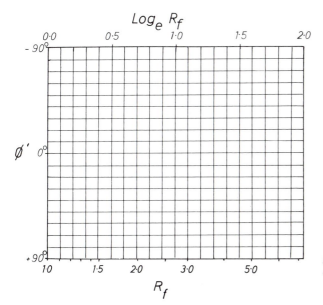

Figure 5.7. Thin section (negative print × 10) of an ironstone oolite, with graph for plotting the recorded data. See Question 5.2.

In our discussion of Figure 5.3 we have already seen that most of the R_f values of deformed, initially elliptical markers are greater than the value of the ellipticity of the finite strain ellipse R_s. The *arithmetic mean of R_f*,

$$\bar{R}_f = \frac{R_{f_1} + R_{f_2} + R_{f_3} \ldots + R_{f_n}}{n}$$

always lies considerably above the true R_s value. Other types of mathematical mean values of R_f and these generally give a better approximation to R_s than the arithmetic mean.

The *geometric mean of R_f data* is given by:

$$G = (R_{f_1} \times R_{f_2} \times R_{f_3} \ldots \times R_{f_n})^{1/n}$$

We have already deduced that where $R_s > R_i$ then the real value of R_s is the geometric mean of the minimum and maximum values of R_f.

The *harmonic mean of R_f data* is given by:

$$H = n/(R_{f_1}^{-1} + R_{f_2}^{-1} + R_{f_3}^{-1} \ldots + R_{f_n}^{-1})$$

There is no logical reason to choose this mean in preference to others, except that the harmonic mean takes a lower value than the arithmetic and geometric means and lies closer to the true value of R_s. Lisle (1977) has examined this problem in detail and his curves showing how the various means differ from the true value of R_s are reproduced in Figure 5.8.

With the R_f data previously acquired in answer to Question 5.2 make arithmetic, geometric and harmonic means of R_f, and compare their values with the R_f/ϕ' results obtained from the previous question.

If you have a programmable calculator, prepare a graph to show how the geometric and harmonic means change as progressively more data are taken in. Can you suggest a minimum number of R_f data values that would give an approximate but useful result for a value of the finite strain R_s? Does the harmonic mean have really significant practical advantages over the geometric mean?

Shapes of original particles

Question 5.4★

Can you suggest a method whereby, when we have established the finite strain ratio R_s, we could calculate the

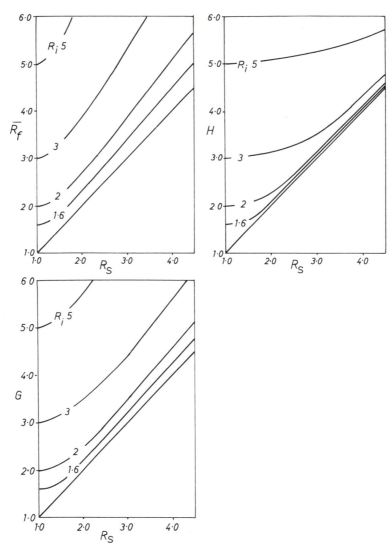

Figure 5.8. *Sets of curves showing how the arithmetic mean $\bar{R}_f$, harmonic mean H and geometric mean G of R_f data give values of R_s which differ from those of the true value. From Lisle (1977).*

undeformed shapes and orientations of the initially elliptical particles?

Question 5.5★

Figure 5.9 shows an outcrop of xenolithic granite from the
Pennine nappe zone of the Swiss Alps. The granite was
intruded during late Hercynian times, and later subjected
to a deformation during the Alpine orogeny. Nearby outcrops show that, where unalpinized, the xenolith long axes
are oriented randomly. Using the harmonic mean of the
shapes of the deformed xenoliths compute a value for the
finite, two-dimensional strain.

ANSWERS AND COMMENTS

Strain determination using initially circular objects

Answer 5.1

Figure 5.10A presents 100 data points derived from the
elliptical sections of the *Scolithus* tubes on the bedding
surface. The points lie about a straight line. Taking an
arithmetic mean of these data and finding the best fit line
(line *a*) passing through the origin $(0, 0)$ we obtain a slope
of 1·6. This value gives the strain ellipse ellipticity R_s. The
orientations of the long axes of all the ellipses are fairly
constant, the main deviations of the shapes from true
ellipses are caused by the interference and overlap of
adjacent tubes. The fact that the ellipses have a rather
constant ellipticity and orientation confirms that these tube
sections were initially circular in shape. The orientation

θ' of the long axes of the ellipses is 3° from the orientation
marker line on Figure 5.1.

A best fit line for the data of Figure 5.10A gives

$$k(1 + e_1) = 1·44\, k(1 + e_2) + 0·18$$

with a coefficient of determination, chi-squared, of 0·73.
This line (line *b*) is located quite close to that obtained
from the arithmetic data mean, but it does not pass through
the origin. This could relate to irregularity in initial shapes
of the sections, or to imperfections in data measurement.

The form of the elliptical section of a *Scolithus* tube
perpendicular to the tube axis is not a true strain ellipse,
because it has been derived from a section that was itself
elliptical (see Figure 5.10B).

Figures 5.1 and 5.2 are from the same locality. Although
the shear strain parallel to the bedding surfaces is variable,
leading to a curvature of the *Scolithus* pipes (Figure 5.2),
the forms of the pipe sections on the bedding surface are
clearly elliptical. It should be clear from these two observations that the deformation here is not a heterogeneous,
simple shear parallel to the bedding planes. Simple shear
parallel to the bedding surfaces would not alter the initial
circular tube sections on the bedding surface. For example,
in our simple card deck models of simple shear no distortion
took place in the plane of the cards.

Strain measurement from initially elliptical objects

Answer 5.2

You will have noted that no firm limit was set on the
amount of data you used to answer this question. An
absolute minimum is about 30, better would be 50, even

Figure 5.9. *Surfaces of deformed granite with sub-elliptical xenoliths, Laghetti, Ticino, Switzerland. See Question 5.5★.*

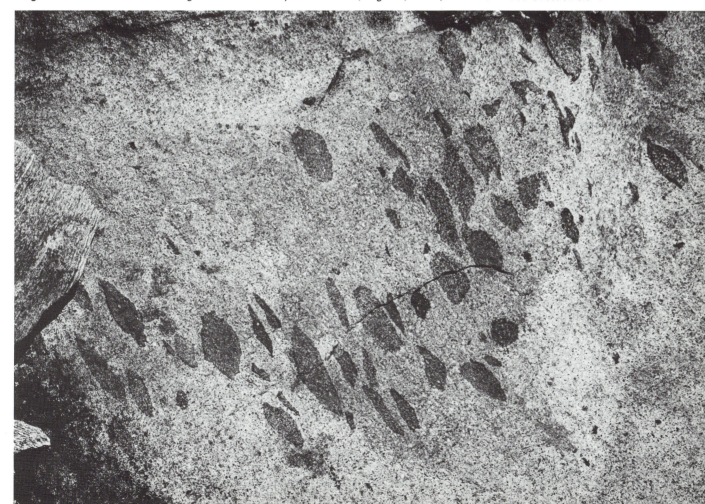

better all 280 possible ooid measurements. However, the data measurement is very time consuming and we always have to consider the problem of whether the result justifies the means. For many investigations the sophistication of an elaborate and highly accurate determination of strain does not justify the amount of work involved. Given a limited time to solve a regional problem, it is generally true that time spent gathering data from more locality points pays off better scientifically than would hours working to obtain very accurate data from fewer locations. One should analyse carefully what is required from the strain data. Does, for example, an R_s value of 2·5 instead of 2·7, or a ϕ value of 65° instead of 67° make a significant difference, especially if the time spent collecting much data from one outcrop precludes data collection from another? In the case of our analysis with a thin section of a deformed oolite the time spent in the field is minimal—collecting and orienting the specimen. However, when we use the R_f/ϕ' technique on larger scale objects (conglomerate pebbles for example) where we have to make measurements directly on the outcrop surface, or make a series of photographic records of the rock surface, we must decide how field time should be spent and critically assess the methods which we employ so as to get maximum effect from our work.

It should be clear from an initial inspection that the ooids vary in shape and that their long axes are variably oriented. The variations in shape and orientation are not obviously systematic across the section; this suggests that the main features of these variations are not likely to be primarily the result of heterogeneous strain through the rock. Although there is a range of long axis orientation, there is a general orientation diagonally across the section. This implies that the tectonic strain ellipse generally has a higher ellipticity than the ellipticity of the initial ooids.

We will now discuss the solution to our particular problem of deformed ooids by accurate R_f/ϕ' technique. Figure 5.11A illustrates the R_f/ϕ' data points from all 280 ooids

of Figure 5.7. If your data sample contains fewer points, it is likely that the odd points in Figure 5.10 indicating a local fluctuation greater than 90° will have been missed. Statistically these are unimportant in defining the strain state; what they do signify is that there were a very few ooids with a marked initial elliptical shape which exceeded that of the strain ellipse.

The average long axis orientation is found by making a frequency diagram of the ϕ' directions at 5° intervals, as shown in Figure 5.11B. The data show a symmetric distribution about a maximum oriented at 42° to the marker reference line. The symmetric nature of this distribution indicates that the initial orientation of the ooid long axes was almost random. If such a frequency diagram shows a marked asymmetry, it is either because not enough data points have been plotted or, if this possibility can be eliminated, it implies that the initial long axis distribution was not random (perhaps the result of some fabric arising during sedimentation or sediment diagenesis).

The next problem is to find the best fit R_f/ϕ' curves which contain and envelope the data points. One has to admit that even with 280 data this is not too easy because there is quite a scatter of data. The problem can be resolved by producing contours of the intensity of data distribution (Figure 5.11C). This contoured map indicates that we have the general conditions $R_s > R_i$, as we deduced by initial inspection. By "eyeballing-in" a best fit general R_f maximum, R_f minimum, and R_f/ϕ' curves we obtain

$$R_s = 1·7$$

maximum $R_i = 1·6$

We now look at the fluctuation. Ignoring the few widely scattered points and concentrating on the contoured data frequency map, the 1·5% contour range gives a fluctuation F of about 60°. From the curves of Figure 5.6, this accords quite well with the numerical deductions given above. It should be clear that the R_f/ϕ' technique is not perfect, and

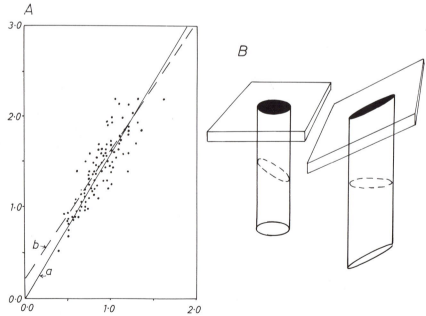

Figure 5.10. A: Data plot from Question 5.1—abscissa and ordinate record lengths of short- and long-axes of elliptical sections of Scolithus tubes. B shows that the shape of a perpendicular section of a deformed tube is not proportional to the shape of the strain ellipse, whereas the ellipse shape on the bedding surface (black ellipse) has the same ellipticity as that of the strain ellipse.

it is not completely easy to arrive at an exact specific value of the tectonic strain. It is, however, one of the best methods available. The principal difficulties in interpreting an R_f/ϕ' diagram arise mainly from the scatter of points, the scatter itself arising from the variable shape of the initial elliptical object population. It should be clear from the R_f/ϕ' curves shown in Figure 5.11D that we have a whole range of initial shapes ranging from circles, to ellipses with R_i of 1·6, and even a few ellipses with initial shape ratios which exceeded that of the tectonic strain ellipse (1·7).

Returning to Figure 5.7 it is interesting to note that there are tectonic extension fissures oriented at an angle of $-38°$ from the marker line. The perpendicular to these extension fissures is therefore oriented at $10°$ from the principal finite tectonic extension. This $10°$ difference in angle could be accounted for in two ways. The discrepancy may have arisen because the extension fissures formed perpendicular to the main extension line in three dimensions, and this line is not necessarily coincident with the maximum two-dimensional extension that we have com-

puted in the section surface. A second possibility is that the fissures may be related to one particular stage of the deformation geometry formed during a particular strain increment. If this is so, their geometry will not be related to the total strain geometry. The geometry of fissures forming as a result of a changing deformation history will be discussed later in Session 13.

Effect of non-random orientation of the initial elliptical markers

The R_f/ϕ' technique assumes that the long axes of the initial elliptical particles are randomly arranged. Where the particles have an initial fabric with a statistically preferred orientation, a simple application of the R_f/ϕ' technique will produce an incorrect answer for the separation of tectonic strain and initial shape. We will look at this problem in more detail so that we are in a position to recognize the presence of an initial fabric and see how we might resolve the problem of separating the components.

If there is a perfect alignment of the initial elliptical

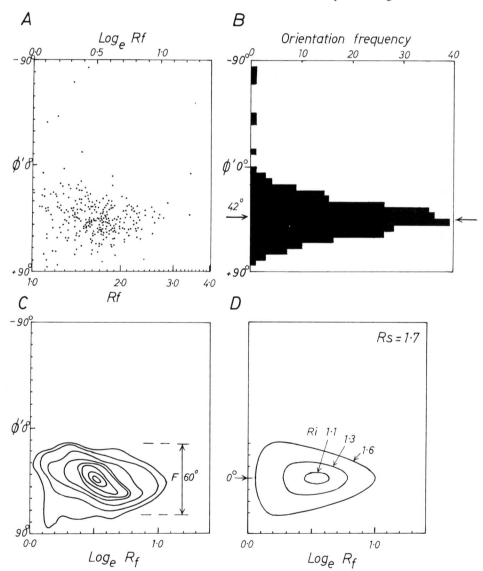

Figure 5.11. A, R_f/ϕ' data points from the deformed oolite of Question 5.2 (Figure 5.7); B, analysis of orientation frequency; C, contoured version of A, contour intervals 1% per area, F is the fluctuation; D, theoretical R_f/ϕ' curves for a strain of $R_s = 1·7$ and for different initial shape values of R_i.

particles and if the particle ellipticities R_i are constant, no solution is possible. The reason is as follows: after a homogeneous strain all the final ellipses will have identical shapes and orientations, and such a geometry is indistinguishable from the deformation of initially circular objects.

The fabric of most rocks with embedded particles is, however, not usually so perfect. There is usually a range of orientation of the particle long axes about a mean value. Figure 5.12A illustrates such an initial fabric before deformation. Twelve elliptical particles (A–L) of differing ellipticities are arranged so that their distribution is symmetrically disposed about a line (which could represent a bedding plane trace) oriented at 45° to the edge of the block. The initial values of ellipticity and the orientation of these ellipses is illustrated as a R_i/ϕ plot in Figure 5.12B. The block was subjected to a homogeneous strain ($R_s = 2 \cdot 0$) with the principal strains aligned parallel to the edges of the block. The resulting ellipse geometry (A'–L') has been calculated mathematically (Figure 5.12C) and represented in an R_f/ϕ' plot in Figure 5.12D. This plot shows a different type of data distribution from that which was developed from the deformation of randomly oriented particles. The data field is no longer symmetric about a particular angular direction, and the distribution of the

points in this field is markedly uneven. Although the initial ellipses were symmetrically oriented about the bedding trace, the data points of the deformed ellipses show an asymmetric grouping across this trace, and show a type of tectonic imbrication. It is no longer valid to take the directional mean of the data field as representative of the principal strain direction. The mean, in fact, lies between the principal extensional strain axis and the bedding trace. It is also not valid to use the maximum and minimum values of R_f to compute the values of the principal strain ratio R_s as can be done with a symmetric R_f/ϕ' diagram.

The problem of separating R_i and R_s is soluble, but the solution is rather complex without a computer. The method requires that the direction of the principal strain be known. The technique consists of "unstraining" the ellipse shapes progressively by small coaxial decrements, and at the end of each unstraining decrement checking if the distribution of the long axes of the elliptical markers is symmetric or asymmetric about the bedding trace. During the unstraining process, the particle long axes gradually become more symmetric, and when the result is perfectly symmetric the unstraining is stopped. When a symmetric distribution has been achieved, the total strain ellipse derived from the decrements gives the reciprocal strain ellipse of the finite

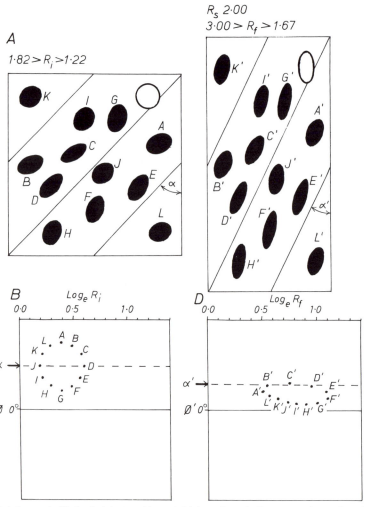

Figure 5.12. Analysis of deformed elliptical objects with an initial preferred alignment about the oblique line in A. R_i/ϕ data plot of the undeformed aggregate shown in B. C shows the deformed block A with a strain ellipse R_s value of 2·0, and D shows an R_f/ϕ' plot of the ellipses in C.

strain. Further details and programs for this technique can be found in a paper by Dunnet and Siddans (1971).

Rapid methods for the determination of tectonic strain

Answer 5.3★

The arithmetic, geometric and harmonic means of the data were analysed by taking the data in groups of 50 R_f values. Table 5.1 shows the results from each of such groups, together with the means for the total data. The values of these means accord with the predictions of Lisle (1977), and for each column

$$\overline{R_f} > G > H$$

The reasons for the variation shown by each group were investigated further by plotting geometric means for the first two data points, then the first three etc. until the mean of the total group of 50 points has been plotted. The results are shown in Figure 5.13. Each curve begins by showing a relatively wide variation, because the differences of each individual datum affects the mean. Then the mean value gradually settles down, so that after about 20 points of each group the mean does not appear to show significant changes. This suggests that, for rapid analysis, the geometric or harmonic mean of 20–30 R_f data measurements gives a strain value to within a few per cent of the correct value. In this problem the differences between the geometric and harmonic means is negligible. For certain problems such a solution, involving much less work than a full R_f/ϕ' analysis, might be considered satisfactory.

The analysis of the various data groups shown in Table 5.1 and Figure 5.12 indicates that the groups show significant variations in their means. When the locations of each group are examined (Figure 5.14), it is seen that those data giving high mean values (Zones 1 and 2) come from regions where the density of ooids is comparatively low.

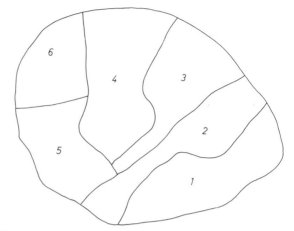

Figure 5.14. Location of Figure 5.7 of the zones used for the data analysis shown in Figure 5.13.

It therefore seems likely either that the proportion of ooid to matrix has influenced the finite strain, or that there is a slight competence contrast between the ooids and matrix material. Examination of Figure 5.7 shows that ooids in contact with each other sometimes show slight mutual indentation, and so the second of these alternatives seems to offer the most likely explanation for the variation in mean value.

Table 5.1

	zone 1 1–50	zone 2 51–100	zone 3 101–150	zone 4 151–200	zone 5 201–250	zone 6 251–280	Total 1–280
Arithmetic mean $\overline{R_f}$	1·76	1·86	1·68	1·55	1·59	1·69	1·67
Geometric mean G	1·72	1·81	1·65	1·51	1·56	1·65	1·64
Harmonic mean H	1·68	1·77	1·62	1·48	1·51	1·60	1·60

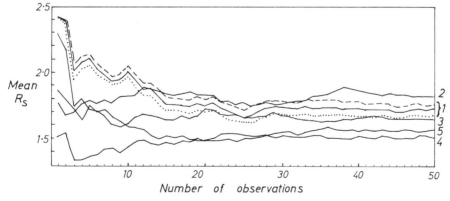

Figure 5.13. Means of tectonic strains calculated from various zones of Figure 5.7. The numbers on the right-hand side of the graph refer to the zones shown in Figure 5.14. Solid lines are geometric means, and dash–dot lines show harmonic and arithmetic means respectively from zone 1.

Shapes of original particles

Answer 5.4★

The removal of tectonic strain from the deformed ellipses requires a computer, particularly if a large number of ellipses needs to be unstrained. The steps in the calculation are as follows:

A. From the final ellipse orientation ϕ' and ellipticity R_f calculate a matrix representing the total shape (see Appendix C, Equation C.16).
B. Knowing the tectonic strain R_s determine the deformation matrix (Equation C.18) and from this determine the reciprocal deformation matrix (reverse the position of the term of the first column first row, with that of the second column second row).
C. Multiply the shape matrix with the reciprocal deformation matrix (Equation C.22).
D. From this matrix product calculate ϕ and R_i of the undeformed particle (Appendix B, Equations B.14 and B.20 respectively).

Answer 5.5★

A harmonic mean of the xenolith shapes gives a value of the finite strain of $R_s = 2 \cdot 6$. The technique of strain measurement in igneous rocks using the shapes of deformed xenoliths is particularly valuable in assessing the significance of the shapes of plutonic bodies. In this example the deformation appears to have arisen by tectonic processes imposed on the rock much later than the original emplacement. In large plutons, however, the emplacement history itself often leads to deformation of previously consolidated parts of the intrusion. Many large batholithic granites show outer zones which appear to have been stretched like an inflating balloon as a result of late stage infilling of the intrusion core.

KEYWORDS AND DEFINITIONS

Fluctuation F   The range of orientation of ellipse long axes in a deformed aggregate of particles with initial (sub-)elliptical shape. The value of F depends upon the initial ellipticity R_i and the ellipticity R_s of the strain ellipse (see Equation 5.1).

KEY REFERENCES

The following articles are recommended as presenting data, ideas and discussions of practical application of the techniques described in this Section.

Cloos, E. (1947). Oolite deformation in the South Mountain Fold, Maryland. *Geol. Soc. Am. Bull.* **58**, 843–918.

This is a classic paper which marked the reinvigoration of structural geology from its between-war doldrums. Its approach to numerical data and the practical problems of deformation in the Appalachian fold belt has played a major role in influencing thought and methodology in strain analysis. Cloos discusses in great detail the practicality and limitations of particle measurement, and the paper presents a large amount of excellent data, with strain variations related to fold profiles and with maps of regional variations.

Dunnet, D. (1969). A technique of finite strain analysis using elliptical particles. *Tectonophysics* **7**, 117–136.

This presents a very detailed analysis of the R_f/ϕ' technique and discusses the practical problems arising in analysing strain in oolitic limestones and clastic sediments using R_f/ϕ' curves as well as arithmetic and geometric data means.

Dunnet, D. and Siddans, A. W. B. (1971). Non random sedimentary fabrics and their modification by strain. *Tectonophysics* **12**, 307–325.

The theory of the deformation of particles with an initial preferred orientation is set out together with practical ways of separating the initial shape and tectonic strain components.

Elliott, D. (1970). Determination of finite strain and initial shape from deformed elliptical objects. *Geol. Soc. Am. Bull.* **81**, 2221–2236.

This introduces a novel graphical method for the solution of strain problems, and discusses a simple method for determining the original shapes and orientations of strained particles.

Lisle, R. J. (1977). Estimation of tectonic strain ratio from the mean shape of deformed elliptical markers. *Geol. Mijnb.* **56**, 140–144.

This paper gives a very useful discussion of the significance of arithmetic, geometric and harmonic data means for rapid strain analysis.

Ramsay, J. G. (1967). "Folding and Fracturing of Rocks", 568 pp. McGraw-Hill, New York.

Pages 185–199 discuss strain analysis of circular and spherical objects, and pp. 202–211 formulate the mathematical background for the geometry of deformed elliptical objects, introducing the R_f/ϕ technique. Using data taken from the work of Cloos, the significance of fluctuation is established and evaluated in mathematical terms. A discussion of the effects of a preferred orientation of the initial particles is given on pp. 216–221.

SESSION 6

Practical Strain Measurement:

2. Lines

Stretched amphibolite dyke, SW Africa

The way the extension e varies with a general two-dimensional displacement is investigated with respect to initial and final positions of a line element. Simple equations relating extensions to positions in the strain ellipse lead to an investigation of how many different extension parameters are required to define the strain ellipse. Methods of solving practical geological problems of strain analysis with one line extensions are developed using the Mohr construction as a simple device for solving the basic strain equations.

INTRODUCTION

In the previous Session we examined the possibility of determining the strain ellipse directly from the measured shapes of initially circular or elliptical markers. In a sense, these initial objects contain a complete spectrum of differently oriented line elements which define the circle or ellipse diameters before deformation. In many deformed rocks we do not have such a complete range of directions of known initial length, but we may find ways in which we can measure extensions in a few different directions. You might see possibilities from our previous discussion of boudinage and ptygmatic folding: for example, we can always measure the final length of a layer and we might be able to measure the initial length by reconstructing in some way its initial geometrical form. This is a good idea, although we will see later that we have to be a little careful in our reconstructions. Our project for this Session is to discover how extension varies with direction in a homogeneously strained material, and how many extensions in different directions must be measured in order to reconstruct the main features of the strain ellipse.

We have already seen from our analysis of length change taking place in simple shear (Sessions 1 and 2) that the extension e along any direction depends on the initial orientation of the line, and the values of the four component terms of the strain matrix. Those readers who have completed Questions 1.6★ and 1.7★ will have arrived at exact mathematical formulae which relate these features in a simple shear system. We now want to establish these relationships in a general case of displacement and a general case of homogeneous strain.

Figure 6.1A illustrates an undeformed unit circle in which the location of a number of diameters (A–R) have been drawn making differing angles ($0°$, $10°$, $20°$ . . . etc.) with the abscissa axis. Figure 6.1B illustrates the resulting geometry after a finite displacement leading to a state of homogeneous strain. The strain matrix is:

$$\begin{bmatrix} 1·2 & 0·2 \\ 0·4 & 0·8 \end{bmatrix}$$

Each line initially located at an angle α to the x-direction is usually deflected to make a new angle α' to the x-axis, and the lines take up new positions A'–R', becoming diameters of the finite strain ellipse. The lines also generally show length changes, becoming either stretched ($e + $ ve) or shortened ($e - $ ve). It should be clear from a study of Figure 6.1 that the changes in angles and lengths are very systematic. Figure 6.2 shows two types of graphical representation of the changes. One shows the extension plotted against the *original orientation* α, the other plots extension against the *final orientation* α'. These two graphs relate to Lagrangian and Eulerian concepts respectively. The first curve (joining A–R) reaches maximum and minimum values situated 90° apart (maximum between C and D, minimum between L and M), and these orientations clearly represent the initially perpendicular directions *which become* the principal axes of the strain ellipse after deformation (semi-axis lengths $1 + e_1$ and $1 + e_2$). The graph of e-variation with α is symmetric about these directions. The second curve (joining A'–R') also reaches maximum and minimum values which have a perpendicular relationship and represent the orientations of the axes of the strain ellipse *after deformation*. The difference between the two sets of mutually perpendicular lines gives the rotational component (ω) of the deformation. The second graph (e-variation with α') is symmetric about the principal strain directions, but it is not the same shape as the first graph. The distribution of the points A', B' etc. shows a clustering around the axis of greatest elongation ($1 + e_1$), and a moving away from the axis of greatest shortening ($1 + e_2$).

It is possible to determine exactly the relations between line extension e and either initial orientation α or final

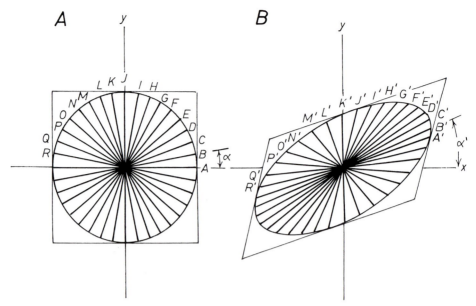

Figure 6.1. *Relationships between orientations of lines and extension in a general linear displacement and homogeneous strain.*

orientation α' in any displacement system leading to a homogeneous strain. We do not wish to interrupt the main flow of this session by developing these equations here. The complete mathematical proofs of these equations are set out in Appendix B.5, B.6 and B.7. With a general linear coordinate transformation with strain matrix

$$\begin{bmatrix} a & b \\ c & d \end{bmatrix}$$

these relationships are

$$(1 + e)^2 = \tfrac{1}{2}(a^2 - b^2 + c^2 - d^2) \cos 2\alpha + (ab + cd) \\ \times \sin 2\alpha + \tfrac{1}{2}(a^2 + b^2 + c^2 + d^2) \quad (6.1)$$

$$(1 + e)^2 = (ad - bc)^2 [\tfrac{1}{2}(d^2 + c^2 - a^2 - b^2) \cos 2\alpha' \\ - (ac + bd) \sin 2\alpha' + \tfrac{1}{2}(a^2 + b^2 + c^2 + d^2)]^{-1} \quad (6.2)$$

$$\tan \alpha' = (c + d \tan \alpha)/(a + b \tan \alpha) \quad (6.3)$$

We might note here how in many of these equations

expressing length change the expression $(1 + e)^2$ appears. This is defined as the **quadratic extension λ** (see Appendix A).

These relationships are of fundamental importance in the theoretical development of strain, but they are not so valuable in the handling of field data. Part of the mathematical complexity arises from the hidden presence of the rotational component of the strain in the equations.

We have previously noted (pp. 21–22) that the rotational component of strain is generally very difficult to measure in naturally deformed rocks. In geological systems we try to calculate the distortional features measured by the strain ellipse principal axes $(1 + e_1, \ 1 + e_2, \ R = (1 + e_1)/(1 + e_2))$ and their orientations. Although it is important to understand fully the geometrical features of a general displacement, such as we recorded in Figures 6.1 and 6.2, it is more important in practical analysis to establish how changes in lines and angles relate to the principal strain

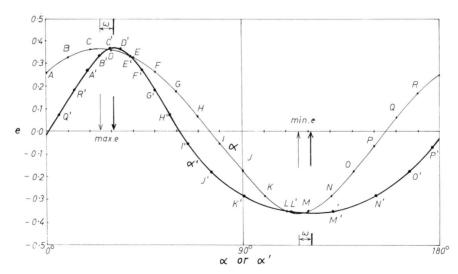

Figure 6.2. *Graphical representation of the relationships between extension e and angle α before deformation, and e and angle α' after deformation of Figure 6.1. ω is the rotational component of the strain.*

directions of the strain ellipse. In such an approach *our simplifications do not imply that we are forgetting components*, all we do is *simplify by allowing our coordinate reference to rotate* with the rotation component of the strain.

The important basic equations relating to length changes in a strain ellipse are established in Appendix D (Equations D.3 and D.7). As we have seen with the general Equations 6.1 to 6.3, two sets of formulae result: one with reference to the orientation ϕ' of a line with respect to the strain ellipse axes, a second with reference to the orientation ϕ of this line with respect to the directions in the undeformed state which will become the strain ellipse axes. The length change equations are most simply expressed in terms of the **quadratic extension** ($\lambda = (1 + e)^2$) or the **reciprocal quadratic extension** ($\lambda' = 1/\lambda$):

$$\lambda = \lambda_1 \cos^2\phi + \lambda_2 \sin^2\phi \qquad (6.4)$$

$$\lambda' = \lambda_1' \cos^2\phi' + \lambda_2' \sin^2\phi' \qquad (6.5)$$

The relationships of the various angles α, α', ϕ and ϕ' are shown in Figure 6.3. Basically what we are doing here is shifting all the points A–R on the curve of Figure 6.2 through the angle of rotation so that the two curves of Figure 6.2 have the same symmetry axes (i.e. we are applying the body rotation component ω to all directions in the surface), then establishing a new coordinate frame (Figures 6.3 and 6.4 "new x" and "new y") along these symmetry axes. The relationships between extensions and orientation expressed by Equations 6.4 and 6.5 are now represented by the two curves and the new x and y axes.

Of these two equations, that expressing length changes in terms of the angles *after* deformation is especially important for practical strain analysis. This is because the absolute value of an extension (or its reciprocal quadratic extension λ') is a function of the two absolute values taken by the principal strains. We can modify these equations to express the proportional stretch S of a line compared with that of the minimum extension.

$$S = (\lambda/\lambda_2)^{1/2} = \left[\frac{R^2}{\cos^2\phi' + R^2 \sin^2\phi'} \right]^{1/2} \qquad (6.6)$$

where R is the ellipticity $(\lambda_1/\lambda_2)^{1/2}$. This function is plotted in Figure 6.5. The curves are markedly "peaked" about the greatest extension direction, and it is of interest to note

that for directions close to the minimum extension the proportional stretch S does not differ much from that of the minimum extension axis. This implies that isolated length strain measurements taken close to the short axis of the strain ellipse are not especially reliable for evaluating the strain ellipse shape. We will return to these curves in the next session when we discuss more general methods of assessing statistically the shape of the strain ellipse from a large number of approximately equal length lines (the centre to centre technique).

How many data for a strain determination?

The first problem we have is to decide *how many* measured extensions we need to establish the strain ellipse. If we know the orientation of the strain ellipse axes (the y-axis of our new coordinate frame) then for two differently oriented line elements (orientations ϕ_A' and ϕ_B') with reciprocal quadratic extensions λ_A' and λ_B' we can formulate two equations of the type of Equation 6.5 to describe how these features are related to the two principal reciprocal quadratic extensions λ_1' and λ_2':

$$\begin{aligned} \lambda_A' &= \lambda_1' \cos^2\phi_A' + \lambda_2' \sin^2\phi_A' \\ \lambda_B' &= \lambda_1' \cos^2\phi_B' + \lambda_2' \sin^2\phi_B' \end{aligned} \qquad (6.7)$$

Mathematically we have an equation system which is easily soluble; these are two equations with two unknowns (λ_1' and λ_2').

If the orientation of our strain ellipse is unknown, we must find three differently directed lines along which we know the elongation, because three different equations are required to solve three unknowns. For example, if the angles between line A and B is α' and between line A and C is β',

$$\begin{aligned} \lambda_A' &= \lambda_1' \cos^2\phi' + \lambda_2' \sin^2\phi' \\ \lambda_B' &= \lambda_1' \cos^2(\phi' + \alpha') + \lambda_2' \sin^2(\phi' + \alpha') \\ \lambda_C' &= \lambda_1' \cos^2(\phi' + \beta') + \lambda_2' \sin^2(\phi' + \beta') \end{aligned} \qquad (6.8)$$

These equations are soluble, but the solution is not easy. We will find later a graphical construction (the Mohr construction) which enables us to make a solution quite rapidly.

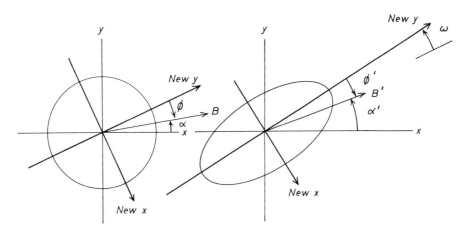

Figure 6.3. Relationships between angles ϕ and ϕ' with respect to principal strains and those α and α' referred to a general coordinate system.

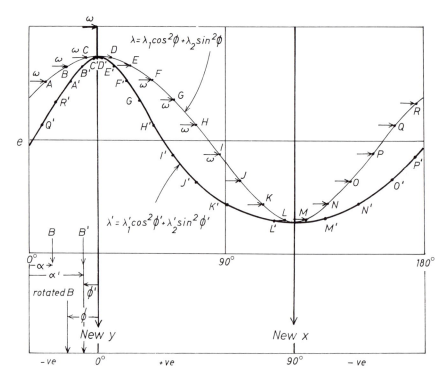

Figure 6.4. Relationships of new x and y coordinate system related to principal strain directions and the general coordinate system.

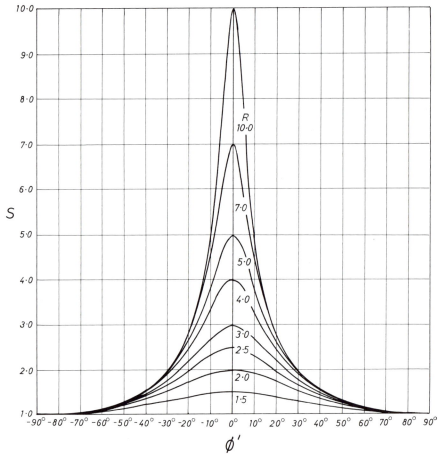

Figure 6.5. Proportional stretch $S = (1 + e)/(1 + e_2)$ of a line of orientation ϕ' with changing ellipticity $R = (1 + e_1)/(1 + e_2)$.

Two known extensions

Question 6.1

Figure 6.6 is a drawing of a hand specimen of deformed limestone containing two stretched belemnites. These fossils are found in strata of Jurassic and Cretaceous age and are the hard parts of a marine animal related to the squids and *Octopus*. They consist of a conical rod (guard) made up of fibrous calcite, with a conical cavity at one end which contained the soft parts of the animal. The calcite fibres are radially arranged around the cone axis and give rise to a mechanical anisotropy and weakness in the fossil when it is stretched along the cone axis. When stretched, the guard breaks into pieces and these pieces separate from each other as the elongation continues. The spaces between the separating fragments are often filled with crystalline material (usually fibrous quartz and calcite) during the deformation, and the fibre axes are aligned parallel to the separation direction of the fossil fragments. The so-called "stretched belemnites" are common in many parts of the Alps and have attracted the attention of many geologists since the earliest tectonic studies were made on these rocks (Heim, Daubrée). These stretched belemnites offer a means of measuring extension. Usually the fossil itself suffers very little internal shape change. To determine extension we measure the total length after deformation, then calculate the original length by reconstructing the separated pieces. Knowing the length before (l_0) and after (l') deformation, we can evaluate the extension $e = (l' - l_0)/l_0$ and the reciprocal quadratic extension $\lambda' = (l_0/l')^2$.

In Figure 6.6 the surface of the specimen shows a lineation produced by the alignment of the long axes of small deformed clastic carbonate grains. This lineation most probably marks the direction of greatest extension in the rock, and we can therefore measure the angle ϕ' between the overall trend of all the belemnite fragments and the strain ellipse long axis for each of two differently oriented fossils.

Using the relationships of Equations 6.7, solve for the values of the two principal strains—express this in the semi-axis lengths of the strain ellipse axes $1 + e_1$ and $1 + e_2$.

What has been the area change in the specimen surface?

The pieces of belemnite are arranged in an en-echelon manner. Can you suggest how this arrangement relates to the principal features of the strain?

The Mohr Diagram

In 1882 the German engineer Otto Mohr developed a very elegant graphical method for analysing stress variation in bodies. The mathematical structure of the equations describing homogeneous stress fields is very similar to that describing homogeneous strain fields, and it therefore allows that the Mohr technique is applicable to the analysis of finite strain. We will discuss stress variation in a later Session, but perhaps here it is worth commenting that, although the mathematical structures of stress and strain are similar, stress and strain are physically very different entities. Stress applied to a body does lead to the development of a strain, but the actual relationships between the two are generally very complex.

The Mohr method develops from an appreciation of the special structure of the equations expressing how extension and shear strain vary with orientation in a strain ellipse. The two basic equations relating these features to orientation ϕ' are:

$$\lambda' = \lambda_1' \cos^2\phi' + \lambda_2' \sin^2\phi'$$

$$\frac{\gamma}{\lambda} = (\lambda_2' - \lambda_1') \sin\phi' \cos\phi'$$

The first of these we have met previously, the full proof of the second will be found in Appendix D (Equation D.16). We could produce another version of the second equation by dividing the right hand side by the value of λ' given by the first equation, but for the purpose of the Mohr construction we keep this form and define a new parameter γ' (**gamma prime**) as $\gamma' = \gamma/\lambda$. It is often a puzzle to beginners to see how we can justify the invention of such a strain parameter combining as it does shear strains and extensions. We justify it because it makes the equations arrive in their most simple form and because such simplicity is vital for our construction.

The next thing we do to this pair of equations is even more surprising to a beginner. We change the trigonometric terms of single angles into those for double angles ($\cos^2\phi' = (1 + \cos 2\phi')/2$, $\sin^2\phi' = (1 - \cos 2\phi')/2$, $\sin\phi' \cos\phi' = \sin 2\phi'/2$). This modifies our original equations into something that at first sight looks very complex.

$$\lambda' = \frac{\lambda_1' + \lambda_2'}{2} - \frac{(\lambda_2' - \lambda_1')}{2} \cos 2\phi' \qquad (6.9)$$

$$\gamma' = \frac{(\lambda_2' - \lambda_1')}{2} \sin 2\phi' \qquad (6.10)$$

Why have we done all this mathematical juggling? The reason is that Equations 6.9 and 6.10 have a very special form.

The parametric equations defining points on a circle of radius r and with centre on the x-axis at distance c from the origin (Figure 6.7A) are:

$$x = c - r \cos \alpha \qquad (6.11)$$

$$y = r \sin \alpha \qquad (6.12)$$

Equations 6.9 and 6.10 have an identical structure to Equations 6.11 and 6.12 if we make the following correspondences:

$$x = \lambda'$$
$$y = \gamma'$$
$$c = (\lambda_1' + \lambda_2')/2$$
$$r = (\lambda_2' - \lambda_2')/2$$
$$\alpha = 2\phi'$$

This implies that the only possible pairs of values of λ' and γ' in a state of homogeneous strain always lie on a circle when plotted according to the graphical system of Figure 6.7B. This is the **Mohr Circle**. The circle intersects the λ' axis at values of λ_1' and λ_2' referring to the maximum and minimum extensions in the system. In such a system of principal strains the values of λ' and γ' indicated by the point X for example are impossible. The Mohr circle always lies to the right of the γ' ordinate axis because the reciprocal quadratic extension is always a real, positive number. The angle $2\phi'$ is always measured from a line joining λ_1' and the circle centre. Positive angles in the body (measured

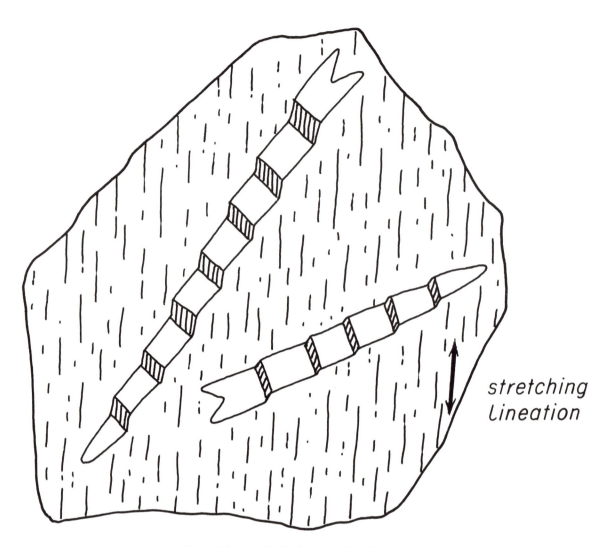

Figure 6.6. Stretched belemnites. See Question 6.1.

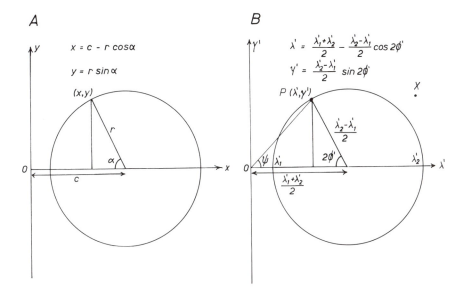

Figure 6.7. A: *The general parametric form of the equations for a circle.* B: *The application of the parametric form of the circle to the equations of strain—the Mohr circle.*

anticlockwise from the λ'_1 direction) are always measured in a clockwise sense in the diagram, negative body angles are measured in an anticlockwise sense from the λ' axis.

The angular shear strain ψ for any orientation ϕ' can be found by joining the point with coordinates (λ', γ') on the circumference with the point $(0, 0)$ (Figure 6.7B, line PO). Because of our definition of gamma prime ($\gamma' = \gamma/\lambda$) it follows that

$$\psi = \tan^{-1}\gamma = \gamma'/\lambda'$$

and that the angle $PO\lambda'$ is the angle ψ for direction ϕ' ($2\phi'$ in the Mohr diagram).

You will probably find it a little difficult to assimilate all the geometric features and implications of the Mohr diagram at first. However, it is a very important construction, and it does offer rapid practical ways of solving strain problems. In several of the following questions we will use the Mohr construction, and by repeated use its significance and potential will become clearer.

We can use the Mohr construction in two ways. First, if we know values of the principal strains e_1 and e_2 in a homogeneous strain field it is possible to use the method to predict rapidly all the possible λ and γ values for every orientation ϕ' in the surface. Second, and more useful in problem solving, from measured deformed objects we can often construct the Mohr circle and use it to evaluate the values of the principal longitudinal strains and their orientations. To understand the geometry of the Mohr diagram we will first examine the prediction techniques, then we will solve particular problems using deformed line elements.

Mohr circle construction

Question 6.2

Draw a Mohr circle representing states of strain given by $e_1 = 0.70$, $e_2 = -0.26$. Locate points on the Mohr circle at intervals of $2\phi' = 20°$ to determine values of λ and γ for physical directions given by $\phi' = 0°$, $10°\ldots90°$ in the deformed body. Plot these values graphically.

Locate directions in the body where the shear strain has maximum and minimum values. Note that zero shear strains ($\gamma = 0$) occur in directions $\phi' = 0°$ and $90°$, i.e. parallel to the principal strain axes. Is there any correspondence between the directions of maximum and minimum shear strain and the directions of no finite extension?

If we know the maximum shear strain in a homogeneously deformed body, what could we determine about the values of the principal strains?

Before going on to the next question it is probably advisable to check the Answers and Comments section.

Two known extensions: Mohr circle solution

Question 6.3

In Question 6.1 we solved the two belemnite problem mathematically. How could you make a solution using a Mohr construction?

With all of these problem-solving constructions we have to find a method of expressing our data in a form that will enable us to locate the Mohr circle. It is always advisable to check first the basic mathematical equations to see that we do have enough information for a solution.

For example, with line elements alone from which we can evaluate the reciprocal quadratic extension the possibilities of solutions are set out in Table 6.1.

Table 6.1

Data	Mathematical structure	Possible computations
λ'_A	1 equation, 3 unknowns	Nothing
$\lambda'_A\phi'_A$	1 equation, 2 unknowns	Nothing
$\lambda'_A\lambda'_B\phi'_A\phi'_B$	2 equations, 2 unknowns	$\lambda_1\lambda_2$ (Questions 6.1, 6.3)
$\lambda'_A\lambda'_B\lambda'_C$ + angles between lines	3 equations, 3 unknowns	$\lambda_1\lambda_2\theta'$ (Question 6.4★)

Return to the data of Question 6.1. Devise a method for plotting these data so as to construct a Mohr circle. Hint: draw the circle first, fit the known data and calculate the scale of the diagram at the end of the computation.

Now return to the Answers and Comments.

STARRED (★) QUESTIONS

Three known extensions: Mohr circle construction 1

Question 6.4★

Figure 6.8 shows an outcrop of calcareous slate of Liassic age from Leytron, Valais, Switzerland, one of the classic localities for stretched belemnites in the Alps. The three belemnites have differing orientations and it is clear, from the discussions above, that we can find the values and orientations of the principal strains. A mathematical solution is not easy: the Mohr construction offers a simpler method. Measure the extensions for fossils A, B and C and calculate the reciprocal quadratic extensions λ'_A, λ'_B, λ'_C. Measure the angles between the fossils $A \angle B$, $B \angle C$, $C \angle A$.

The method we employ here is to place the measured angular data into a circle, then arrange the orientation of the circle until the points take up their correct relationship relative to the measured reciprocal quadratic extensions.

1. Draw a circle of any radius on a piece of tracing paper, then construct three radii with angular relationships twice those of the measured angles (because the physical angle ϕ' always appears as a $2\phi'$ angle in the Mohr construction, Figure 6.9, step 1).
2. Place the circle centre on a horizontal (λ') axis drawn on a piece of graph paper (Figure 6.9, step 2). Calculate the value of the proportion $(\lambda'_B - \lambda'_A)$ $(\lambda'_C - \lambda'_A)$ from the measured data. Rotate the tracing paper about a fixed centre and measure the proportions of the perpendicular from points A, B and C on to the λ' axis. By trial and error rotate the circle until the measured proportion is achieved (Figure 6.9, step 3). This will occur at one point only, and this represents the unique solution to the problem.
3. Calculate the position of the origin. Now read off values of λ'_1, λ'_2 and determine the angle $2\phi'$ between the principal λ'_1 axis and one of the three directions (Figure 6.9, step 4).

Three known extensions: Mohr circle construction 2

Question 6.5★

A second method for solving this type of problem uses a secondary construction for the determination of the shear strain for each direction and establishing three points on the periphery of the Mohr circle (by calculating coordinates, e.g. (λ'_A, γ'_A)).

It may at first seem surprising that we can calculate shear strains from measured length data alone. But if you remember that the three lines enable us to calculate the complete form and orientation of the strain ellipse, and that all the strain features are contained in this ellipse, perhaps it is not such a surprise.

1. Draw on a sheet of paper a triangle, the sides of which have the same angular orientations as the measured lines (Figure 6.10, step 1). The size of this triangle is unimportant. This triangle represents a deformed version of some other undeformed triangle. Measure the lengths of its sides x', y', z'.
2. Calculate the lengths of the sides of the undeformed equivalent triangle x, y, z using the relationships such as $x = x'/(1 + e_A)$ and construct this triangle (Figure 6.10, step 2).
3. In the reconstructed triangle, construct perpendiculars from the three corners on to their opposite sides. These will be our reference lines to determine shear strain. Each perpendicular will intersect the opposite side at a point which divides that side in a characteristic ratio (Figure 6.10, step 3, x_1/x_2).
4. Return to the original (deformed) triangle and divide the corresponding side in the same ratio (Figure 6.10, step 4, $x'_1/x'_2 = x_1/x_2$). This is a valid construction because the strain along this direction is homogeneous. Now draw the line from this division point to the opposite corner (Figure 6.10, step 4). This line will not normally be perpendicular to the reference direction, and its deviation gives a measure of the angular shear strain ψ_A. Calculate ψ_B and ψ_C in the same way. Calculate $\gamma'_A = \tan \psi_A / \lambda_A$ etc.
5. We have now obtained three pairs of coordinates for a Mohr circle. Plot these on a graph with fixed λ', γ' scales. Find the centre of the circle where the right bisector of each coordinate pair crosses the λ' axis (Figure 6.10, step 5). There are three ways of obtaining the centre, and the coincidence (or otherwise) of these intersections gives a good check on the accuracy of the constructions (NB it does not check the accuracy of the primary λ'_A, λ'_B and λ'_C measures).
6. Read off from the completed Mohr circle values for λ'_1, λ'_2 and ϕ'_A (Figure 6.10, step 6).

ANSWERS AND COMMENTS

Two unknown extensions

Answer 6.1

By reconstructing the original form of the belemnites, and determining the angle ϕ' between the line joining the centres of the fragments of each belemnite and the stretching lineation we obtain:

$$\lambda'_A = 0.62 \qquad \phi'_A = 33°$$

$$\lambda'_B = 0.75 \qquad \phi'_B = 62°$$

The two simultaneous equations we solve are

$$0.62 = 0.70\lambda'_1 + 0.30\lambda'_2$$

$$0.75 = 0.22\lambda'_1 + 0.78\lambda'_2$$

which give

$$\lambda'_1 = 0.54 \ (e_1 = 0.36)$$

$$\lambda'_2 = 0.81 \ (e_2 = 0.11)$$

Figure 6.8. Three deformed belemnites. Leytron, Switzerland.

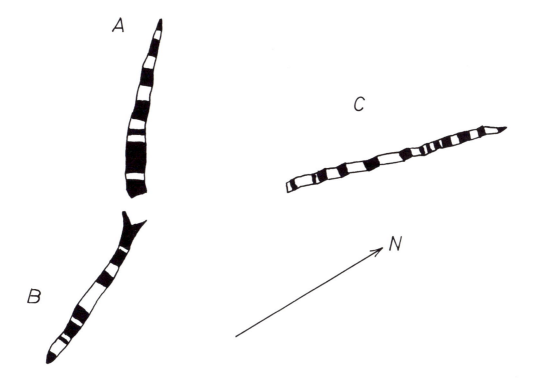

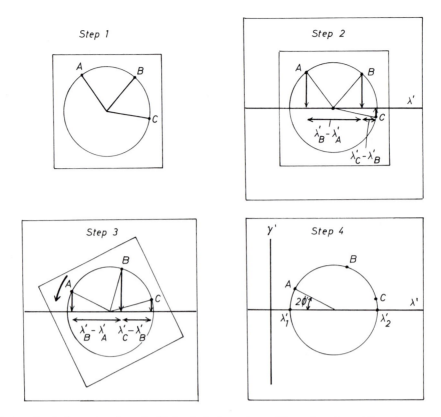

Figure 6.9. First method of construction of a Mohr circle from three differently oriented line elements. See Question 6.4★.

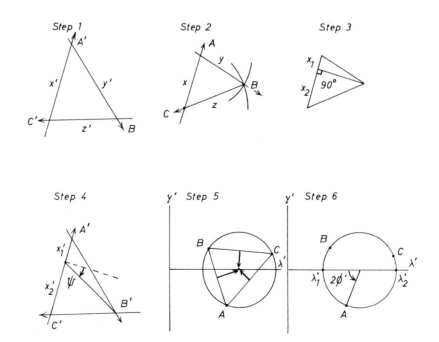

Figure 6.10. Second method of constructing of a Mohr circle from three differently oriented line elements. See Question 6.5★.

The area change Δ_A is given from

$$1 + \Delta_A = (1 + e_1)(1 + e_2)$$

$$\Delta_A = 0.51 \ (51\% \text{ area increase})$$

Although we have not yet introduced strain in three dimensions, it is perhaps worth commenting on the significance of this area change. The surface of our specimen contain the maximum (e_1) and intermediate (e_2) strains in three dimensions, and is perpendicular to the maximum shortening (e_3). The area increase arises from the strong compression normal to the surface leading to a situation where both the principal surface two-dimensional extensions are positive. It should be clear that in some other surfaces the two-dimensional strain ellipse is one of decrease in surface area (e.g. the section containing e_2 and e_3).

The en-echelon arrangement of the fragments, which arises because of the rigid nature of the belemnite guard, is analogous to the en-echelon boudins we discussed in Session 2. During the deformation, the rotational couple exerted on the belemnite fragments by the deforming matrix is insufficient to cause the fragment to rotate at the same rate as that of the *overall* line of the belemnite. The en-echelon sense is right-handed on one side of the maximum principal extension, and left-handed on the other, and the en-echelon effect is at a maximum where the belemnite axis makes an angle of about 30° with the maximum extension direction. Figure 6.11 is a drawing from the Leytron quarries of a block showing this effect very clearly. It should be stressed, however, that the rotation of rigid particles in a ductile matrix is not just a function of finite strain, but it is also a function of strain history. The same finite strain can be built up in many different ways. The behaviour of rigid particles shows asymmetric differential rotation patterns about the maximum extension direction which reflect these various deformation paths. We will return to this topic later when we discuss progressive deformation.

As the belemnite fragments separate, spaces (or perhaps "potential" spaces) open up between the fragments. In deforming rocks such spaces are probably immediately

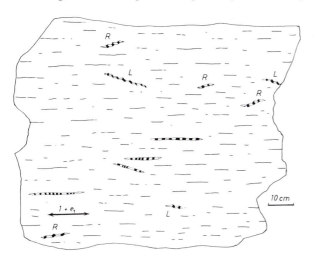

Figure 6.11. En-echelon arrangements of stretched belemnites. R and L refer to right-handed and left-handed en-echelon patterns respectively. Belemnites parallel to the maximum elongation show no en-echelon pattern.

filled with fluid from which crystalline material is deposited. Thus opening and filling go on simultaneously. This is born out by the structure of the quartz and calcite crystals – they are always fibrous and never show such good development of crystallographic faces as characterizes a crystal growing freely into a large, fluid filled cavity. The fibres always link points which were once in contact and their form is related to the movement separation path traced out by the moving rigid fragments. Fibre geometry contains much information about the progressive movement and deformation history. We will return to this topic again in Sections 13 and 14. Where the minimum, two-dimensional extension e_2 is close to zero, the fibres are often sub-parallel to the greatest extension direction because fragment separation is generally more or less in that direction. Where the e_2 strains have positive values, the fibres show a more complex geometric form, and differently oriented belemnites have differently aligned fibres.

Methods of measuring extension

Many geological situations offer potential data for determination of values of extension in particular directions these will be discussed below. Our main problem will be to determine whether the object itself is internally deformed. If it is there will be difficulties in determining the original length necessary for an accurate computation of extension.

1. Belemnites

Belemnites offer potentially good material for measurement of extension. Where they are preserved in a shale or marl matrix, cross sections generally show little or no deformation, so the restored fragments do generally give a reasonably good approximation to the original length (as we have already seen in Question 6.1). If they are embedded in limestone, or if they are in an environment where deformation has been accompanied by low or high grade metamorphism, they may show significant internal deformation. Such an effect can generally be spotted if the originally circular cross sections of the guard cone have been modified into an elliptical shape. The principal feature which can lead to errors arises when the belemnite has not taken up the full extent of the elongation, owing to the matrix at the ends of the guard having slid past the tip without generating breaks in the guard. It is a general rule with belemnites (as with many deformed prismatic objects used for extension measurement) that the greater the length–breadth ratio, the more accurate the calculation.

Professor Badoux of Lausanne University made some interesting studies of deformed belemnites from the Helvetic nappes of the Alps. His research has a bearing on the accuracy of extension parameters derived from stretched belemnites. He showed that the extensions calculated at one locality from many differently oriented belemnite guards showed a considerable variation (Figure 6.12), and it seems likely that at least part of this variation is the result of the belemnites' taking up a less than appropriate full stretch. An average of these data for different sector orientations produced a convincing strain ellipse geometry, but it seems likely that this ellipse is probably an underestimate of the strain at this locality.

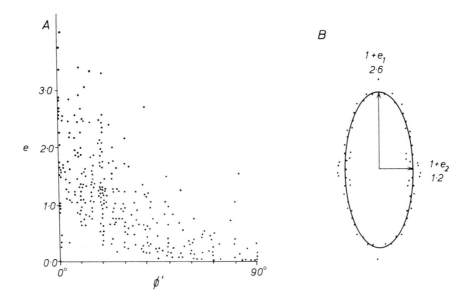

Figure 6.12. A: *Extension e determined from individual deformed belemnites as a function of angle* φ' *from the maximum extension direction.* B: *Strain ellipse constructed from A by averaging the extension data over 5° intervals. Data from Badoux (1963).*

2. Crinoids

Crinoid columns offer excellent material for computing extensions. This is because the articulating surfaces between the ossicles provide mechanically weak junctions through the column enabling the ossicles to separate when subjected to a positive extension. It is sometimes possible to determine negative extensions along crinoid columns. The articulating surfaces frequently undergo localized pressure solution. Providing that the crinoid dimension parallel to the column axis is reasonably constant, as is the case with many species, the reduced spacing after pressure solution can be used to calculate the extent of contraction along the columns.

3. Other fossils

Many fossil species have characteristic dimensions when they reach maturity. If information on the original dimensions is available, it may be possible to use direct measurements of the deformed fossils to calculate extensions in different directions (Wright and Platt, 1982). The dangers of this technique come about because immature and mature forms may be mixed. The thecal spacing of graptolites, for example, have been used to define extension in deformed graptolitic shales.

4. Crystals

In metamorphic terrains which have been subjected to deformation after crystallization it is not uncommon to find broken crystals with the parts separated in a similar way to the features we have seen in belemnites and crinoids. Such separations can be used to calculate values for the late strains. Particularly useful are those crystals which show a marked prismatic habit, especially those with a well developed cleavage normal to the prismatic elongation. The commonest and most useful for these measurements

are tourmaline (see Figure 6.13A), epidote, kyanite and amphiboles of the tremolite–actinolite group (see Figure 6.13B).

5. Boudins

Boudinaged competent layers in an incompetent matrix can provide good markers for positive extension. The calculated stretch value will often be less than the true value because there is often some plastic flow in the boudin which makes the reconstructed boudin assemblage exceed the length of the original layer. Boudins with high competence relative to their matrix give the best results (see Figure 1.9A, layer a). At low layer parallel extensions, such boudins show rectangular sections. At high extensions they may show "fish mouth" forms, and the original length is then measured in the boudin from "throat" to "throat". Pinch and swell structure is not of general use for extension measurement, because such structure develops where the layer–matrix competence contrast is low, and therefore much stretching takes place inside the competent layer. Chocolate tablet boudinage offers interesting possibilities for reconstructing the tablets to produce a full two-dimensional strain analysis in a single competent layer and we will return to this possibility in Session 13.

6. Folds

Buckle folds developed in isolated competent layers are useful for computing layer shortenings, but only where the competence contrast between layer and matrix is high. Folds of this type are recognized by a well developed ptygmatic form (the so-called elastica, see Figures 1.14 and 2.15). Folds that have short wavelengths relative to layer thickness, cusp-shaped contacts and signs of internal deformation within the competent layer (e.g. the presence of cleavage) should not be used for extension calculations.

A

Figure 6.13. A: *Stretched tourmaline crystals in a metasedimentary gneiss. Pizzo Molare, Ticino, Switzerland. B: Thin section of a stretched amphibole porphyroblast in a mica schist. Otta, Norway (×50).*

B

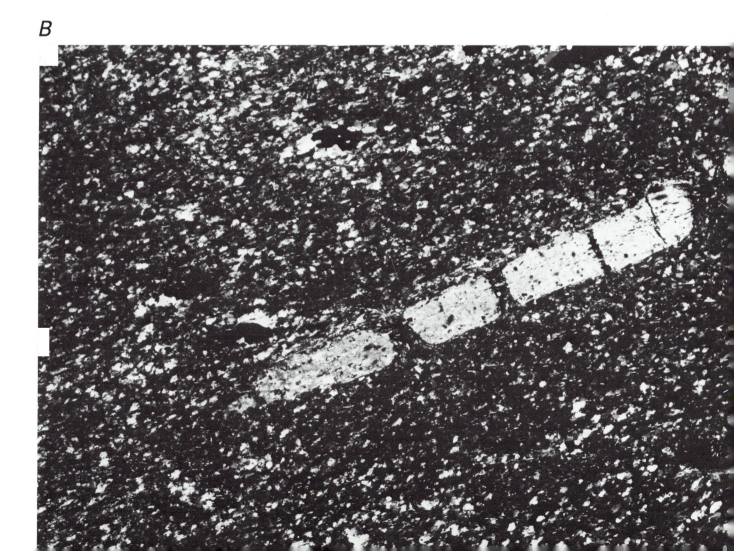

A

Figure 6.14. A: *Combination of boudin structure and ptygmatic folds in competent aplite dykes cutting less competent granitic gneiss. Hoggar, Central Sahara. B: Ptygmatic folds in pegmatite veins cutting deformed granitic gneisses. Monte Rosa nappe, Ascona, Switzerland. The pegmatite sheets were intruded at different times during the deformation and therefore took up different amounts of longitudinal strain.*

B

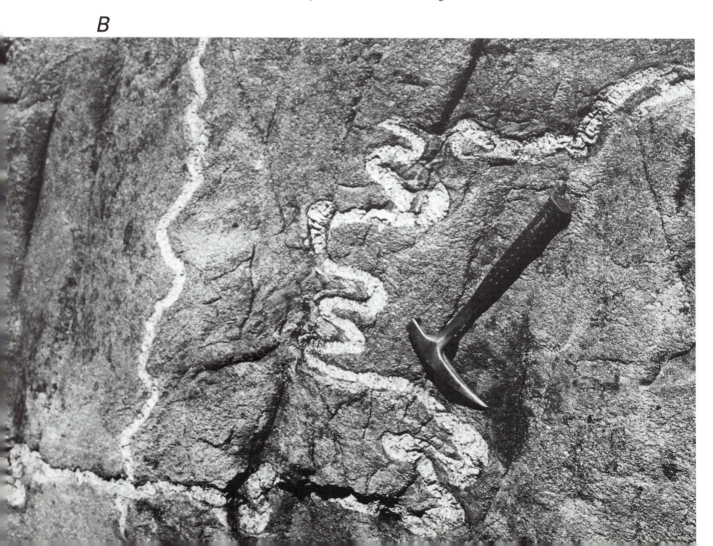

These folds develop in situations of low competence contrast (see Figure 1.9A, layers b and c) where the competent layer suffers considerable shortening parallel to the layering before developing the sideways buckling instability. This means that the measured fold arc length is much shorter than that of the original layer.

Ptygmatic folds often occur in several directions in gneissic complexes where they may be combined with boudinaged veins (Figure 6.14A). These structures can offer a method of computing the strain ellipse (and three-dimensional strain ellipsoid). In some instances, however, the veins may be emplaced at different times during the progress of the deformation, each vein taking up different proportions of the total strain and producing geometric relationships that cannot be analysed with a single strain ellipse (Figure 6.14B). The work of Talbot (1970) provides a good example of strain investigations in gneissic rock complexes in Africa using measurements on folded and boudinaged veins to compute the strain.

The Mohr circle construction

Answer 6.2

Figure 6.15A illustrates the Mohr circle for principal strains $e_1 = 0.70$, $e_2 = -0.26$, and Figure 6.15B shows the two graphs of values of λ and γ derived from it. In this example the lines of no finite extension are symmetrically arranged about the principal strain axes at angles of $\pm42°$ from $1 + e_1$. The ellipse is a Field 2 ellipse ($e_1 +$ ve, $e_2 -$ ve, see Session 4, p. 65).

Shear strains reach a maximum value ($\gamma = +0.93$) at 23° clockwise from the greatest extension axis, and a minimum value ($\gamma = -0.93$) at 23° anticlockwise from this axis. There is no correspondence between the directions of no finite elongation and the directions of maximum and minimum shear strain. In fact, not all ellipses have directions of no finite elongation (e.g. Field 1 and Field 3 ellipses, see Session 4) but all ellipses have symmetrically paired directions of maximum and minimum shear strain.

For a given maximum shear strain it is possible to construct many different Mohr circles. These ellipses all have differing λ'_1 and λ'_2 values, yet the ratio λ'_2/λ'_1 (equivalent to the square of the ellipticity $= R^2$) is constant for all circles. The values of the maximum shear strain is therefore a function of the ellipticity of the strain ellipse.

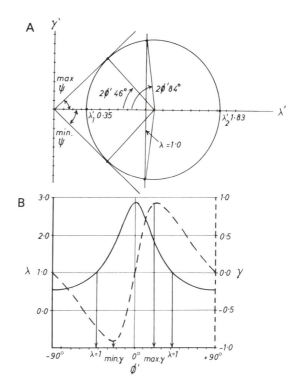

Figure 6.15. *Mohr circle construction for Question 6.2. The graph shows the variation of quadratic extension λ and shear strain γ along lines making different angles ϕ' to the maximum elongation direction.*

Two known extensions: Mohr circle solution

Answer 6.3

The solution to this problem using the Mohr construction is shown in Figure 6.16. The construction proceeds by first drawing the Mohr circle with points A and B positioned on its circumference using the $2\phi'$ relationships. These two points establish the scale along the λ' abscissa from the measured value of $\lambda'_B - \lambda'_A$. This scaling factor is then used to fix the origin of the coordinate system. Once the origin is established, the values of λ'_1 and λ'_2 can be read off the abscissa scale.

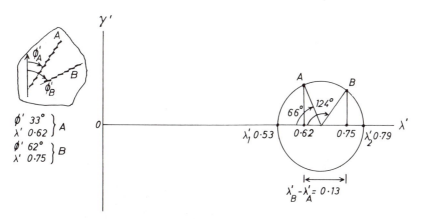

Figure 6.16. *Mohr circle solution to Question 6.3.*

Three known extensions: Mohr circle construction 1

Answer 6.4★

The solution to the three-belemnite problem using the first technique is set out in Figure 6.17. This particular example gives a good clean solution because the angles between the three objects are moderately large. It should be clear that a less precise solution results if any two of the objects are closely oriented. If this occurs, the establishment of the ratio $(\lambda'_A - \lambda'_B)/(\lambda'_B - \lambda'_C)$ is subject to great error where there are small errors in the extension measurement data.

Three known extensions: Mohr circle construction 2

Answer 6.5★

Using the constructions described for the second technique the following pairs of (λ', γ') coordinates are found

$$\lambda'_A = 0.41 \; \gamma'_A = 0.08$$

$$\lambda'_B = 0.30 \; \gamma'_B = 0.13$$

$$\lambda'_C = 0.18 \; \gamma'_C = 0.01$$

These coordinates are close to those illustrated for the points A, B and C in the Mohr circle in Figure 6.17.

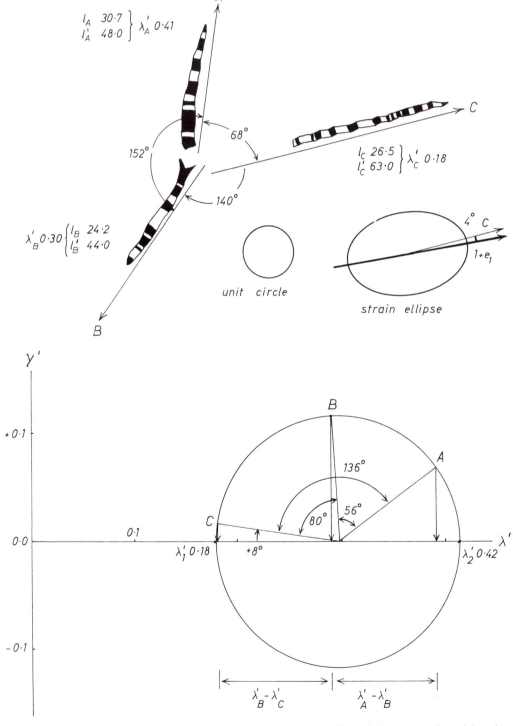

Figure 6.17. Data from the belemnites of Question 6.4★ and 6.5★, and the Mohr circle construction giving the solution to the problem.

KEYWORDS AND DEFINITIONS

Mohr diagram   A graphical method for representing variations of longitudinal and shear strains for a field of homogeneous strain. The Mohr circle represents the values of all appropriate pairs of reciprocal quadratic extension ($\lambda' = 1/(1 + e)^2$ (plotted as abscissa) and gamma prime values ($\gamma' = \gamma/\lambda$) (plotted as ordinate) for given directions in the deformed plane.

KEY REFERENCES

Badoux, H. (1963). Les bélemnites tronçonées de Leytron (Valais). *Bull. Lab. Géol. Min. Géoph. Musée Géol. Univ. Lausanne* **138**, 1–7.

This paper presents data from the locality of Question 6.1 and 6.4★ and provides a useful commentary on the variations of longitudinal strains in the region.

Beach, A. (1979). The analysis of deformed belemnites. *J. Struct. Geol.* **1**, 127–135.

A detailed study in which extension measurements are used to establish the finite strains in a region of the W. Alps, and the influence of original preferred orientations of the belemnite guards on the computation are discussed.

Ramsay, J. G. (1967). "Folding and Fracturing of Rocks", 568 pp. McGraw-Hill, New York.

This reference is useful to give additional background information about the diagram and the methods for determining strain state using Mohr circle constructions (pp. 69–81). Methods of strain analysis using length changes are described on pp. 247–251.

Sanderson, D. J. and Meneilly, A. W. (1981). Analysis of three dimensional strain modified uniform distributions: andalusite fabrics from a granite aureole. *J. Struct. Geol.* **3**, 109–116.

A strain analysis using stretched prismatic crystals of andalusite in the hornfels aureole of the Ardara granite, Eire, shows that the dominant deformation is of three dimensional flattening type, supporting the idea of pluton growth by balloon-like expansion.

Talbot, C. J. (1970). The minimum strain ellipsoid using deformed quartz veins. *Tectonophysics* **9**, 47–76.

This work shows how an analysis of the orientations of boudinaged and folded veins in gneissic complexes can be used to determine the finite strain state.

Thakur, V. C. (1972). Computation of the values of the finite strains in the Molare region, Ticino, Switzerland, using stretched tourmaline crystals. *Geol. Mag.* **109**, 445–450.

A short but very useful account shows how boudinaged prismatic crystals in the Penninic nappes of the Alps can be used to evaluate the late stage strains. The analysis shows that the greatest extensions occur parallel to the fold axes and was of a constrictional type.

Wright, T. O. and Platt, L. B. (1982). Pressure dissolution and cleavage in the Martinsburg shale. *Am. J. Sci.* **282**, 122–135.

This paper describes how finite strains in the Appalachian slate belt were analysed by the change in lengths of graptolites and concludes that the strains are consistent with a material volume loss of about 50%.

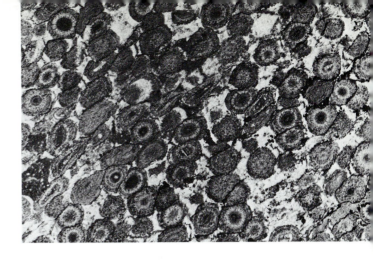

SESSION 7

Practical Strain Measurement

3. The Centre to Centre Technique and Strain Partitioning

Under certain conditions the distribution of particle centres of rock components making up an aggregate (e.g. oolites, conglomerates) can be used to determine the state of finite strain. The method involves determining particle centres which were originally statistically nearest neighbours, finding nearest neighbour tie lines, and analysing the changes of length of these tie lines. Different analytical procedures for strain determination applied to a rock may give different values of strain. These differences can be real, and related to the activity of different rock deformation mechanisms. The possibility of apportioning the total strain between different active mechanisms (pressure solution, grain boundary sliding, plastic flow), termed strain partitioning, is investigated.

INTRODUCTION

The aim of this Session is two-fold. In the last Session we developed techniques for determining strain from extension measurement with individual line elements. In the first part of this Session we analyse the possibility of measuring strain by determining the redistribution of locatable points using the way the joins or tie lines between the points change length. The so-called *centre to centre technique* is very important and enables us to determine the strain from the overall redistribution of rock particles such as ooids, pebbles and even clastic sediment grains. The second part of this Session discusses *strain partitioning*, that is the way we can separate the differing mechanisms of deformation that have led to the overall bulk strain. Many naturally deformed rocks deform by several mechanisms (crystal plasticity, grain boundary sliding, pressure solution, cataclasis etc.) sometimes these mechanisms acting synchronously, sometimes successively with pressure and temperature change. In strain partitioning studies we discover what role each of the various mechanisms play in making up the total deformation. The centre to centre technique often plays an important part in enabling us to separate the various components.

THE CENTRE TO CENTRE METHOD

We will now discuss methods of determining the strain ellipse when we have a rock containing points which we can locate (ooid centres, pebble centres, sand grain centres, porphyroblasts, porphyroclasts), but where we cannot use the object shape to unambiguously determine the strain. The pebbles in deformed conglomerates, for example, are often of a markedly different competence from each other

and from their matrix (Figure 7.1). The pebble shapes are often sub-elliptical, but these forms do not necessarily relate directly to the shape of the strain ellipse of the rock as a whole. The relationship between pebble shape after deformation is a very complicated function of original shape, competence contrast between the pebble and its matrix and the influence of neighbouring pebbles. Figure 7.1 shows these various factors very clearly. The light coloured pebbles consist of quartz and felspar (they are granites, pegmatites and aplites) and are much less elliptical than the dark pebbles of mica schist. The large, competent, centrally located pebble has developed a complex heterogeneously strained field in its contact zone, and these strain heterogeneities have influenced the shapes of nearby pebbles as well as the orientation of the schistosity. The problem of determining the *bulk strain* or *average strain* for a rock such as this is quite difficult.

A very marked competence contrast between pebbles of differing composition is illustrated in Figure 7.2. This conglomerate is made up of dark coloured limestone pebbles and light coloured dolomite pebbles in a sparse matrix of fine grained calcite mudstone. The limestone pebbles are clearly much less competent than the dolomite pebbles and appear to have deformed by some mechanism of plastic flow. In contrast the dolomite pebbles are either undeformed or undergo break up (cataclasis) with the formation of extension and shear fractures. It is clearly not possible to record the bulk strain of this rock using the methods of analysis described in Session 6.

Another example of difficulties arising from competence contrasts of the particles is illustrated in Figure 7.3, and the scale of this example is very different from that discussed above. Figure 7.3 is a thin section of oolitic lime-

107

Figure 7.1. *Polymict conglomerate: pale fragments of aplite, granite and granite gneiss; dark fragments of mica schist, deformed by different amounts as a result of strong regional tectonic strain. Lebendun nappe, Cristallina, Ticino, Switzerland.*

Figure 7.2. *Deformed carbonate conglomerate. The dark fragments are highly deformed limestone clasts, the pale fragments are little deformed dolomite clasts. Naukluft Mountains, Namibia.*

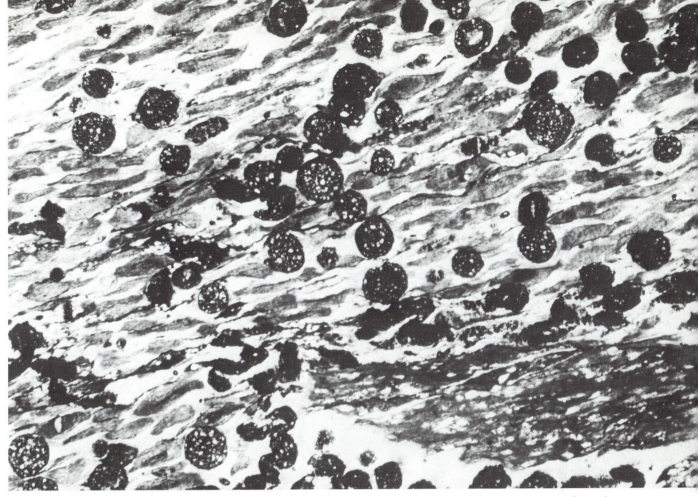

Figure 7.3. *Thin section of a deformed oolitic limestone, SW Sardinia. The circular dark areas are almost undeformed dolomite ooids, the paler elongated light grey areas are highly deformed calcite ooids,* × *15.*

stone which has been partially dolomitized. Some of the oolitic particles have been completely replaced by dolomite, whereas others consist of the unaltered calcite. The rock bulk has been strongly deformed: the calcite ooids are strongly deformed, yet those ooids made up of dolomite are almost undeformed. We could not make measurements of the shapes of the dolomite and calcite ooids to arrive at any sensible computation of total rock strain.

A further example of strain measurement difficulties using particle shapes is illustrated in the frontispiece of this Session. This is an oolitic limestone from Devonian limestones in southwest England where strong pressure solution has taken place at the ooid–ooid contacts. Neither the external form of the ooids nor the shape of the elliptical internal banding gives a correct value for the total strain, because local solution and migration of material has played an important part in giving the bulk shape change of this rock.

Let us first analyse what happens to the spatial distribution of a series of points as a result of a displacement which leads to a homogeneous strain, if we deform a set of points which is completely random.

The centre to centre method of strain determination is a technique which enables the bulk rock strain to be calculated using the redistribution of points in the deformed rock and the distances between these points as extended line elements.

First, let us consider what happens to the spatial distribution of a series of points as a result of displacement and strain. Figure 7.4A illustrates a *completely random initial set of points* (known as a Poisson distribution) in a surface,

the coordinates of the points selected from a random number list. In such an aggregate one finds clusters of points as well as regions relatively empty of points. If we homogeneously deform such a body and observe the changed positions of the points we find that the resulting distribution remains random (Figure 7.4B). The points regroup into new clusters and gaps, but the overall random pattern has the same properties as the original field. In such a redistribution of points it is clearly *impossible to use the point distribution for strain calculations unless we can recognize tie line joins between the points*. One geological possibility for recognizing tie lines might be found using desiccation cracks in sediments. The polygonal cracks in a sediment could provide a random net of tie lines which might be used to identify the original nearest neighbour triple points of crack intersections (see Roder, 1977; Harvey and Ferguson, 1981).

In many geological situations, the initial distribution of points is not random in the broadest sense. The points may be scattered but distributed in such a way that the distances between them is more or less constant (Figure 7.5A). In such an aggregate we do not see the clustering and anti-clustering as in a truly random distribution (cf. Figures 7.4A and 7.5A). Such a distribution (termed a *statistically uniform distribution*) generally arises because the objects (pebbles, ooids) have a characteristic initial size, and this means that the centre to centre separations of nearest neighbours is controlled by the geometric constraints of the packing. An aggregate of points with such a statistically uniform distribution has special geometrical properties when it is deformed. Figure 7.5B illustrates the effect of

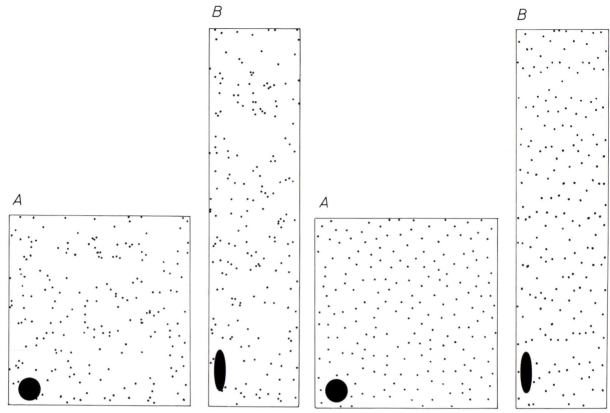

Figure 7.4. A *is an original* Poisson random distribution *of points showing characteristic clustering and anticlustering effect.* B *shows the result of deforming this point distribution. Although the points are displaced the resultant distribution still shows a Poisson random distribution.*

Figure 7.5. A *is an originally* statistically uniform distribution *of points, and* B *shows the result of deformation on the point distribution. After deformation the point distribution is no longer uniform: distances between points are greatest in the direction of maximum longitudinal strain and least in the direction of minimum longitudinal strain.*

deforming the uniform anticlustered distribution of Figure 7.5A. The spatial distribution is no longer the same as that of the initial distribution: there is a greater separation of points parallel to the long axis of the strain ellipse, and a maximum closing up of distances between points parallel to the short axis of the strain ellipse. The spacing distances between points in the deformed aggregate is a function of the elongation in the direction joining the points. Such a deformed aggregate *does offer the possibility of strain determination.* Two basic techniques are suggested.

Method 1 "Nearest neighbour" centre to centre technique

This technique can be used where we have an aggregate of objects which was initially fairly uniform in the sense that the distances between particle to particle centres were initially statistically constant, and where the particle centres are relatively easy to define (ooids, conglomerates, sand grains). Figure 7.6A illustrates an example of an aggregate of circular objects. Each particle has near neighbours and far neighbours. For example, particle A has near neighbours B, C and D and far neighbours E, F, G, H and I. We define what is a near neighbour by drawing a tie line between particle centres. The centre tie lines AB, AC and AD cross no other particle whereas AE–AI all cross other circular particles. Figure 7.6B illustrates the geometry of the deformed aggregate. The distances between the particle centres are modified by strain but the tie lines between the

particle centres retain their non-crossing or crossing properties of the initial aggregate, and can therefore be used to establish what were initially near neighbours.

If we examine the variations in lengths (d) of the tie lines of particles that were initially near neighbours we find that tie line length is independent of initial direction α, and shows a statistical length related to the particle size

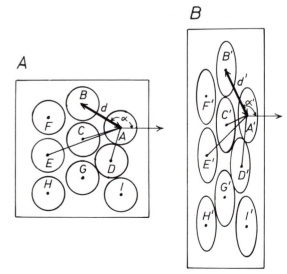

Figure 7.6. *Method for determining nearest neighbours in an aggregate of particles before and after deformation*

and packing. After deformation, however, the tie lines change orientation (α') and their lengths (d') vary as diameters of the strain ellipse. The centre to centre tie lines can therefore be used to calculate the shape and orientation of the strain ellipse.

Question 7.1

Figure 7.7 shows a negative photograph of a thin section of a deformed oolitic limestone from the Silurian strata of central Wales. From the geometric form of the ooids it is clear that the rock has been deformed by a combination of two mechanisms. The concentric structure of the original ooid has been modified in part by some mechanism of internal flow and the form of outer boundaries of the original concentric shells of the ooids has been truncated by a pressure solution chemical transfer process (Figure 7.8). The ooid shells have been solved along pressure solution seams, and this material has been precipitated in spaces opened up by the pulling apart and local detachment of the ooid shells. Clearly the sub-elliptical form of the ooid shell records only a part of the deformation. To determine the bulk strain of this rock we must use the centre to centre technique.

On a transparent overlay on Figure 7.7 locate the central point of each ooid. Examine the point distribution on this overlay carefully. Could you tell from this distribution that the rock has been strained? Draw tie lines between these points and select original near neighbours by rejecting all tie lines which cross another ooid. Measure the distance (d') between near neighbour tie lines and measure the orientation of each tie line (α') from the base azimuth. Plot these data on a graph with α' as abscissa and d' as ordinate.

In many problems of strain measurement, the main problem is to decide how many data to use. Data collection is time consuming, and after a while more data do not appreciably increase the accuracy of the computation. For this strain analysis we need to find a best fit curve, to discover the maximum and minimum values of this curve and its symmetry axes. The most accurate method is to determine the arithmetic means of α' for a certain angle interval (90°–80°, 85°–75°, 80°–70° etc.) and to draw the curve which best satisfies these averages. The best fit curve should be of the form of Equation 6.5,

$$kd' = \left(\lambda_1' \cos^2(\theta' + \beta') + \lambda_2' \sin^2(\theta' + \beta')\right)^{-1/2} - 1 \quad (7.1)$$

where λ_1' and λ_2' are the principal reciprocal quadratic extensions, β' is the orientation of the principal strains from the base line azimuth, and θ' is the orientation of a measured direction from the principal strain, and $\alpha' = \theta' + \beta'$. k is a constant which depends upon the initial spacing of the particle centres.

The maximum and minimum values of this curve will not represent the absolute values of $1 + e_1$ and $1 + e_2$, but from these values the ellipticity R of the strain ellipse can be determined: $R = d'_{max}/d'_{min}$.

Method 2 The Fry method

This technique, devised by Norman Fry (1979), provides an excellent practical method for finding the best fit solution to the strain ellipse. Current techniques of strain measure-

ment can sometimes be criticized as being too sophisticated and too time consuming. The advantage of the Fry method is that it provides a graphical solution to the centre to centre method which is both rapid and accurate, and that one sees from the developing graph when enough data have been plotted to provide an answer of sufficient accuracy for the investigation in hand.

The basic idea is very simple. Consider an aggregate of statistically uniformly sized particles producing a set of near neighbour centres (Figure 7.9A). Relative to centre point A all the surrounding points have a spatial distribution reflecting the average particle size and type of packing. For simplicity Figure 7.9 has been constructed with a two-dimensional close packing of circles of equal radius. The distances between centres AB, AC etc. cannot get closer than twice the particle radius ($2r$), and for each point like A there will be six near neighbours at distance $2r$. The surrounding particles are situated at distances of $2\sqrt{3}r$ like AD, AE etc., and around A there are six of these. The pattern of neighbours continues with six centres situated at distances of $4r$ from A (e.g. AF, AG) etc. The type of packing therefore sets up certain characteristic distances of the centres of neighbouring particles from any one chosen particle. If the particles have identical radii, and the packing is perfect, this zonal repetition of the distances of centres from any one centre will show regular periodic variations. In natural aggregates however, the regularity decreases with distance from any one chosen point. Figure 7.9B illustrates the geometry of a homogeneously deformed aggregate of 7.9A. The distances between particle centres become modified in proportion to the value of the longitudinal strains along that direction, that is in proportion to the strain ellipse diameters. Because the values of the periodic distance spacings become systematically changed according to the strain ellipse shape, this geometry can be used to establish the shape and orientation of this ellipse.

If one has a series of redistributed originally uniform anticlustered points the Fry construction is as follows:

1. On a sheet of paper mark the centres of all particles. Number these points.
2. Take a transparent overlay and mark a central reference point. Place this reference point over one of the central points (1). Trace the positions of all other points (2, 3 . . . etc.) on the overlay (Figure 7.10A).
3. Move the overlay keeping a constant azimuth so that the overlay reference point lies over point (2). Trace the positions of all other points (1, 3, 4, 5 . . . etc.) on the overlay (Figure 7.10B).
4. Repeat for all other points on the base sheet.

The many points that accumulate on the overlay are not uniformly distributed (Figure 7.10C).

1. Around the central overlay reference point there is a point vacancy, or a region which shows a very low point concentration. This can have a circular or elliptical form. The vacancy arises from the fact that any two original particles cannot come to lie closer than the sum of their radii. A circular vacancy field implies no strain, an elliptical vacancy field implies that the rock has suffered strain. The shape and orientation of the strain ellipse are directly recorded by the elliptical form of the field. If no vacancy field is seen after

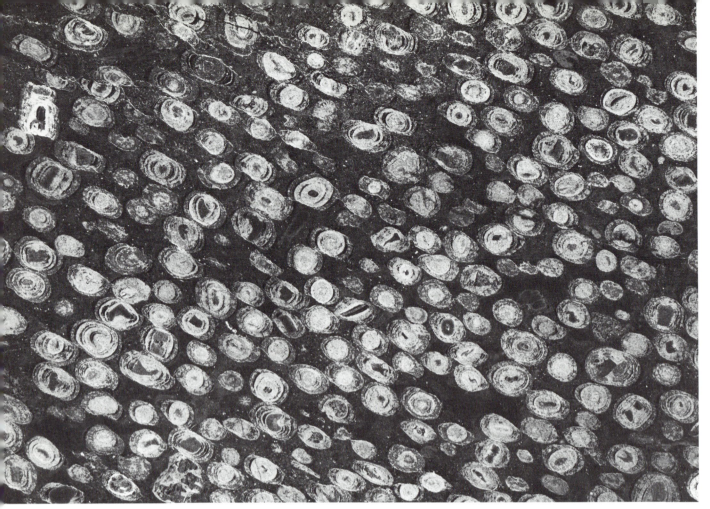

Figure 7.7. *Deformed ironstone oolite in thin section × 5. See Question 7.1.*

Figure 7.8. *Enlargement (× 100) of an ooid in Figure 7.7 showing pressure solution surfaces. The ooid shells have been detached along the direction of maximum longutidinal strain and fibrous quartz and calcite crystallized in the extension fissure.*

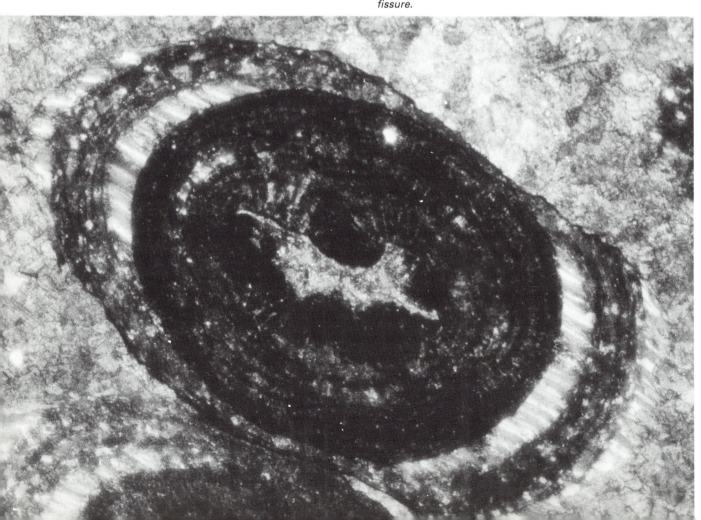

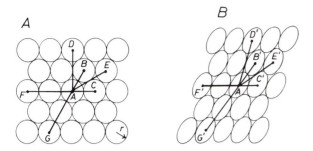

Figure 7.9. *Geometric features of an undeformed* (A) *and deformed* (B) *aggregate of regular hexagonally packed circular disks of radius r. Around centre A there are six neighbours with spacing 2r, six at distance $2\sqrt{3}r$, six at distance 4r etc.*

a plot of some 50 separate moves of the overlay have been made this implies that the initial aggregate of points possessed a completely random arrangement. Under these circumstances no final solution can be found. The rock could be either unstrained or strained if its initial geometric configuration was controlled by a Poisson random distribution (Figure 7.4).

2. Around the vacancy field high concentration of points will occur. This concentration can have either a circular or elliptical form like that of the central vacancy field (Figure 7.10C). This concentration relates to the commonest closest packing distance between the particles (Figure 7.9A *AB*, *AC*, or Figure 7.9B, *A'B'*, *A'C'*). If the initial particle size was fairly constant, and the particle packing a close one, this concentration will be very well marked. If the initial particles had a varied shape, or if the packing was not tight, this concentration will be weak. The elliptical or circular shape, like that of the vacancy area, reflects the form of the strain ellipse.

3. Around the strong concentration is a weak concentration of points reflecting the general lack of packing distances in this range (caused by the distance gap between *AC* and *AE* in Figure 7.9A).

4. If the particle size was initially very constant and the packing close and perfect, other concentric circular or elliptical zones of high and low point densities will occur. For most geological clastic aggregates such geometrically perfect initial packing distributions are unlikely.

Question 7.2

Determine the finite strain of the deformed oolitic limestone of Figure 7.7 using the Fry method described above. Estimate the ellipticity and orientation of the strain ellipse, and compare the result with that from Question 7.1. Compare the time spent on solving the strain problem by the measurement techniques of Question 7.1, and the Fry method.

STRAIN PARTITIONING

The deformation of a rock can be accomplished by various differing mechanisms acting in the crystalline aggregate of which it is comprised. Different crystal species react in different ways to the stresses causing deformation depending upon the environmental conditions operative during deformation. Figure 7.11A shows an initial undeformed aggregate, and B, C and D show the same overall or **bulk strain** produced by three quite different mechanisms.

Deformation by *plastic flow* (B) occurs when individual crystals change shape as a result of **crystal plasticity**, the component atoms or molecular groups inside the crystal move from one stable orientation to another. As a result of the physical rearrangement of the crystallographic lattice by the migration of dislocations, the crystals may show special strain effects (deformation bands, strain polarisation (Figure 7.13), deformation twinning (Figure 7.14)), and the crystals making up the aggregate often develop a shape orientation (fabric (Figure 7.13)) and a preferred orientation of their lattices (texture).

Grain boundary sliding (Figure 7.11C) is another mechanism whereby an aggregate can change shape, but without profound changes of forms of the individual crystals. The crystals slide past one another and, for example, two crystals lying next to one another in the line of extension move apart, the intervening space being filled by crystals moving in from the side. A good analogy for this process is the way the shape of a bag of ball bearings or sand grains can be altered by compressing two sides of the bag. With this type of deformation it is common for the grain fabric and texture of the deformation product to look almost identical to those of the starting material.

Another important mechanism of rock deformation,

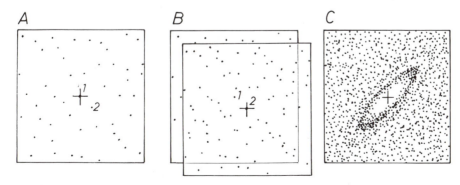

Figure 7.10. *The construction of strain state using the Fry method. A shows the overlay centred at point 1 and all other points located. B shows the overlay recentred over point 2, and all points relative to 2 located. C shows the features of the final diagram. The elliptical distribution of point densities gives the shape and orientation of the strain ellipse.*

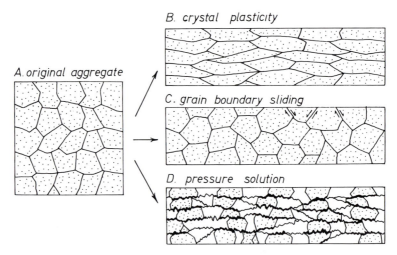

Figure 7.11. Deformation of a crystal aggregate A by different types of deformation mechanisms. In D the residual parts of original crystals are stippled, and the newly crystallized material is left unstippled.

particularly at low metamorphic grades is **pressure solution** (Figure 7.11, D, Session 7 frontispiece): individual grains undergo solution on the sides which face the greatest shortening direction, and this material is transfered to and crystallized on the sides which face the greatest extension direction. This mechanism can be identified by the presence of pressure solution seams or stylolites which cut off preexisting structural features of the grains (Figures 7.8, 7.12). Although the regrowth of the material is often in crystallographic continuity with the parent grain, there is often some recognizable physical difference between the new and old parts of each grain. For example, the new growth may show different types of inclusion or different inclusion density from that of the original parent and, in contrast to the original grain, it may show a fibrous morphology. Although the shapes of individual grains will be modified by the chemical transfer of material, in contrast to shape change by crystal plasticity, the crystallographic orientation of the original parent grains remains unaltered.

Figure 7.11 shows schematically deformation by one or other of these mechanisms, but in naturally deformed rocks it is common to find combinations of these mechanisms in differing proportions which acted together or at different times in the deformation history to build up the finite strain. By making a careful geometric analysis of the characteristic features of individual crystals, crystal groups and bulk strain it is sometimes possible to separate the different componental mechanisms. Such a procedure is known as **strain partitioning**. If we look again at Figure 7.11 in

terms of the centre to centre technique discussed above it will be seen that with crystal plasticity and pressure solution the distribution of grain centres is rearranged as a result of strain whereas grain boundary sliding leads to a grain centre redistribution that is geometrically indistinguishable from that of the original aggregate. Although the modification of the midpoints of the original component crystals in the processes of crystal plasticity and pressure solution may be similar, the grain truncation and grain growth phenomena of pressure solution enable the two processes to be distinguished. It should be pointed out that the geometry of crystal plasticity indicated in Figure 7.11B is often modified by the recrystallization of the deformed original grains into new subgrains, and the original grain structure may be totally obliterated. If this is the case the centre to centre grain analysis will not reveal information about the extent of the strains.

Strain partitioning

Question 7.3

Return to Figure 7.7 and observe carefully the forms of the ooids. The original concentric shells have been systematically truncated along those edges which are oriented perpendicular to the direction of maximum shortening. The innermost shells making the ooid cores have subelliptical forms, but the aspect ratios are less than the form of the strain ellipse. In an enlargement of the field of view (Figure 7.8) it can be seen that the ooid shells have been pulled apart, and the spaces between them have been filled with fibrous crystals of quartz and calcite, the long axes of the fibres being oriented parallel to the direction of the strain ellipse long axis. It appears that we have at least two separate mechanisms of deformation operating in the ooids. One is a relative movement of the grains comprising the ooid (by crystal plasticity or grain boundary sliding), the other is a shape change by pressure solution. The problem is to numerically separate the physical and chemical mechanisms and partition the total strain into its components.

In Figure 7.15B measure the shapes of the central ooid

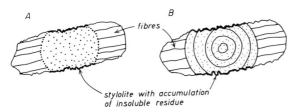

Figure 7.12. Features of clastic grains (A) and ooids (B) characteristic of deformation by a mechanism of pressure solution.

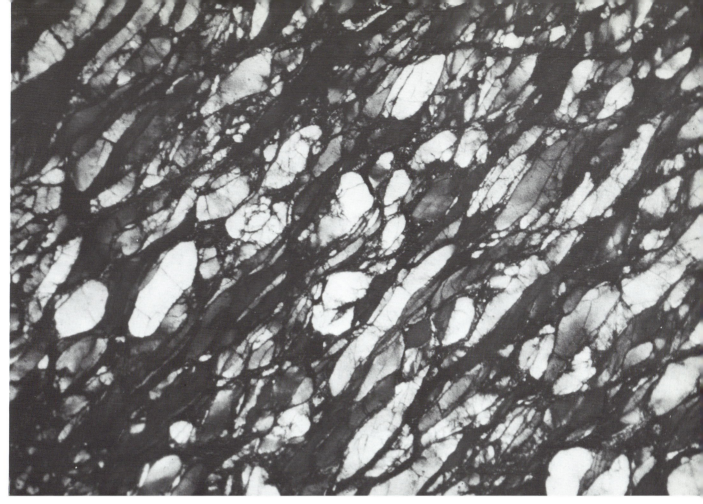

Figure 7.13. Thin section (crossed polars) of clastic grains in a quartzite deformed by crystal plasticity (cf. Figure 7.11B). Note the strongly elongated form of the grains and the presence of undulatory extinction as a result of internal deformation. South Mountain, Maryland, USA, × 30.

Figure 7.14. Thin section (crossed polars) of experimentally deformed Carrara marble (600°C, 1·4 kb, strain rate 10^{-14} s^{-1}, 15·5% shortening). The crystal deformation has taken place with the development of mechanical twinning. From an experiment performed by Stefan Schmid, × 150.

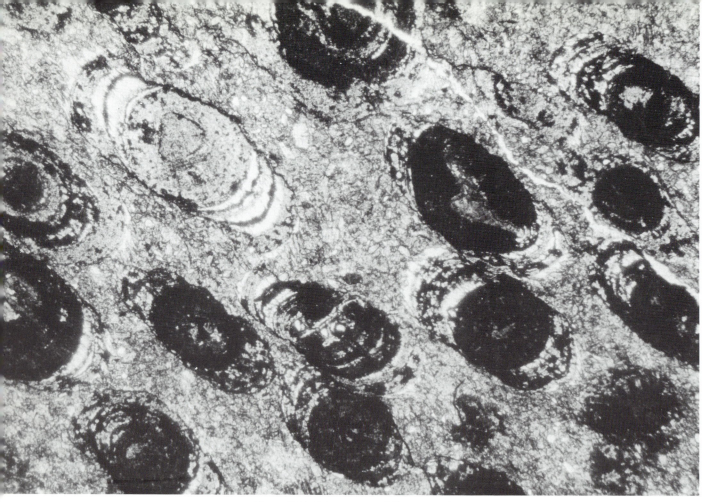

Figure 7.15. Enlargement of Figure 7.7. See Question 7.3.

shells where these appear unaffected by pressure solution truncations, and determine the strain by finding the geometric or harmonic means. Determine the ratios of maximum and minimum dimensions of the ooids, and find the mean of the total shape change. Knowing these two ratios partition the strain of the ooids into physical and chemical components. Relate these measurements to those obtained from the results of Question 7.1 and 7.2.

Now proceed to the Answers and Comments section, then to Session 8 or to the starred question below.

Strain partitioning; grain boundary sliding versus crystal plasticity

Question 7.4★

Figure 7.16 illustrates a thin section of the Cambrian quartzite shown in Figure 5.1, the section being parallel to the bedding surface. The quartz grains appear to represent the modified forms of the initial sand grains and their diagenetic quartz overgrowth. In undeformed rocks these cemented grains are statistically equi-dimensional and of constant size. The quartz shows undulose extinction, but no development of small grains or subgrain recrystallization. There is no evidence of pressure solution (truncated original grain contacts, local concentrations of insoluble mineral components). The two mechanisms that could have dominated the deformation are crystal plasticity and grain boundary sliding. In this example an analysis of the mid-points of grains might be used to partition the strain between the two mechanisms. The elongated grain shapes are the result of crystal plasticity, and the extent of this can be compared with the value of bulk strain obtained previously from an analysis of the shapes of the sections of the worm tubes. Determine the proportions of crystal plasticity and grain boundary sliding.

ANSWERS AND COMMENTS

"Nearest neighbour" technique

Answer 7.1

The distribution of ooid centres (Figure 7.17) does not show any pronounced clustering and resembles the point distribution of Figure 7.5 rather than that of Figure 7.4. This rather uniform distribution probably reflects the original packing of the ooids, the physical proximity of adjacent ooids being influenced by the average size of the ooid spheres. The distribution of points in Figure 7.17 shows that the separation distance is greatest along a line running from the top left to bottom right, and smallest in a direction at right angles, but this geometry is not easy to detect by eye alone.

Lines were drawn joining ooid centres and those crossing any ooid other than those belonging to the centres were rejected as not joining originally nearest neighbours (Figure 7.17). The lengths d' and azimuths α' of 400 original nearest neighbour tie lines were measured and the data graphed (Figure 7.18). The measurement of these 400 data points took several hours of work and we do not expect that the student has amassed so much data. The

data was analysed in groups of 100 points to see if there were significant differences from different parts of the field, but none were detected.

The next analytical problem consists of finding the best fit for the three unknown variables λ_1', λ_2' and β' of Equation 7.1. The maximum and minimum values of d' occur at positions $d'_{max} = k\lambda_1^{1/2}$, $d'_{min} = k\lambda_2^{1/2}$ (where k is an unknown constant the value of which depends upon the average length of the centre tie lines before deformation, and β' is the orientation of these maximum and minimum values). Averages of d' were obtained by finding arithmetic data means over 10° intervals, and these are shown as circles in Figure 7.18. The best fitting curve to these means gave $d'_{max} = 17 \cdot 0$, $d'_{min} = 9 \cdot 5$, and the ellipticity of the finite strain ellipse is $R = 1 \cdot 79$. The principal axes are situated at $-25°$ and $+65°$ from the azimuth direction.

Clearly much time was involved in this computation to obtain a high degree of accuracy. The student will probably have used fewer points and the resulting strain calculation will probably diverge slightly from the numbers given above. However, for many purposes computations of lower degree of accuracy are quite acceptable. As with all strain measurement techniques it is always best to consider what is the point of making the strain analysis, as the answer to this question will inevitably guide the choice of technique.

The Fry method

Answer 7.2

Figure 7.19 illustrates the result of the analysis using the distributions of centres around 50 chosen ooids taken from Figure 7.17. The point vacancy at the centre of the diagram is sub-elliptical indicating that the distribution of the centres in different directions is not uniform. The long axis of this empty area makes an angle of about $-30°$ with the reference azimuth. The sub-elliptical area has the same shape as the strain ellipse and has a measured aspect ratio of $2 \cdot 0 : 1 \cdot 2$, giving the ellipticity of the strain ellipse $R = 1 \cdot 7$. This result is not so precise as that determined from the centre to centre technique because it has been derived using fewer ooid centres. However, it was acquired much more rapidly, and involved no time consuming direction and distance measurements. For many geological investigations the result would be perfectly adequate. If greater accuracy is required it is not difficult to construct a computer program which, using digitized x, y-coordinates of each centre point it would be possible to plot and statistically analyse the resulting point distribution.

The central vacancy of the Fry diagram of Figure 7.19 is bounded by a region of relatively high point concentrations, and this region is itself surrounded by a region of lower than average point concentration. This variation in point density results from the constraints of initial distances between the centre points arising because of the packing distances of the initial ooids.

Strain measurement in clastic rocks

We began this Session by discussing some of the geometric features of deformed aggregates where the particles making up the aggregate (e.g. pebbles, ooids) have different

Figure 7.16. Thin section of quartzite (crossed polars × 30) parallel to the bedding surface of Figure 5.1. See Question 7.4★.

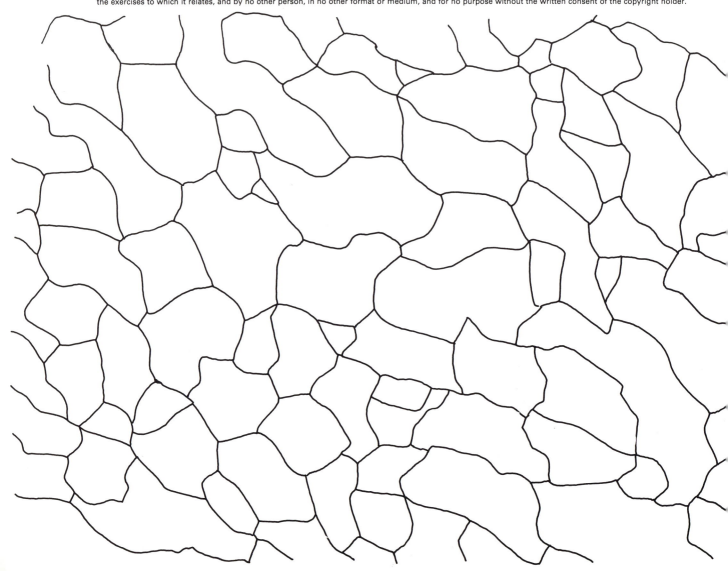

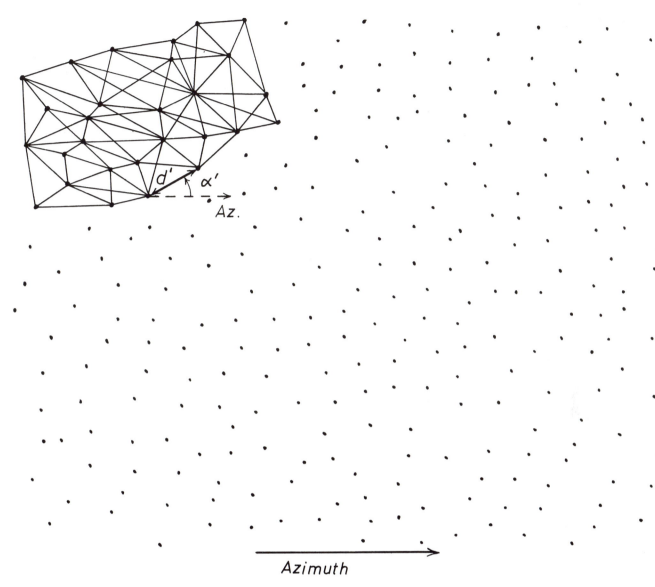

Figure 7.17. Ooid centres from Figure 7.7 plotted as points: note the lack of clustering. The lines joining centres in the top left hand corner are nearest neighbour tie lines determined according to the principles shown in Figure 7.6. For each tie line the distance d′ and orientation α′ were recorded.

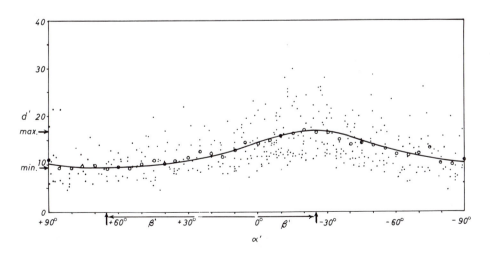

Figure 7.18. Graph of lengths of tie lines d′ against orientation α′. The circles represent the data means of d′ taken over 36 ten degree orientation sectors.

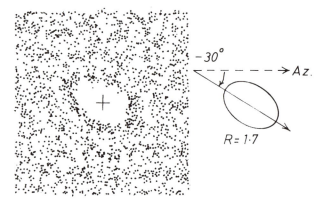

Figure 7.19. *Fry diagram of distances between 50 centre points of ooids from Figure 7.7.*

ductilities from their matrix. It has to be admitted that the calculation of strains from such rocks poses complex problems. However, such rocks are of great importance for calculating strains for a number of reasons. For example, the distribution of conglomerates is widespread through the whole geological column, and even after quite intense deformation and metamorphism the pebbles can still be recognized. Furthermore, it is clear that strains determined from deformed pebbles do relate to the total strain since the conglomerate was deposited. Another important feature of deformed conglomerates is that the final pebble shapes vary with pebble composition, and therefore we have the potential for comparing the ductilities of the several different rock types making up the pebbles. The principle features of a particle aggregate which control the shapes of the deformed particles are:

1. The rheological properties of the particle and of the matrix (competence contrast).
2. The original shape of the particle (ellipsoidal or non-ellipsoidal, centre- or non-centre-symmetric).
3. The original orientation of the particle.
4. The anisotropy of the particle (internal bedding, banding, effects of ooid cores etc.)
5. Closeness of particle packing—is the particle free to move without being influenced by its neighbours?
6. Strain history—whether increments are coaxial or non-coaxial.

Clearly there are a great number of factors here, and we will consider them in more detail in Volume 3. Because of the complexity of analysing the shapes of the deformed particles the strength of the techniques developed in this Session can be appreciated. Providing that the particle distribution in the aggregate was fairly uniform and the initial particle centres had an initial characteristic spacing distance, an analysis of the distribution of the particle centres can be used to determine the bulk strain, even though the individual particles may change shape in a very complex manner.

Strain partitioning

Answer 7.3

The cores of the ooids were reconstructed where pressure solution surfaces had removed part of the central zone. Analysis of the ellipticity of the central cores of the 13

ooids gave a mean value $R = 1.22$. The shapes of the external boundaries of the ooids were also determined, and gave a mean ellipticity $R = 2.24$. The first ratio gives the extent of the deformation by physical rearrangement of the crystal components (by crystal plasticity and grain boundary sliding) and the second gives the total strain in the ooids. The pressure solution component can therefore be obtained from the quotient $R = 2.24/1.22 = 1.84$.

It is interesting to note that the total deformation of the ooids ($R = 2.24$) is significantly higher than the total rock strain derived from the centre to centre technique ($R = 1.79$). How can this be explained? Returning to Figure 7.7 it will be seen that zones of intense pressure solution (pressure solution seams, marked by concentrations of insoluble phylosilicates) are concentrated at the edges of the ooids. It seems likely that the strain in the ooids is higher than that of the matrix because the ooids were more competent than the matrix and acted as "stress risers". Abnormally high pressure solution was induced along the ooid boundaries and the shape change in the ooid boundary was greater than the aspect ratio of the bulk strain ellipse.

One of the most outstanding pieces of work on strain partitioning was carried out by Shankar Mitra in the quartzites of the Appalachian fold belt (Mitra, 1978). We have briefly met the results of the work of Ernst Cloos in this area in which he determined strains in folded Palaeozoic carbonates using ooid shapes (Figure 2.8). Mitra investigated the deformation in the Cambrian quartzites which lie beneath these carbonates and which make an envelope around the volcanic rocks forming the core of the South Mountain anticlinorium. The deformation took place under greenschist facies metamorphic conditions (temperature 350°C, pressure 3.5 kb). Mitra found that the range of strain axial ratios ($X:Z$ from 1.7 to 7.0) was similar to those finite strains recorded by Cloos (1.6 to 6.9) and that the strain distributions in the folds were also similar: low strains on the normal fold limbs, high strains on the overturned limbs. He showed that, in thin section, the original clastic grains could be identified by concentrations of inclusions along the grain boundary, and that, at many localities, these original grain boundaries were cut on two opposite sides by pressure solution surfaces and overgrown on other sides by new quartz (cf. Figure 7.12A). Clearly pressure solution was an important deformation mechanism and, by making measurements of the grain solution and grain growth areas, he was able to put values on the extent of deformation by the pressure solution process. He also observed the presence of fine hair-like crystals of rutile inside some of the original quartz grains. In fact, rutilated quartz sand grains are not uncommon in many other quartzites derived from the erosion of medium to high grade metamorphic source areas (Figure 7.20). What made his observations especially pertinent to rock deformation studies however, was that he showed that the rutile needles possessed structures indicative of the distortion of the quartz crystal enclosing them. The needles were either stretched and showed micro-boudinage (Figure 7.22B), or contracted with the development of micro-folds (Figure 7.22A). The quartz grain had been deformed by processes of crystal plasticity (he termed this dislocation creep) and, by making a geometric analysis of the forms of differently oriented rutile needles before and after deformation, he showed how it was possible to determine the extent of the

Figure 7.20. Thin section (dark field illumination) of needles of rutile embedded in a sand grain in a quartzite. The needles are oriented randomly and form continuous grains because the quartz is undeformed (× 60).

Figure 7.21. Highly deformed magnetite bearing quartzite, South Mountain, Maryland, USA. This rock has suffered deformation both by crystal plasticity and pressure solution, and the extent of the pressure solution has been enhanced by the presence of deformation resistant magnetite grains, (× 30).

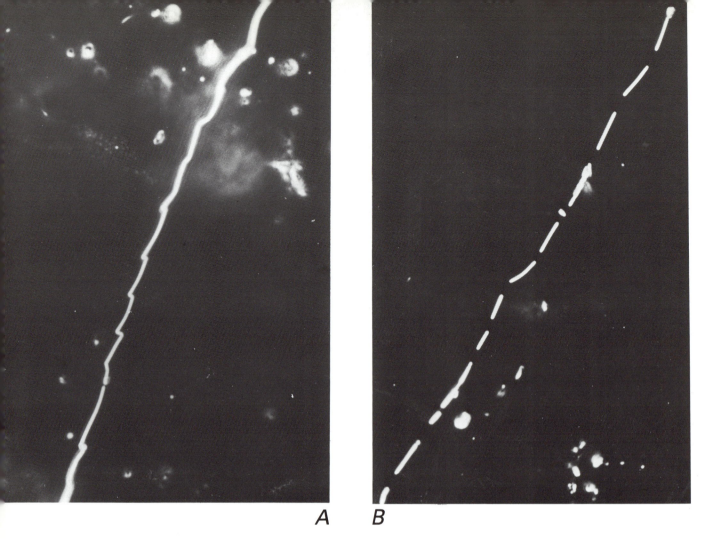

A   B

Figure 7.22. Photo micrographs (dark field illumination × 1500) of individual rutile needles embedded in quartz: A, shortened and folded; B, elongated and boudinaged. From Mitra (1978).

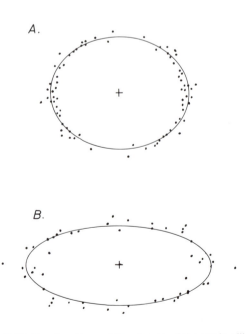

Figure 7.23. Constructions of the shapes of the strain ellipse for the component of crystal plasticity derived from an analysis of longitudinal strain suffered by rutile needles (from Mitra, 1978). Each point represents a calculated longitudinal strain. A, data from normal fold limb; B, from overturned limb.

crystal–plastic part of the deformation (Figure 7.23). Mitra developed methods of accurately assessing the extent of deformation by pressure solution and by plastic flow of quartz, and he examined the partitioning of these mechanisms with other geological factors: position of the layer within a fold (and relationship to total strain), lithology and temperature.

He showed that, in a given lithology, the partitioning varied with location in the folds. Crystal plasticity was often insignificant on the normal limbs, but its relative magnitude increased sharply on the overturned fold limbs, locally becoming the dominant deformation mechanism. Because the total strains on the fold limbs were geometrically related he suggested that the two mechanisms probably acted more or less at the same time, and that the rock deformation obeyed a composite flow law.

One of the quartzite units was poorly sorted and contained significant amounts of mica and magnetite. The effect of these impurities was quite spectacular and they appear to have led to an intense activation of pressure solution deformation mechanisms (Figure 7.21).

By comparing the partitioning of strain in rock specimens selected from similar lithologies and structural sites, but differing temperature regimes he was able to demonstrate that an increase in temperature led to an enhancement in the proportion of deformation by crystal plasticity over pressure solution.

The very informative study by Mitra showed, for the first time, how observations of the geometrical features of

Figure 7.24. Deformed pebbles from the Morcles fold nappe of the Helvetic Alps: A, inverted fold limb with deformation by plastic flow; B, normal fold limb with deformation predominantly by pressure solution.

naturally deformed rocks can be used to separate, or partition, the contributions of differing mechanisms operating on the scale of the rock crystals in the total finite strain state. We are of the opinion that studies along these lines will greatly assist our understanding of crystal deformation.

We would also like to emphasize that Mitra's conclusions have important correlations with other terrains and how observations made at different observational scales can be useful in revealing strain partitioning. In the Swiss Alps the Morcles nappe is recognized as a large scale recumbent anticline. The normal limb shows generally rather weak strains, whereas the overturned limb is intensely deformed (see profile section Figure 11.10). These total strains have been assessed by analysing the shapes and centre to centre distribution of pebbles found at the Tertiary–Cretaceous contact. In detail, the geometric forms of adjacent pebbles on the normal and inverted limb are quite contrasted. On the lower, inverted limb of the Morcles anticline the pebbles are extremely elongated parallel to the X-direction and strongly shortened in the Z-direction. The deformation appears to be rather homogeneous through the whole conglomerate mass and the pebble matrix, and the crystallographic orientation of the component calcite grains shows a strongly defined fabric and texture (Figure 7.24A, see previous page). In contrast, the same conglomeratic horizon on the normal limb of the Morcles fold shows geometric features which suggest deformations by a combination of plastic flow and pressure solution mechanisms (Figure 7.24B). The pebbles are mutually impressed into each other along stylolitic pressure solution contacts, and there is also a shape change of individual pebbles that cannot be simply related to the pressure solution process. The predominance of pressure solution over plastic flow on the upper limb of the Morcles fold is clear and it is associated with the strong development of extension vein systems which act as the sink for the material removed on pressure solution surfaces.

The advanced student would be advised to think carefully about the implications of the geometric modifications that arise by pressure solution. For example, grain to grain solution in a rock might lead to very local deposition of the pressure solved material in local extension vein systems. How would these redistributions of material affect local computations of strain? They might give rise to very different strains calculated from the local strain geometry as compared with the bulk strain of the overall rock mass.

Answer 7.4★

A centre point analysis of the grains of Figure 7.16 made by the Fry method is illustrated in Figure 7.25. It is not

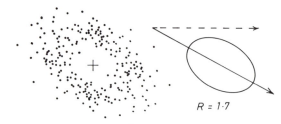

Figure 7.25. *Central part of a Fry diagram prepared from the midpoints of quartz grains of Figure 7.16.*

an easy matter to identify with confidence the dimensions of the elliptical form of the point data. The ellipse fit we have made gives a strain ratio $R = 1.7$. This value is very close to that determined from the deformed cross sectional shapes of *Scolithus* pipes (see Answer 5.1, $R = 1.6$), and we deduce that the rock deformation took place predominantly by crystal plasticity with little or no grain boundary sliding. The work of Mitra also led him to conclude that grain boundary sliding was probably insignificant as a deformation mechanism in the Cambrian quartzites of South Mountain fold.

KEYWORDS AND DEFINITIONS

Bulk strain	The average strain in a heterogeneously deformed material.
Crystal plasticity	A deformation taking place inside individual crystals as a result of glide and climb of dislocations. It leads to change of shape of the crystals and generally to a change in the crystallographic orientations (Figure 7.11B).
Grain boundary sliding	A deformation which takes place by sliding of individual grains one past the other along crystallographic grain boundaries (Figure 7.11C).
Pressure solution	See p. 53 and Figure 7.11D.
Strain partitioning	The separation of the different componential rock deformation mechanisms which have acted together or in sequence to produce the finite strain observed in the rock.

KEY REFERENCES

Fry, N. (1979). Random point distributions and strain measurement in rocks. *Tectonophysics* **60**, 89–105.

This paper marks a major technical advance in the understanding of the geometry of deformed particle aggregates found in rocks, and develops the method, termed in this book the "Fry method", for analysing strains in naturally deformed rocks. This would be a good paper to read in conjunction with Question 7.2.

Lisle, R. J. (1979). Strain analysis using deformed pebbles: the influence of initial pebble shape. *Tectonophysics* **60**, 263–277.

A good introduction to some of the complex geometric features of deformed conglomerates.

Mitra, S. (1978). Microscopic deformation mechanisms and flow laws in quartzites within the South Mountain anticline. *J. Geol.* **86**, 129–152.

Mitra, S. and Tullis, J. (1979). A comparison of intracrystalline deformation in naturally and experimentally deformed quartzites. *Tectonophysics* **53**, T21–T27.

The first of these two related papers describes the microscopic strain analysis of quartzites which enables a major advance in the techniques of strain computation, namely the possibility of separating the various mechanisms in a rock which led to the finite strain state we observe today. The second paper describes laboratory experiments, which are a follow up to the analysis of naturally deformed rocks, and which confirms the validity of using rutile needles as markers to record the extent of crystal plasticity.

Ramsay, J. G. (1967). "Folding and Fracturing of Rocks", 568 pp. McGraw-Hill, New York.

Pages 195–197 set out, for the first time, the methods of analysing strain using centre to centre measurements of particles making up rock aggregates. This technique was developed for analysing strain in aggregates where pressure solution has played a dominant role, and where particle shapes cannot be used to define finite strain.

SESSION 8

Practical Strain Measurement:

4. Angles

The way that shear strain γ varies with a general two-dimensional displacement is investigated with respect to initial and final orientations of a line element. The basic equations are used to solve for the values of ratio and orientations of the principal strains from data derived from initially perpendicular lines. The methods are applied to fossils with initial bilateral symmetry. Solution of the equations using graphical methods and the Mohr construction are described. Statistical analysis of preferred orientation of initially randomly oriented lines is used to evaluate strain. Strain determination from non-perpendicular angles and spiral forms is presented.

INTRODUCTION

In Session 6 we investigated how extension e varied with direction in a homogeneously strained body, and we arrived at methods whereby individual measurements of values of e could be used to determine the form and orientation of the strain ellipse. In the following session we shall investigate how angular changes between lines vary in a homogeneously deformed surface, and how such variations can be used to calculate the principal features of the strain ellipse.

We shall first re-examine the geometric features of a general displacement which leads to a state of homogeneous two-dimensional strain, particularly those features relating to the changes in angles between two initially perpendicular lines. In several of the previous Sessions we saw that, in general, the angle between any two perpendicular lines becomes deflected from the initial 90° by an amount termed the angular shear strain ψ, or the shear strain gamma ($\gamma = \tan \psi$). On a surface there are two initially perpendicular directions where the angular shear strain is zero, and these directions become the two principal strains parallel to the major and minor axes of the strain ellipse. The only exception to this general rule is where the strain sets up a uniform dilation (i.e. the strain ellipse has a circular form). In this special case of deformation all initially perpendicular lines remain perpendicular. Let us return to a more general strain where the principal strains have different values. In order to study the way shear strain varies with direction, we should go back to the general displacement illustrated in Figure 6.1. The various lines A–R were selected at 10° intervals so that the angles $A \angle J, B \angle K, C \angle L \ldots I \angle R$ were initially 90°. After displacement the angles between the displaced lines $A' \angle J', B' \angle K'$ etc. are generally not perpendicular. We can demonstrate how shear strain varies with orientation in two graphs (Figure 8.1): the first graph relates shear strain γ to orientation α before displacement and strain; the second relates γ to the orientation α' after straining. These two graphs are equivalent to the two graphs we constructed to demonstrate variation of length change with orientation α and α' (Figure 6.2). Both shear strain graphs intersect the $\gamma = 0$ axis at two points and for each of the graphs these intersections are perpendicular. These two sets of perpendicular intersections clearly relate to the lines of no shear strain, and are parallel to the directions of principal strains before and after deformation. They coincide with the directions of maximum and minimum extension shown in Figure 6.2, and the differences in positions on the α–γ and α'–γ graphs relate to the rotational component ω of the strain. The α–γ graph reaches maximum and minimum values at points located at angles of ±45° from the axis of greatest extension. This relationship is true for all types of homogeneous strain (except the special case of uniform dilatation mentioned above). The two lines which eventually take up the maximum and minimum shear strains are always initially perpendicular and lie symmetrically at ±45° to the perpendicular lines which will become the axes of the strain ellipse. After deformation the final positions of the directions of maximum and minimum shear strain are always located closer to the axis of greatest stretch at symmetric positions which depend on the ellipticity R of the strain ellipse. In the case of the displacement investigated here they are oriented at angles of ±25·4° from the axis of greatest extension.

For all the intermediate points on the two curves it is possible to express how shear strain is a function of both orientation (α or α') and components of the deformation matrix a, b, c and d. These are

$$\gamma_\alpha = \frac{2\cos 2\alpha(ab + cd) - \sin 2\alpha(a^2 - b^2 + c^2 - d^2)}{2(ad - bc)} \qquad (8.1)$$

$$\gamma_{\alpha'} = \frac{[(bm' - d)^2 - (c - am')^2](ab + cd) \quad - (a^2 - b^2 + c^2 - d^2)(c - am')(bm' - d)}{[(c - am')^2 + (bm' - d)^2](ad - bc)} \qquad (8.2)$$

where $m' = \tan \alpha'$.

The proofs of these equations are set out in Appendix B (Equations B.25, B.26). As with the equations we discussed earlier relating elongation to orientation, the complexity of mathematical structures arises because of the presence of factors relating to the rotational component of the strain. In our discussions of the variations of extension of line elements we found that the relationships of e with orientation were considerably simplified if they were related to the directions of principal strain, and that these simpler equations became of great practical importance in determining strain in deformed rocks. The same holds true for the shear strain equations. We can refer shear strains to directions measured with respect to the principal direction either before (ϕ) or after (ϕ') deformation.

If we choose the directions of our coordinate system to coincide with the principal strains after deformation (Figure 8.2 "new x" and "new y") and shift all the γ–α points through the rotation component ω, the two sets of curves become symmetric about the new coordinate axes. At this stage in our discussion it might be advisable to return briefly to the comments we made on the validity of this process on pp. 90–91.

With these new coordinate directions we can relate the relationships of ϕ and ϕ' and the variations of shear strain γ with initial (ϕ) or final orientation (ϕ'). The full mathematical proofs are set out in Appendix D:

$$\tan \phi' = R^{-1} \tan \phi \qquad (8.3)$$

$$\gamma = (R - 1/R)\cos \phi \sin \phi \qquad (8.4)$$

$$\gamma = \frac{(R^2 - 1) \tan \phi'}{1 + R^2 \tan^2 \phi'} \qquad (8.5)$$

where R is the ellipticity of the strain ellipse.

Strain determination using measurements of angular shear strain

In structural geology it is often possible to find practical methods for measuring angular shear strain. Features such as originally bilaterally symmetric fossils, desiccation cracks initially perpendicular to bedding planes, *Scolithus* tubes oriented perpendicularly to the bedding, cooling cracks in dykes and sills initially perpendicular to the intrusion walls are examples of line features which, in deformed rocks, might be used to determine shear strain. With such initially perpendicular line elements we measure deflections (ψ) from the perpendicular, and we must relate the sign of the shear strain to the reference direction. In Figure 8.3 two initially perpendicular lines A and B are transformed to new orientations A' and B'. The shear strain with reference to direction A' is $+\psi$ because line B' has moved relatively anticlockwise, whereas the shear strain with reference to line B' is $-\psi$ because line A' has moved clockwise relative to A'. In the computations which follow it is most important to specify the positive or negative sign of the shear.

It should be clear from the structure of the equations set out above that by making measurements of shear strain alone we can never determine absolute values of the principal strains, but only the value of the ellipticity $R = (1 + e_1)/(1 + e_2)$. The reason is that angular variations in a deformed body are independent of dilatation (see Figure 8.4).

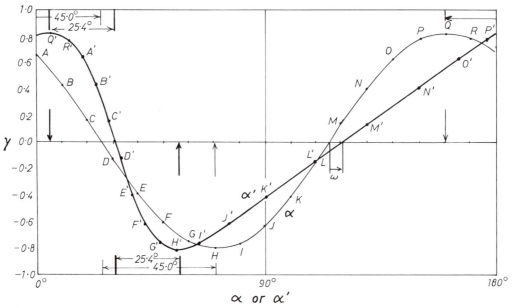

Figure 8.1. Graphs showing variation of shear strain γ with orientation before (α) and after (α') displacement, as developed in the displacement scheme shown in Figure 6.1. Maximum and minimum values of γ occur at $\alpha = \pm 45°$ and $\alpha' = \pm 25.4°$ from the greatest principal extension direction.

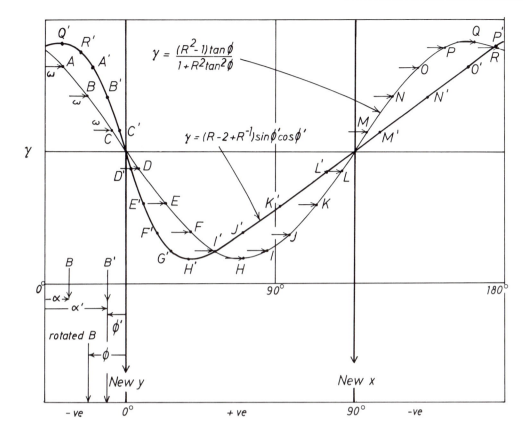

$$\gamma = \frac{(R^2-1)\tan\phi}{1+R^2\tan^2\phi}$$

$$\gamma = (R-2+R^{-1})\sin\phi'\cos\phi'$$

Figure 8.2. *Relationships of shear strain γ and angles between the principal extension direction before (φ) and after (φ') deformation. Points on the original α' curve of Figure 8.1 have been shifted sideways by an amount equal to the rotational component (ω).*

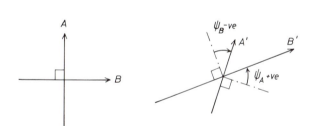

Figure 8.3. *Sign convention for angular shear strain.*

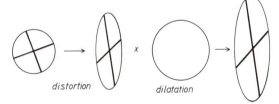

Figure 8.4. *The values of angular shear strain ψ are independent of the absolute values of the principal strains because they are independent of any dilation component $(1 + \Delta_A)$.*

From the form of Equation 8.5 it will be apparent that if we can make one measurement of shear strain $\gamma = \tan\psi$, and if we know the orientation of the principal strains from this reference direction (ϕ'), one equation is sufficient to determine the one unknown principal strain ratio R. There are two practical methods for solving for R. Both are simple and convenient graphical methods: the first uses graphs of the function of Equation 8.5 and was first suggested by the German structural geologist Hans Breddin; the second uses a Mohr circle construction to solve the strain equation.

Breddin graphs

For any particular ellipticity R it is possible to graph how angular shear strain ψ varies with orientation ϕ' that the line makes with the axes of the strain ellipse. Families of curves for differing R values are plotted in Figure 8.5. Any two values of ψ and ϕ' (NB +ve and −ve sign convention) from, for example, a single deformed fossil can be used to directly evaluate R.

Mohr construction

In Session 6 (p. 95 and Figure 6.7), we discussed the relationships of angular shear strain ψ and ϕ' as they appear in a Mohr construction. It might be an idea to return briefly to revise this. To derive a Mohr circle so that we can solve the strain problem posed by single values of ψ and ϕ' we proceed as follows:

1. Draw an orthogonal λ' (abscissa), γ' (ordinate) coordinate frame.
2. From the origin O construct a line making an angle of ψ with the λ' axis—see Figure 8.6A, NB sign convention.
3. At *any point P* along this line construct a line parallel to the λ' axis, and set off an angle of $2\phi'$ to intersect the abscissa at some point C (Figure 8.6A).
4. With centre C and radius CP construct a circle (Figure 8.6B). This circle is the Mohr circle for the strain state—note that the angle PCO is $2\phi'$, and that the point P must represent a state of strain in this circle

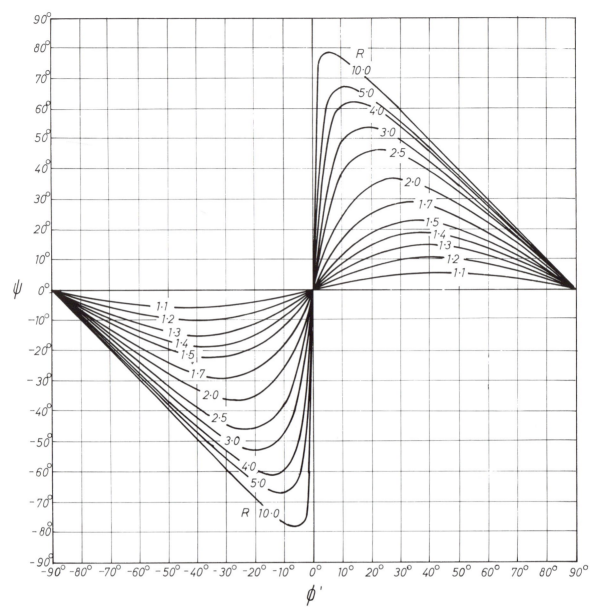

Figure 8.5. Breddin curves showing variations in angular shear strain ψ with orientation φ' of the greatest principal extension direction.

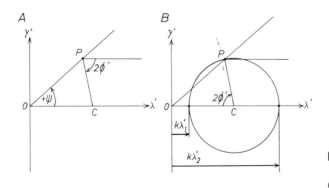

Figure 8.6. The Mohr construction for the solution of strain ratio from a single observation of angular shear strain.

with angular shear strain at an angle φ' from the greatest extension.

5. The two points where the Mohr circle intersects the λ' axis are equivalent to the principal strains. It is clear that we have no absolute scale fixed along the λ' axis, but we can measure the distances $k\lambda'_1$ and $k\lambda'_2$ and hence derive the strain ratio $R =$ where k is a constant of unknown value.

Deformed trilobites

Question 8.1

Figure 8.7 illustrates 2 specimens (A and B) of the trilobite *Angelina* from a locality in the Ordovician slates of Lleyn, North Wales. The fossils originally had bilateral symmetry,

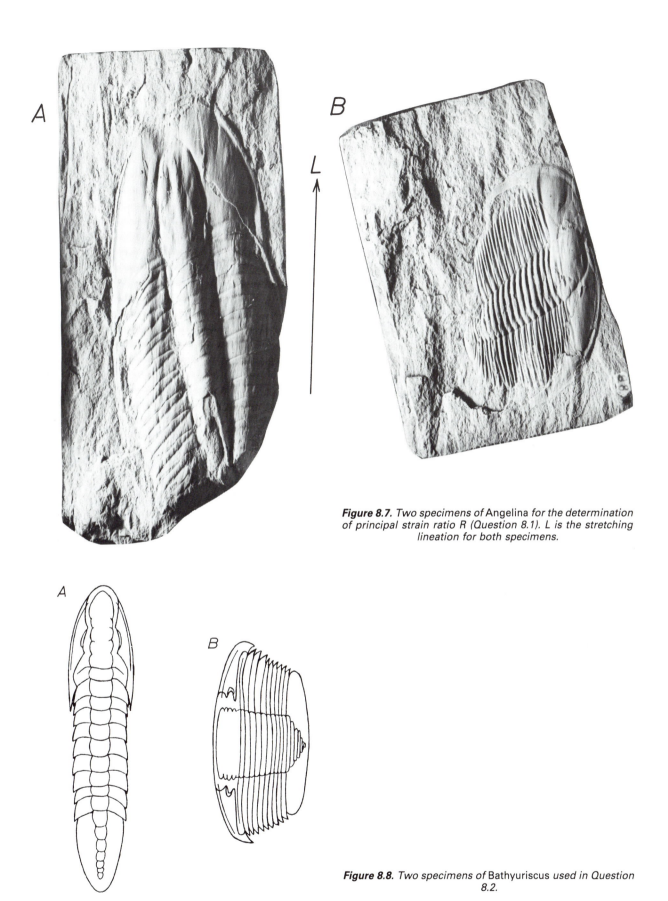

Figure 8.7. *Two specimens of* Angelina *for the determination of principal strain ratio R (Question 8.1). L is the stretching lineation for both specimens.*

Figure 8.8. *Two specimens of* Bathyuriscus *used in Question 8.2.*

but this symmetry has been lost on account of tectonic deformation during the Caledonian orogeny. Determine the angular shear strain along the axis of each fossil (ψ_A and ψ_B). A fine linear stretching fabric runs obliquely across the fossils. Assuming that this lineation L is parallel to the maximum extension direction in the surface, determine the angles ϕ'_A and ϕ'_B of the fossil axis from the maximum extension.

Using the Breddin graphs of Figure 8.5, determine the value of the strain ratio R for each fossil.

Construct a Mohr circle for each of the fossils according to the method described above and again determine two values for R.

Construct an orthogonal grid over fossil B and determine the original form of the fossil by reconstructing this grid in the undeformed state.

Deformed trilobites

Question 8.2

Figure 8.8 is a sketch of two specimens of the deformed trilobite *Bathyuriscus* taken from the same locality. Although these fossils are deformed, they still show bilateral symmetry. What does this imply? Can the Breddin or Mohr techniques be used to establish the values of the two-dimensional strains? If not why not? Derive a technique to determine the strain ellipse shape R. Hint: return to those methods used for evaluating strains from elliptically shaped particles and check how the shape of an initial ellipse changes when it is coaxial with the strain ellipse.

Deformed crinoid and brachiopod

Question 8.3

Figure 8.9 is a sketch of a block of deformed marly limestone from North Devon, England which contains a stretched crinoid stem and a deformed brachiopod. Fibrous quartz and calcite have been deposited between the separated ossicles of the crinoid.

Calculate the following strain parameters in the direction of the hinge line of the brachiopod—e, λ, λ', ψ, γ, γ'. Can you deduce anything about the orientations and values of the principal strains in the surface? Hint: assume that the radial ribs were arranged initially in an equiangular fashion, and use this to find the directions of the principal longitudinal strains.

Deformed plants

Question 8.4

A sketch of a bedding surface of a Carboniferous plant-bearing shale is illustrated in Figure 8.10. On this surface are three fragments of fern A, B and C. For each fragment calculate the angular shear strain along the stem axis. What can you deduce about the orientations of the principal strain axis in the bedding surface and the nature of the strain field? Calculate the angles that the pinnules originally made with the stem axis.

Before proceeding to the next question check the Answers and Comments section.

Strain ratio and orientation determined from two or more measurements of shear strain

If measurements of shear strain (ψ_A and ψ_B) can be made along two different directions A and B of known angular relationships (angle α' between the two directions) then it is possible to calculate both the orientation and the ellipticity of the strain ellipse. This possibility results from the mathematical structure of the equation relating strain ratio R, orientation of the principal axes (ϕ' from direction A) and shear strain γ. For directions A and B respectively we can write:

$$\gamma_A = \frac{(R^2 - 1) \tan \phi'}{1 + R^2 \tan^2 \phi'} \tag{8.6}$$

$$\gamma_B = \frac{(R^2 - 1) \tan(\phi' + \alpha')}{1 + R^2 \tan^2(\phi' + \alpha')} \tag{8.7}$$

Providing that the strain is homogeneous (R and ϕ' both constant), these two equations are soluble for the two unknown quantities R and ϕ'. The solution can be made algebraically, but answers can be obtained more rapidly using the Breddin graphs or Mohr construction.

Breddin graph method

The basic method consists of plotting data relating to orientation and angular shear strain on to a piece of transparent paper, and then adjusting the orientation of this data as an overlay on the set of strain curves of Figure 8.5 so that the data satisfy a curve of constant R value.

Using the same scales for abscissa and ordinate as Figure 8.5 plot points representing two (or more) directions (see Figure 8.12). It is most important to set down correctly the signs of the angles between the various directions and the angular shear strains (clockwise angular relationships are negative, anticlockwise are positive). The transparent overlay with these data is then placed over the reference curves of Figure 8.5 with the abscissa axes coinciding. The overlay is then moved sideways (keeping the abscissa coincident) until the datum points lie on (or about) a single R curve. The R value is directly read off this position, and the orientations ϕ' of the principal axes of strain from one of the reference directions are determined using the reference scale of the abscissa.

Mohr construction

This method, like that using the Breddin graphs, also uses an overlay technique to relate primary data to the location of a Mohr circle.

1. The axes for a Mohr diagram are constructed (abscissa λ', ordinate γ') and the primary angular shear strain data (ψ_A, ψ_B) are plotted as straight lines drawn through the origin making angles of ψ_A and ψ_B with the λ' axis (Figure 8.13A)

2. The second part of the construction is made on a transparent overlay. A circle with centre c of any convenient radius r is drawn and two points A and B representing the strain states of directions A and B are located on the circumference such that the angle AcB is twice the measured angle between the two directions (i.e. $2\alpha'$) (Figure 8.13B).

3. The transparent overlay is placed over the axes of the

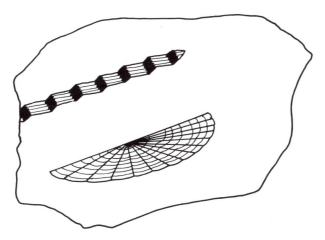

Figure 8.9. Block containing a stretched crinoid stem and brachiopod. See Question 8.3.

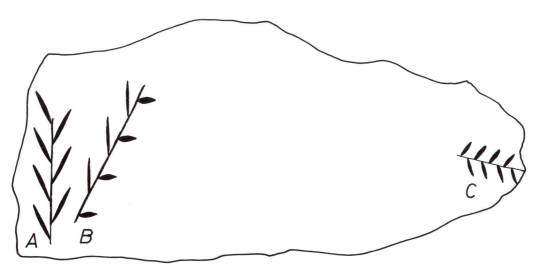

Figure 8.10. Bedding surface containing deformed plant fragments. See Question 8.4.

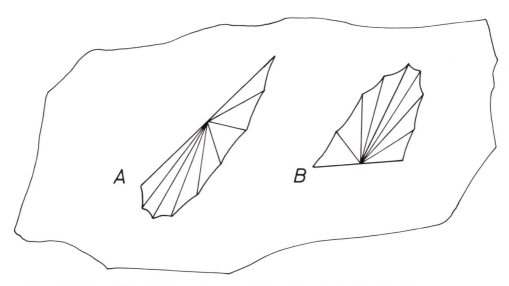

Figure 8.11. Two originally bilaterally symmetric brachiopods. See Question 8.5.

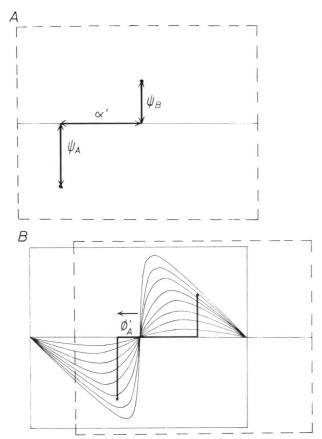

Figure 8.12. *The Breddin graph method for the determination of R and φ' from two measurements of angular shear strain.*

Mohr diagram so that the centre of the circle lies on the λ' axis. The overlay is moved sideways (the circle centre c is kept on the λ' axis) and rotated about the centre c until the points A and B on the periphery of the Mohr circle lie on their respective ψ_A and ψ_B lines. There is only one position where all the data are consistent.

4. The values of $k\lambda_1'$ and $k\lambda_2'$ can now be read off the abscissa axis, and the ellipticity of the strain ellipse calculated from $R = (k\lambda_2'/k\lambda_1')^{1/2}$. The orientation ϕ_B' of the greatest extension from direction B can be found from the angular relationships in the Mohr diagram where the angle $BcO = 2\phi_B'$.

Deformed brachiopods

Question 8.5

Figure 8.11 shows a sketch of two deformed brachiopods lying in a strained bedding surface. Assuming that the strain is homogeneous, determine, using first the Breddin graphs and second the Mohr construction, the value of the principal strain ratio R and the orientations of the principal strains.

Now pass on to the Answers and Comments section; then either answer the starred advanced questions below or move on to Session 9.

STARRED (★) QUESTIONS

Deformed coral and crinoids

Question 8.6★

In Question 8.4 we saw that although some fossils do not have an initial bilateral symmetry it may be possible to make certain geometric constructions which enable angular shear strains in certain directions to be measured, and hence the principal strain ratios and orientations to be computed. We always need a minimum of two angular shear strain values for this calculation.

Figure 8.14A illustrates a specimen of the coral *Favosites* from deformed Siluran slates of Pembrokeshire, South Wales. When undeformed, the polygonal coral polyps have the form of a perfectly regular hexagonal honeycomb. Determine the orientation and ratio of the principal strains.

Figure 8.14B shows a bedding surface in a deformed Devonian limestone from North Spain containing ossicles of the crinoid *Pentacrinus*. Undeformed crinoids show a perfectly regular pentagonal symmetry, with the rays oriented at 72° to each other. Determine the orientation and ratio of the principal strains.

The change in orientation of passive line markers as a result of strain

Question 8.7★

In our discussion of the geometry of a deformed spectrum of differently oriented line elements we saw that strain leads to a change of orientation of these lines with a

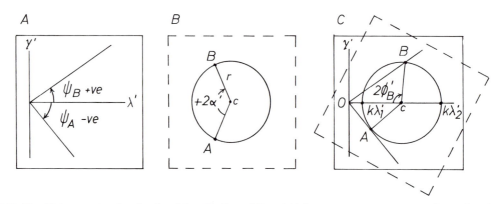

Figure 8.13. *The Mohr construction for the determination of R and φ' from two measurements of angular shear strain.*

A B

Figure 8.14. A, *deformed* Favosites; B, *deformed* Pentacrinus. *See Question 8.6★.*

concentration of the lines towards the direction of maximum extension. Any line initially making an angle of ϕ with the direction which will become the greatest elongation is displaced to make an angle ϕ' with the long axis of the strain ellipse. The relations between ϕ and ϕ' depend only on the ratio R of the principal strains (Equation 8.3). The concentration effect as an assemblage of passive line markers can be used as a measure of R. Figure 8.15 illustrates the change in frequency f of initially randomly oriented lines in a deformed surface.

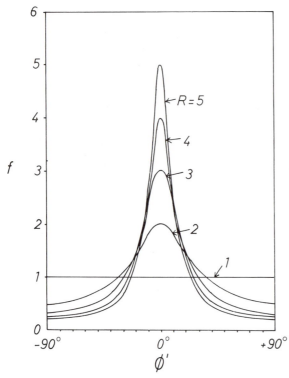

Figure 8.15. *Graphical plot of frequency of concentration F of passive line elements with respect to strain ratio R and orientation ϕ' from the maximum extension direction.*

To use this geometry to compute the strains we must be sure that the following conditions are met:

1. The line elements must be *truly passive*. There should be no competence contrast between the linear objects and their matrix. If there is a competence contrast, the relationships between ϕ and ϕ' are complicated by additional rigid-body rotations which depend upon the object shape and the way the finite strain is built up (i.e. the deformation path). Rotational finite strains will produce a different line element geometry from that of an irrotational strain. If the line elements are mechanically active, their maximum final concentration will not necessarily be symmetric about the maximum extension axis as occurs with passive elements. It follows that line elements such as prismatic crystals and belemnite fragments are therefore inapplicable to this type of analysis. Objects which may be used for this analysis are certain trace fossils (feeding trails, bioturbation lines, worm burrows etc.) or very thin elongate fossils such as leaves or plant stems.
2. The method depends on a statistical analysis of the frequency of certain directions. For such an analysis

to be accurate we must have *enough line elements* for statistical significance. To define "enough" is not easy: probably 50 lines is a minimum value.
3. We must be sure that the lines either had a *truly random initial orientation*, or, if this is not the case, we must have some knowledge of the initial orientation distribution. For example, certain trace fossils may have an initial preferred orientation (feeding burrows either perpendicular to the bedding or controlled by sea floor slope orientation), and elongate fossils may have been oriented by current activity during the deposition of their enclosing sediment.

Figure 8.16A shows the bedding surface of a Liassic marl from the Grandes Rousses massif of the western Alps of France. The surface shows small, randomly oriented, silt-filled, tubular trace fossils. There appears to be little or no alignment of the long axes of the tubes, and the rock appears to be essentially undeformed. Figure 8.16B shows a rock of identical lithology from another locality in the region. Here the trace fossils show a strong, statistically preferred orientation. The rock also contains a stretched belemnite (right hand side) and a number of calcite filled pressure shadow areas formed around small deformation-resistant particles of pyrite.

Analyse the degree of preferred orientation in Figure 8.16B and make a histogram plot of direction frequency against orientation. Compare the result with the theoretical analysis of passive line elements presented in Figure 8.15 and compute the orientation and strain ratio R of the principal strains. Determine the absolute values of the principal strains.

Strains determined from initially non-perpendicular angles

Question 8.8★

So far we have concentrated on evaluating strains using measurements of initially perpendicular line elements because these are relatively common relationships in natural situations. However, it is not necessary to have initially perpendicular angles in order to compute strain, *any angle* can be used to compute the strain ratio and the orientation of the strain ellipse.

If we know the values of angles α, β before deformation, and α' and β' after deformation, where ϕ' refer to a knowledge of the principal strain directions, Table 8.1 sets out the possibilities of strain computation.

The relationships of angles of known or unknown initial values are probably best resolved graphically or by computer analysis of the various sets of equations. To apply

Table 8.1

Data	Mathematical structure	Possible computations
$\alpha\alpha'$	1 equation, 2 unknowns	Nothing
$\alpha\alpha'\phi'$	1 equation, 1 unknown	R (see Ramsay, 1967, pp. 243–244)
$\alpha\alpha'\beta\beta'$	2 equations, 2 unknowns	$R\phi$ (see Ramsay, 1967, pp. 244–245)

A

Figure 8.16. *Silt filled trace fossils (probably feeding burrows) in a marly sediment.* A, *undeformed;* B, *deformed. See Question 8.7★.*

B

the basic analysis applicable to all initial angles consider Figure 8.17. Figure 8.17A illustrates two line elements initially making an angle α with each other, and with one of the elements making an angle ϕ with the direction which will become the principal maximum extension. After deformation these angles become α' and ϕ' respectively. Then

$$\tan \phi' = R^{-1} \tan \phi$$
$$\tan (\alpha' - \phi') = R^{-1} \tan(\alpha - \phi)$$
$$\alpha' = \tan^{-1} \phi' + \tan^{-1}(\alpha' - \phi')$$

or

$$\alpha' = \tan^{-1}(R^{-1} \tan \phi) + \tan^{-1}(R^{-1} \tan(\alpha - \phi)) \quad (8.8)$$

Thus knowing α it is possible to evaluate a series of graphs showing how α' varies with R and ϕ'. The Breddin graphs we have discussed earlier are one special set of these families of curves (for $\alpha = 90°$). Figure 8.18 illustrates the effect of differing strains on initial angles of $30°$ and $60°$, and Figure 8.19 shows graphically the functions derived from Equation 8.8. If we know any initial angle α, we can use these graphs in the same way as we used the Breddin graphs to calculate R and ϕ'—two observed values of α' are necessary for a solution.

If the initial value of α is unknown, we have to have available sets of α' curves for a number of initial α values. Then, having three values of α', it is technically possible to find a unique solution (the three points should fall on one curve) to evaluate R, ϕ' and α.

If we have many values of α' then it is best to try and evalute the maximum and minimum values of α' because these values can be used to establish the strain ratio R, the initial angle α and the orientation ϕ' of the principal strains. It should be clear from a study of the geometric features shown in Figure 8.18 that maximum and minimum values of α' occur where α' is bisected by the minimum and maximum strain axes respectively. Figure 8.17B illustrates the geometric features of these special positions to the initial positions of the lines initially intersecting at an angle of α. The minimum and maximum values of α' are

$$\min \alpha' = 2 \tan^{-1}(R^{-1} \tan \alpha/2) \quad (8.9)$$

$$\max \alpha' = 2 \cot^{-1}(R^{-1} \cot \alpha/2) \quad (8.10)$$

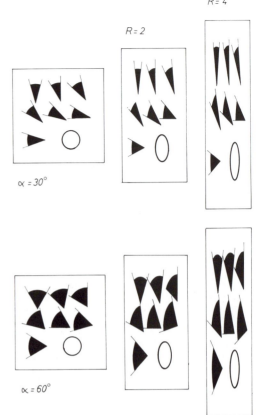

Figure 8.18. *Change in initial angles $\alpha = 30°$ and $\alpha = 60°$ as a result of strain with principal strain ratio R.*

From these relations we find that

$$R = \left(\frac{\tan(\max \alpha'/2)}{\tan(\min \alpha'/2)}\right)^{1/2} \quad (8.11)$$

and

$$\alpha = 2 \tan^{-1}\left(\tan\left(\frac{\max \alpha'}{2}\right) \tan\left(\frac{\min \alpha'}{2}\right)\right)^{1/2} \quad (8.12)$$

Figure 8.20 shows a bedding surface containing several examples of the chitinous Ordovician graptolite *Didymograptus*, also known as the tuning-fork graptolite from the

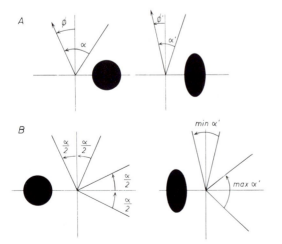

Figure 8.17. A, *Relationship of any initial angle α and its deformed product α'; B, Relationship of maximum and minimum α' angles to principal strains.*

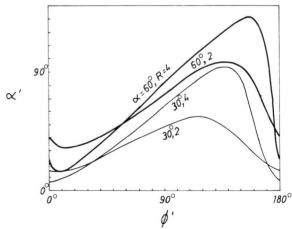

Figure 8.19. *Graphs showing variation in angle α' between two line elements with orientation as illustrated in Figure 8.18.*

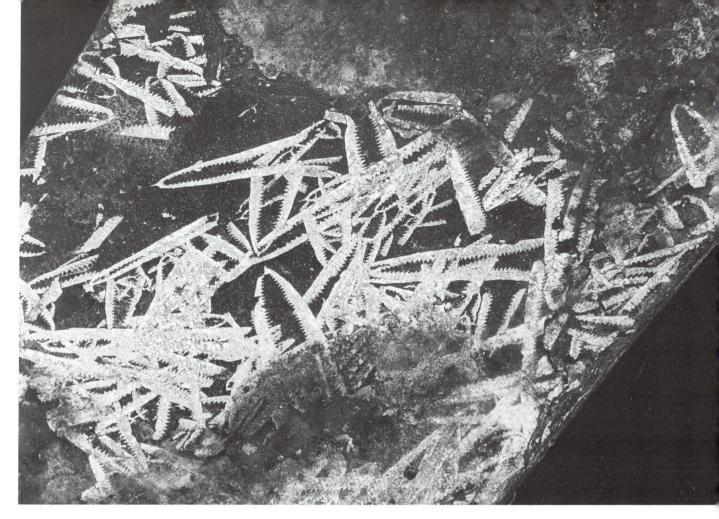

Figure 8.20. *Bedding surface with deformed* Didymograptus. *See Question 8.8★.*

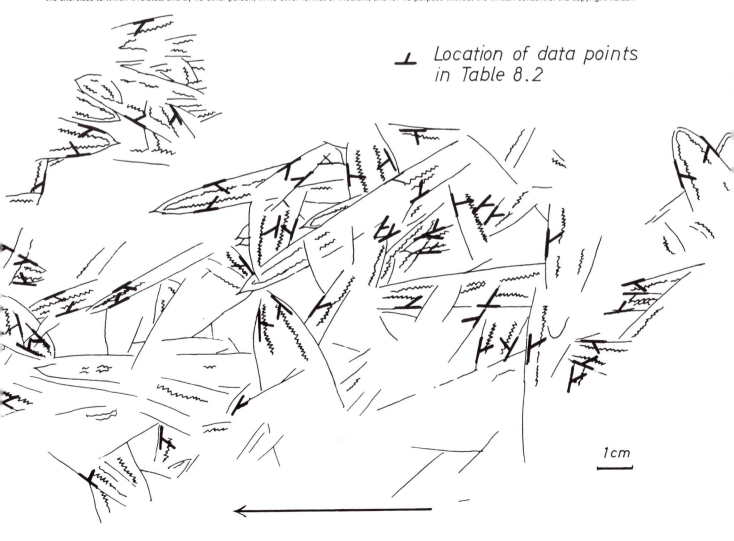

⊥ Location of data points
in Table 8.2

1cm

shape of the two branches (stipes) hanging down from a joining point like the prongs of a tuning fork. Along these stipes are the so-called thecae, originally tube like containers of the actual living parts of the organism. In this deformed rock the axes of the thecae make variable angles α' with the stipe depending on the orientation of the stipes with the axes of the strain ellipse.

Using the angular relationships of α with line orientation (NB one must systematically record the orientation of one of the lines to the reference orientation line on the specimen surface) determine the maximum and minimum values of α'. From a graphical plot of α' against orientation, determine (using Equations 8.11 and 8.12) the strain ratio R, initial angle α and orientation of the principal strains with respect to the reference orientation line. In order to speed the solution of this problem the measurements derived from Figure 8.20 are set out in Table 8.2.

Table 8.2

Observation point	Orientation of line	α'	Observation point	Orientation of line	α'
1	+18	45	31	−1	81
2	−81	68	32	−61	78
3	−29	83	33	+32	45
4	+13	60	34	+49	37
5	−3	66	35	+68	35
6	+64	51	36	+74	42
7	+26	53	37	+68	42
8	+20	44	38	+38	34
9	+49	34	39	+6	59
10	+81	45	40	+56	35
11	+9	65	41	+36	30
12	+29	47	42	+24	32
13	+52	37	43	+89	59
14	−32	60	44	−1	82
15	−53	71	45	+12	86
16	+27	39	46	+42	43
17	+70	46	47	+50	46
18	−34	85	48	+55	43
19	−35	71	49	+78	50
20	+69	47	50	−47	73
21	−9	74	51	+85	60
22	−50	70	52	+87	52
23	+9	60	53	+83	55
24	−88	60	54	+30	45
25	+46	39	55	−1	68
26	+16	50	56	+57	40
27	+76	44	57	+45	40
28	+45	42	58	+17	50
29	−28	82	59	−7	70
30	+53	32	60	−47	60

Strains from deformed spirals

Question 8.9★

A number of animals (ammonites, goniatites, gastropods, some pelecipods) build hard parts in a spiral form, often very close to that known mathematically as a logarithmic spiral (Mosely, 1838; Thompson, 1942). The form of such a spiral can be expressed most generally in a parametric equation

$$r = ke^{\theta \cot \alpha} \tag{8.13}$$

where r is the radius vector from the origin, k a constant, e the exponential constant, θ the angle through which the curve has evolved and α the **spiral angle** between the tangent to the spiral and the radius vector (Figure 8.21). The value of the spiral angle is related to the tightness of the spiral coil—values close to 90° give tightly coiled spirals, whereas lower angles give more open spirals.

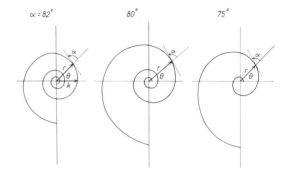

Figure 8.21. Geometric forms of various logarithmic spirals with differing spiral angles α.

When such a logarithmic spiral becomes deformed we can define the shape in terms of (x', y') coordinates and the strain ratio R. If the greatest extension is parallel to the y-axis, the coordinates of points on the spiral become

$$x' = R^{-1/2} \cos \theta k e^{\theta \cot \alpha} \tag{8.14}$$
$$y' = R^{-1/2} \sin \theta k e^{\theta \cot \alpha}$$

The angle between the tangent to the deformed spiral and the radius vector is now modified to α', and varies with position of the tangent with respect to the principal strain axes. The relationships of α' to the strain ellipse are therefore exactly identical to those we discussed in the previous question. Tan pointed out in 1973 a method for evaluating strain using variations in α'. He showed that

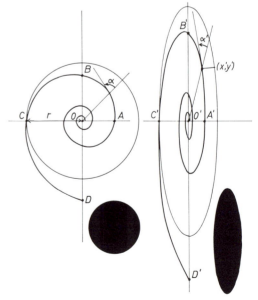

Figure 8.22. Relationship of the geometry of deformed to an undeformed logarithmic spiral.

certain Jurassic ammonites showed a rather restricted initial spiral angle from 80° to 84°. From curves he had plotted showing variations in α' with variation in R, he developed a method whereby two observations of α' were sufficient to specify which of the various R curves in a certain α family satisfied the measured data, a technique like that discussed in the last question.

Figure 8.23 illustrates a deformed ammonite from the black marls (the so-called *terres noires*) of the western Alps. The initial spiral angle α of this species is 86°. Curves showing variation of α' with direction ϕ' of the tangent to the deformed whorl for different values of R are illustrated in Figure 8.24. Use these curves to determine the principal strain ratio of the ammonite illustrated in Figure 8.23.

Deformed ammonite, Blake's method

Question 8.10★

If one has a well preserved, deformed logarithmic spiral it is possible to determine strain very rapidly by a method proposed by Blake in 1878. *Blake's method* utilizes certain special features of distances from the pole of the spiral to points on the spiral. Returning to the equation for the spiral, and considering values of θ of 0, $\pi/2$, π, and $3\pi/2$ we find the lengths of OA, OB, OC and OD respectively in Figure 8.22.

$$OA = ke^0 = k$$
$$OB = ke^{\pi\cot\alpha)/2}$$
$$OC = ke^{\pi\cot\alpha} \qquad (8.15)$$
$$OD = ke^{(3\pi\cot\alpha)/2}$$

From these relationships it follows that

$$OC = (OB \cdot OD)^{1/2} \qquad (8.16)$$

If we were to draw a circle centre O, radius $r = OC$ in the undeformed spiral of Figure 8.22, then the value of r in the direction OB and OD would be $r = (OB \cdot OD)^{1/2}$. If the spiral were now homogeneously strained such that the principal strains are oriented parallel to OB and OC respectively the circle would become strained to an ellipse with axes of lengths $r(1 + e_1)$ and $r(1 + e_2)$. Because of the relationships of Equation 8.16 then it follows that

$$R = \frac{(O'B' \cdot O'D')^{1/2}}{O'C'}$$

Thus, if we can establish the axial directions of the strain ellipse by inspection, measurements of the principal diameters of the deformed spiral lead to a rapid determination of the strain ratio.

The spiral angle of the undeformed spiral can be obtained from the equation

$$\cot\alpha = \pi^{-1}\log_e(OC/k)$$

or

$$\cot\alpha = \pi^{-1}\log_e(O'C'/O'A') \qquad (8.17)$$

The term in the brackets is the ratio of the length of *any* radius vector with direction θ to that with direction $\theta-\pi$ even in deformed spirals. This offers a very convenient and statistically sound way of finding the initial spiral angle of for example a deformed ammonite species, by measuring

whorl thicknesses at several different orientations of θ and finding the arithmetic mean.

Use the Blake method to evaluate the strain from Figure 8.23.

ANSWERS AND COMMENTS

Deformed trilobites

Answer 8.1

With reference to the long axis of each trilobite, the measured data are as follows

Fossil A $\phi'_A = +11°$ $\psi_A = +19°$

Fossil B $\phi'_B = -55°$ $\psi_B = -22\cdot5°$

The Breddin graphs give values of R of 1·75 and 1·72 and the Mohr diagrams (Figure 8.25) give very close answers. It is possible to rearrange Equation 8.5 into the form

$$R^2 = \frac{\tan\psi + \tan\phi'}{\tan\phi' - \tan^2\phi'\tan\psi} \qquad (8.18)$$

and obtain direct numerical solutions for the data. These solutions are $R = 1\cdot72$ and $R = 1\cdot78$ for fossils A and B respectively.

In order to restore the deformed fossil to its original shape we place a unit grid over the deformed specimen B so that one of the sets of the grid lines is parallel to the stretching lineation. We then redraw the specimen shape but with an orthogonal grid with spacing intervals in the ratio of the reciprocal strain ellipse 1·8 : 1·0. The resulting reconstruction (Figure 8.26) is now bilaterally symmetric.

It should be clear that the deformation of fossils can lead to considerable modifications of their original form. In 1846 Daniel Sharpe described how deformed fossils could be used to compute the finite strains in the Devonian slates of southwest England. He noted that sometimes the deformation was so strong that it was difficult to determine the species, sometimes even the genus of a fossil. In a beautifully illustrated account, he showed how the brachiopod *Spirifer* was distorted without its shell being broken in any way, and he reconstructed the initial form of each shell in the way we reconstructed the form of the trilobite *Angelina* showing that several forms previously considered to be different were, in fact, all the same species. Another excellent piece of work along similar lines is that of Fanck (1929). Fanck described deformed fossil shells from the upper Tertiary Molasse sediments of the northern Swiss Alps, showing that the number of species of lamellibranchs was not 426 as originally monographed, but only 62. Clearly, palaeontologists working with deformed rocks should bear in mind the geometric effects of strain before categorizing new species.

Deformed trilobites

Answer 8.2

The technique of using values of angular shear strain to compute the shape of the strain ellipse is not equally applicable for all fossil orientations. It should be clear from

Figure 8.23. Deformed ammonite. See Question 8.9★.

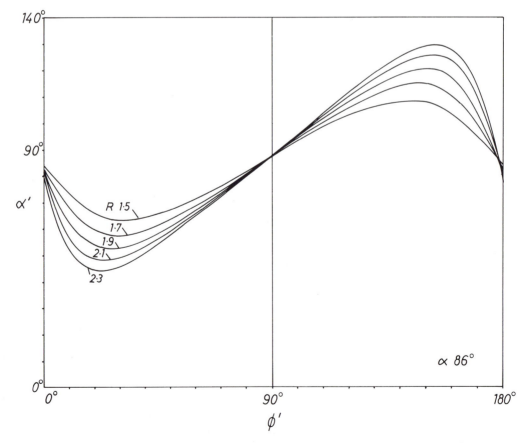

Figure 8.24. Curves showing variations in spiral angle α' with orientation φ' and strain ratio R, with α = 86°.

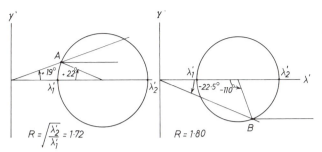

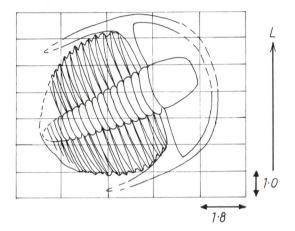

$$R = \sqrt{\frac{\lambda_2'}{\lambda_1'}} = 1\cdot72 \qquad\qquad R = 1\cdot80$$

Figure 8.25. *Mohr diagram solutions to Question 8.1.*

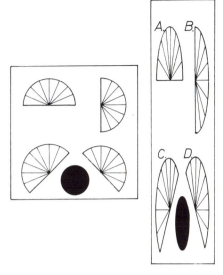

Figure 8.27. *The shape of deformed fossils. A, narrow symmetric form; B, broad symmetric form; C, right-handed oblique form; D, left-handed oblique form.*

L

$1\cdot0$

$1\cdot8$

Figure 8.26. *Reconstruction of the undeformed shape of the Angelina specimen of Figure 7.8B.*

the bunching together of lines on the Breddin graphs at positions close to the principal strain directions that orientations in these regions do not give distinctive values of the strain ratio R. Slight errors in evaluating the angular shear strain result in large errors in calculating the strain ratio. The angular shear strain method gives best results where the numerical value of the angular shear strain is high.

If the fossil lies with its symmetry axis parallel to a strain direction the angular shear strain is zero and the shear strain method cannot be used to determine the strain. Such fossils are termed **symmetric forms** in contrast to those termed **oblique forms**. The latter can be either right handed or left handed depending upon the deflection of the symmetry axis (Figure 8.27C and D). Symmetric forms can be directed into **narrow** (Figure 8.27A) or **broad forms** (Figure 8.27B) depending on whether the fossil symmetry axis is parallel to the long or short axis of the strain ellipse. If we knew the original shape of the fossil in terms of the ratio of its length to breadth $l_0/b_0 = r_0$, then we can use the ratio of the final length to breadth ratio to determine the strain because in the narrow form:

$$r_n = \frac{l_n}{b_n} = \frac{l_0 R}{b_0} \qquad (8.19)$$

and in the broad form

$$r_b = \frac{l_b}{b_b} = \frac{l_0}{R b_0} \qquad (8.20)$$

This is the method that was used by Samuel Haughton in

a paper published in 1856 when he was investigating the significance of slaty cleavage using deformed fossils. This paper is one of the classic works analysing deformed fossil shapes. Incidentally, it appears to contain the earliest reference to the geological use of the basic equations for length changes arising from strain (Equations 6.4, 6.5).

In Question 8.2 we assume that the original shape ratio r_0 is unknown. Then, providing that this ratio is constant in the species being investigated, and that we can measure the ratios r_n and r_b of the narrow and broad symmetric forms respectively, it follows from the relationships of Equations 8.19 and 8.20 that

$$R = (r_n/r_b)^{1/2} \qquad (8.21)$$

$$r_0 = (r_b r_n)^{1/2} \qquad (8.22)$$

The technique is analogous to that we used when analysing randomly oriented, deformed, elliptical objects in Session 5. The first account of this method can be found in a truly remarkable paper on slaty cleavage by Alfred Harker (1885, p. 15). The technique has subsequently been extensively used, especially by Hans Breddin (1956) and many of his co-workers investigating both the deformed Palaeozoic rocks of North Germany and Belgium, and in the Alpine molasse.

In our problem we obtain the following data

$$r_n = 4\cdot60, \quad r_b = 0\cdot49$$

giving solutions for the strain ratio $R = 3\cdot05$, and the original length to breadth ratio of the trilobite $r_0 = 1\cdot50$.

Deformed crinoid and brachiopod

Answer 8.3

By measuring the stretched and disarticulated crinoid, the length parameters can be calculated. It is advisable to measure the final length (l) from the middle of the first ossicle to the middle of the last, and in computing the original length (l_0) it is best to use only the half lengths of the first and last ossicle in the calculation together with the lengths of the five intervening ossicles. The following data are obtained:

$$e = (l - l_0)/l_0 = 1 \cdot 81 \quad \lambda = (l/l_0)^2 = 7 \cdot 90$$

$$\lambda' = (l_0/l)^2 = 0 \cdot 13 \quad \psi = +54°$$

$$\gamma = \tan \psi = 1 \cdot 38 \quad \gamma' = \gamma/\lambda = 0 \cdot 17$$

Without further information about the direction of the axes of principal strain we can go no further with our calculation. It is possible that the crystal fibres between the separated crinoid fragments are aligned in the direction of maximum extension. The orientations of the third and tenth ribs of the brachiopod (located clockwise from the hinge line) are sub-perpendicular. If the rib ornament on the brachiopod surface was originally radially symmetric, these directions would have been initially perpendicular. It follows that the third and tenth ribs could be parallel to the principal extension directions. The direction of the third rib is parallel to the fibre crystals. Although neither of these characteristics give certain proof of the identification of the principal strains, they are highly suggestive that these directions are close to the principal strains. These observations imply that the brachiopod hinge line is oriented at $\phi' = 11°$ anticlockwise from the greatest extension. If this is correct we are ncw in a position to compute both the strain ratio R (from the value of angular shear strain ψ) and the absolute value of the principal strains. The Mohr construction for the solution is illustrated in Figure 8.28. The results are $R = 2 \cdot 95$, $1 + e_1 = 3 \cdot 10$, $1 + e_2 = 1 \cdot 05$. The short axis of the strain ellipse is of almost unaltered length. We have previously discussed (p. 99) the characteristic geometric features of a strain system such as this on separating rigid particles. Because the displacements taking place in the surface are essentially all parallel to the maximum extension of the strain ellipse, the crystal fibres will themselves be parallel to this principal extension direction. This supports one of the lines of reasoning we used in this problem to establish the orientations of the principal strains.

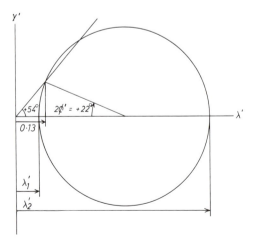

Figure 8.28. Mohr construction giving solution to Question 8.3.

Deformed plants

Answer 8.4

The fossil plants cannot be used directly to evaluate angular shear strain because we have no lines which were initially perpendicular. We must make a construction to provide some initially perpendicular directions. To do this we note that the leaf spacing along each stem is rather regular, and we interpolate identically shaped "leaf elements" between the actual leaves (Figure 8.29). Joining the tips of a real leaf on one side of the stem with a "constructed" leaf on the other side we construct a line which was initially perpendicular to the stem axis. We find the following shear strains: $\psi_A = 0°$, $\psi_B = -38°$, $\psi_C = 0°$.

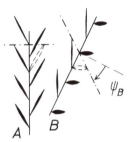

Figure 8.29. Construction used in Question 8.4 to determine the angular shear strain ψ.

It follows that the stems of fossil A and fossil C are parallel to the local principal strain axis. They are not parallel so we must infer that the strain is heterogeneous across the specimen. Fossil A and fossil B lie close together, probably sufficiently close for the strain to be approximately homogeneous. The leaves of B are parallel to the principal strains deduced from A, so we can use the leaf lengths to evaluate the strain ratio $R = d_1/d_2 = 2 \cdot 08$.

Because the final angle between the leaves of fossil B is 90°, and because this was initially 90° (parallel to the principal strains), it follows that the leaves were originally symmetrically oriented at $90/2 = 45°$ to the stem.

Deformed brachipods

Answer 8.5

The solutions to this problem following the Mohr construction and Breddin graph methods are given in Figure 8.30. The basic data are as follows: $\psi_A = -31°$, $\psi_B = -52°$ and the angle between the hinge line of fossil A to that of fossil B is $-39°$, being clockwise. The two methods give an answer for the principal strain ratio $R = 3 \cdot 0$, and the long axis of the ellipse is oriented 15·5° anticlockwise from the hinge line of fossil A (because ϕ'_A is negative, line A has a clockwise sense relative to the principal strain direction).

For this method to give a satisfactory answer any two measurements of angular shear strain are possible, provided that the measures are not related in the way $+\psi_A = -\psi_B$. If this occurs the data from one fossil will not improve the data from the other and the problem is insoluble. Good results will be obtained where the numerical values of the two shear strains are different, and best results occur where one of the shear strains is close to a zero value, with the value of the second close to the maximum (or minimum) shear strain.

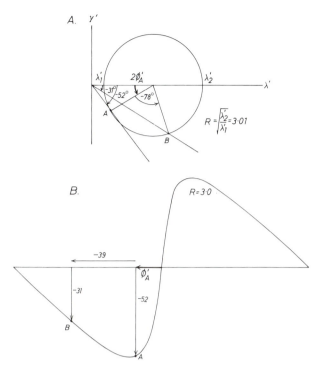

Figure 8.30. Solutions to Question 8.5.

Deformed coral and crinoids

Answer 8.6★

Because the initial geometric forms of the individual corallites making up the compound coral were perfectly regular hexagons, it follows that certain join lines between the apices of the hexagons can be used to construct initially perpendicular sets of lines (see Figure 8.31). From these initial directions we can establish angular shear strain for three separate directions, A, B, and C, and we obtain

$$\psi_A = +28° \quad A \angle B = -47°$$

$$\psi_B = +8° \quad B \angle C = -42°$$

$$\psi_C = -32°$$

We then proceed to construct a Mohr circle using the principles discussed on p. 132 and in Figure 8.13. In the Mohr circle construction remember that positive physical angles between lines (measured anticlockwise in the surface) are recorded in a clockwise sense and vice versa. When the circle overlay containing the orientation data from the three directions is placed on the Mohr graph with its three lines representing the shear strain data, a unique solution is obtained. The data will not be completely consistent on account of inevitable inaccuracies in measurement. From the best fit we read of values of $k\lambda'_1$ and $k\lambda'_2$ and hence determine $R = (k\lambda'_2/k\lambda'_1)^{1/2} = 1.95$, and the graph gives a value of $2\phi'_A$ from direction A of 101°, implying that the long axis of the strain ellipse is situated 50·5° anticlockwise from the direction of line A (see Figure 8.13).

The construction of initially perpendicular sets of lines in the deformed *Pentacrinus* like those of the previous problem depends on the intial angular relationships of the pentagonal star (Figure 8.32). There are five separate directions A to E along which we can compute the angular shear strain, and we clearly have an excess of data for the solution of the problem. The data are as follows:

ψ_A	$-17°$	$A \angle B$	$+50°$
ψ_B	$+27°$	$B \angle C$	$+103°$
ψ_C	$-28°$	$C \angle D$	$+51°$
ψ_D	$+19°$	$D \angle E$	$+81°$
ψ_E	$-4°$	$E \angle A$	$+74°$

359° total gives a check
for accuracy.

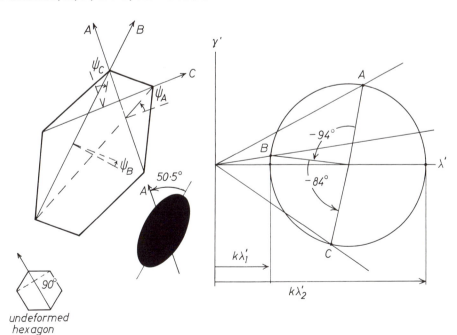

Figure 8.31. Constructions made on the corallites of the deformed fossil Favosites to compute angular shear strains, together with the Mohr diagram solutions to Question 8.6★.

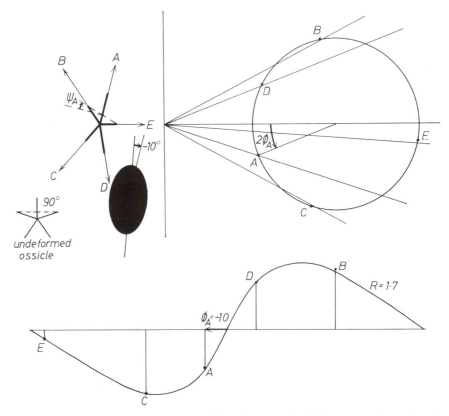

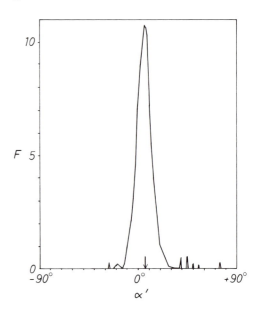

Figure 8.32. *Solutions to the deformed* Pentacrinus *of Question 8.6★ by Mohr construction and using Breddin graphs.*

The Mohr construction is shown in Figure 8.32. The five points on the circle are overlaid on the five lines representing the angular shear data, giving a strain ratio of $R = 1.7$ and a value of θ'_A of $-10°$.

The Breddin graph best fit solution is also shown in Figure 8.32 and gives identical results to those from the Mohr construction.

Deformed passive line markers

Answer 8.7★

It should be clear from the very strong preferred directions of the deformed line elements of Figure 8.16 that the value of the ratio of the principal strains is high. The orientations of 534 lines were determined and Figure 8.33 is a graphical presentation of the data. The concentration factor for each degree of orientation has been plotted, the data being averaged over 2° on either side of the specific orientation to give a smoothing effect. The concentration factor gives a principal strain ratio of $R = 10$, and the orientation of the maximum extension is $+6°$ from the reference direction. It is possible to determine the absolute value of the extension from the stretched belemnite in the right-hand corner of Figure 8.16. Knowing this extension, it is possible to compute absolute values of the principal extensions. The belemnite is oriented at an angle of $+12°$ from the greatest extension direction, and the value of $1 + e$ is 4.06. From Equation 6.5 and the definition of $R = (1 + e_1)/(1 + e_2)$ we find

$$1 + e_1 = ((\cos^2 \phi' + R^2 \sin^2 \phi') \lambda)^{1/2} = 9.33$$

$$1 + e_2 = (1 + e_1)/R = 0.93$$

We will see later when we come to discuss the development of rock fabrics during deformation that the principles of line reorientation discussed above can be used to account for cleavage and certain types of linear structure. In 1853 Henry Sorby pointed out how deformation can lead to reorientation of particles, and he described laboratory experiments to verify this conclusion when he deformed flakes of iron oxide embedded in a clay matrix. Later, in 1856, he used Equation 8.3 to evaluate in detail the degree of concentration of lines as a result of strain.

Figure 8.33. *Concentration factor F of the deformed trace fossils of Figure 8.16 with variation in angle α'.*

Deformed graptolites

Answer 8.8★

Figure 8.34 shows a graph showing variation in α' with the orientation of line elements of the deformed *Didymograptus* of Figure 8.20. It should be noted that the primary orientation data were tabulated very carefully and systematically so that the reference orientation lines always lie on the positive side of the acute angle α'. Sometimes this direction is a stipe direction, at other times a theca direction (see Figure 8.20).

From this graph maximum and minimum values of α' of 85° and 35° respectively are obtained. These are not the *absolute* maximum and minimum values of the data, but represent an eyeball-in of the data points. Using Equations 8.11 and 8.12 these values give a strain value of $R = 1\cdot7$, and the initial angle α between stipe and thecae of 56°.

The orientation of the principal strains are found from the orientations of the maximum and minimum α' values on the best-fit graph calculated from Equation 8.8 (with values of R and α as above). The minimum and maximum points on the graph make angles of $\tan^{-1}((\tan\alpha/2)/R)$ and $-\tan^{-1}((\cot\alpha/2)/R)$ with the direction of maximum principal strain respectively, and make angles of $+22°$ and $-68°$ with the direction of the orientation base line.

You will probably have found that the time spent on data collection, graphical plotting and data analysis has been quite considerable. It is possible to obtain an approximate answer for the strain ratio R by another quicker method. The spacing distance between the graptolite thecae is generally initially rather constant. It follows that by making measurements of the distances between say five thecae along differently oriented stipes it is possible to obtain information about the proportional elongation in the surface. If one had information about the values of these undeformed fossils it would be possible to obtain an absolute value of extension. Knowing proportional values of extension, it is possible to obtain the maximum and minimum values and hence obtain a value for R. It is then possible to restore the shapes of the fossils using the grid method (see Question 8.1) to determine the initial angle α between thecae and stipe, or by using Equation 8.8.

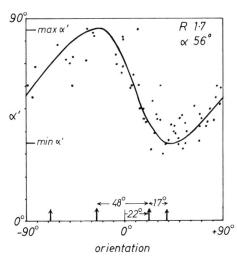

Figure 8.34. Graph showing variation in angle α' with orientation derived from the deformed graptolites of Figure 8.20, with the best fitting R, α curves.

Deformed ammonite, analysis of spiral angle

Answer 8.9★

The plots of deformed spiral angle α' against orientation of the direction of the tangent to the fossil whorl is presented in Figure 8.35. In making this construction the pole of the fossil (Figure 8.22, point O') was determined by finding the intersection point of lines drawn through the contact points of parallel tangents on successive whorls (i.e. tangent isogons). The position of the deformed radius vectors could then be easily constructed and the angle α' between radius vector and tangent measured. The data give a strain ratio of $R = 1\cdot7$, and a direction of the maximum principal strain of $+41°$ from the orientation base line.

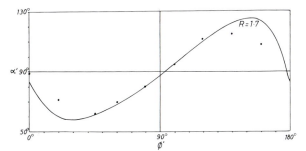

Figure 8.35. Graphical analysis of the data from Figure 8.23.

It is worth noting that the initial spiral angle of some fossils changes with growth of the organism. This possibility should be investigated by plotting successive spiral angles in different whorls using Equation 8.17. If the angle does vary, then the data for successive whorls must be analysed using separate α graphs.

Deformed ammonite: Blake's method

Answer 8.10★

Application of the Blake method on the inner whorls of the ammonite gives a strain ratio of $R = 1\cdot7$. A different value is obtained if the measurements are taken from the periphery of the specimen because differential erosion has produced shape irregularities unrelated to the primary form of the fossil.

Concluding remarks concerning strain analysis from deformed fossils

Although the measurements of strain from deformed fossils can produce excellent and accurate results, it is possible to run into certain problems with this analysis. If the fossils are thin, such as thin-shelled brachiopods or pelecypods, chitinous trilobites or graptolites, then the form of the deformed fossil gives a very good measure of the strain. However, some fossils are quite large, perhaps thick shelled, or with an internal filling of sediment or crystals of different composition or grain size from the surrounding host rock. In such examples there may be marked competence differences between the fossil and its surrounding matrix, and in this case it may not be possible to determine the rock strain from the fossil shape.

Many fossils represent parts of organisms that were living on, or sedimented on a sub-horizontal depositional bedding surface. They therefore frequently show a special preferred orientation relative to the surface, and may lie with their shortest dimension perpendicular to the bedding planes. Such fossils are excellent for measuring two-dimensional strains in the bedding, but their use is limited when it comes to strain determination in surfaces with other orientations. Fossils are distorted for the whole period of their history since interment in the sediment. This means that they may show distortion components which are the result of diagenetic compaction. Normally such diagenetic shortening occurs sub-perpendicularly to the bedding and does not lead to shape modifications of objects lying on the bedding plane. Should a fossil be obliquely inclined to the bedding surface, however, it is possible that such diagenetic compaction could lead to deformations of non-tectonic origin.

KEYWORDS AND DEFINITIONS

Deformed fossils with oblique form Fossils which have their original axis of bilateral symmetry oriented obliquely to the principal strain directions. They may be **right-handed oblique** or **left-handed oblique** depending on whether the symmetry line is deflected clockwise or anticlockwise respectively (Figure 8.27B and C).

Deformed fossils with symmetric form Fossils which have deformed in such a manner that their original axis of bilateral symmetry coincides with one of the directions of principal strain. They are termed **narrow forms** when the symmetry line coincides with the maximum extension, and **broad forms** when this line coincides with the minimum extension (see Figure 8.27A and B).

Spiral angles The angle in a spiral shaped fossil between the tangent to the whorl and the radius vector of the spiral (see Figure 8.22, α).

KEY REFERENCES

There are very many publications describing the angular changes in deformed rocks and especially the use of deformed fossils to compute strain. A detailed list of these will be found in the source list. We recommend below a few publications, those which set out in most detail the main methods of analysis.

Breddin, H. (1956). Die tektonische Deformation der Fossilien im Rheinischen Schiefergebirge. *Z. dt. geol. Ges.* **106**: 227–305.

Several excellent papers on fossil deformation have been published by Hans Breddin in German, setting out the basic equations for practical strain analysis and giving interesting data on regional strain analysis. This paper methodically describes the techniques and introduces the method we have termed the Breddin graph method. It should be noted, however, that the graphs published in this original paper do contain some computational errors.

Haughton, S. (1856). On slaty cleavage and the distortion of fossils. *Phil. Mag.* Ser. 4 **12**, 1–13.

This paper set out for the first time the mathematical methods for strain analysis from the shape analysis of deformed fossils. Most of the analysis proceeds by using changes in lengths of fossils with initially known shapes and sizes. It extends the discussion of two-dimensional strain to an analysis of three-dimensional strain, and gives one of the first accounts of numerical data to estimate the compression necessary to produce slaty cleavage.

Ramsay, J. G. (1967). "Folding and Fracturing of Rocks", 568 pp. McGraw-Hill, New York.

Pages 230–247 summarize the methods for analysing changes in the angular relationships of deformed objects and fossils. The Breddin graphs published in this book (Figure 5.56) contain the same errors as those of the initial paper of Hans Breddin (1956).

Sharpe, D. (1847). On slaty cleavage. *Q. J. Geol. Soc.* **3**, 74–105.

This classic paper sets out the results of the first detailed analysis of deformed fossils and provides clear descriptions of observations and shape analysis. It was the first publication to point out that several previously described separate fossil species were in reality deformed examples of a single species.

Tan, B. K. (1973). Determination of strain ellipses from deformed ammonites. *Tectonophysics* **16**, 89–101.

The method of describing strains from deformed spirals originally presented by Blake is described, and a new method of determining strain from fragments of deformed ammonoids is set out.

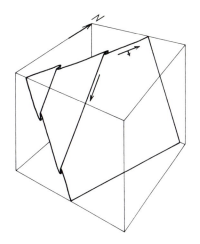

SESSION 9

Orientation Analysis

The stereographic or Wulff net and equal area Schmidt and Billings nets are developed. The basic techniques for plotting lines and planes are developed. The method for measuring angles between lines and angles between planes are described. Geological examples to exploit these techniques are presented. The methods for the body-rotation of projection data about horizontal and inclined rotation axes are set up as an advanced technique and geological examples to problems involving rotation are given.

INTRODUCTION

In the next Sessions we wish to investigate three-dimensional geometric problems. Before we do this we must be able to record and analyse the spatial orientation of planes and lines. Three-dimensional coordinate geometry can be applied but it is not always the method best suited to geological investigations. The methods that we are to develop in this Session are especially apt for describing the angular relations of planar and linear features. They come under the general heading of projection techniques.

Development of the stereographic and equal area projections

The basic concept of a projection can be directly compared with methods of representing the surface of the Earth and the various angular relationships of objects on the surface in the style of a map projection. To represent a plane and a line located on this plane (stippled surface and line *l*, Figure 9.1A) we envisage these forms to be projected downwards into the lower part of a sphere so that they pass through the centre of the sphere and intersect its lower surface (Figure 9.1B). The *spherical projection* of the plane is the trace of the plane on the lower hemisphere, and this curving line is known as a *great circle*. The spherical projection of a line is the point of intersection of the line and the hemisphere (*l* in Figure 9.1B).

To produce a **stereographic projection** from the spherical projection, we project the spherical projection traces on to the horizontal surface of the hemisphere from a point *x* situated at the apex of the upper hemisphere (Figure 9.1B). The stereographic projection of the data therefore appears as lines and points on the circular horizontal section of the sphere, a curved line (great circle trace) representing the original plane and a point or pole representing the original line (Figure 9.1C, point *l*). The complete graph known as a **stereographic** or **Wulff net**, is developed by

projecting sets of differently inclined planes through a N–S diameter and is directly analogous to an equatorial map projection of half the Earth's surface. The projected great circles which pass through the N–S horizontal point are the equivalents of meridians of longitude, and the set of curves cutting the great circles at right angles are projected *small circles* equivalent to parallels of latitude. The small circles are projections of cones that have axes parallel to the N–S axis and varying apical angles.

A convenient method of recording geological data uses this grid to translate orientations of lines and planes into projection form (Figure 9.2). The centre of the net records directions which are vertical, and the ends of the N–S and E–W diameters represent horizontal N–S and E–W lines respectively.

The stereographic net, like the stereographic map projection, is an equiangular—but not an equal area—projection. Angles measured on the net are identical to those measured on the hemisphere surface, but areas lying between two selected adjacent 10° parallels of longitude and cut off by 10° lines of latitude vary across the net. On the hemisphere surface the areas are equal but on the stereographic projection they are not equal. In many problems of structural analysis this area distortion is most inconvenient, particularly when we wish to compare the frequency distribution of data, and it is much easier to interpret points on a constant area net than on a constant angle net. As a result, the geometry of the stereographic net is modified to correct for area distortion. The result is the **Schmidt** or **Lambert net** which will be used throughout this book (Figure 9.3). Although the great and small circles have slightly differing positions from those in the Wulff net, the geometric significance is the same. The Schmidt net is always used as a graph underlying a transparent or semi-transparent sheet attached to the net by a drawing pin placed through the centre point. The various

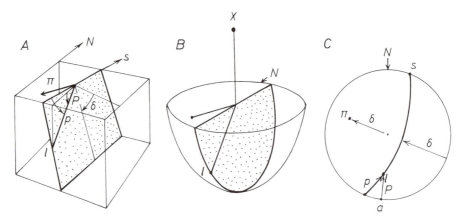

Figure 9.1. *Development of the stereographic projection from a plane (strike s, angle of dip δ), and its normal (π), and from a lineation lying on the plane (azimuth a, angle of plunge P) with a pitch angle of p in the plane. B illustrates the spherical projection of the data, and C shows the stereographic projection.*

construction lines of great and small circles are traced from the underlying net on to the overlay, and it is generally necessary to make rotations of the overlay paper about the net centre to incorporate all the orientations we require.

One variant of the Schmidt net, known as the equal area **Polar** or **Billings net**, is especially convenient for plotting linear data (Figure 9.4). Since it covers the same orientation field as the Schmidt net, diagrams made with Schmidt and Billings nets of the same diameter are interchangeable.

Readers may have previously met the projection technique as a method in mineralogy for determining the angular relations of crystal faces, edges and axes. In mineralogy the stereographic net is generally used, because

its equal angle properties reveal most clearly the angular properties of crystals. In crystallography it is general to use the *upper* hemisphere for projection, whereas in structural geology it is normal to use the *lower* hemisphere.

The main features of geometric analysis using projection techniques are as follows:

1. The net is easy to use in the laboratory and in the field. Very few basic constructions are necessary in order to solve most problems.

2. The degree of accuracy of the resulting computation is within 0·5°, and is well within the range of the accuracy of field measurement.

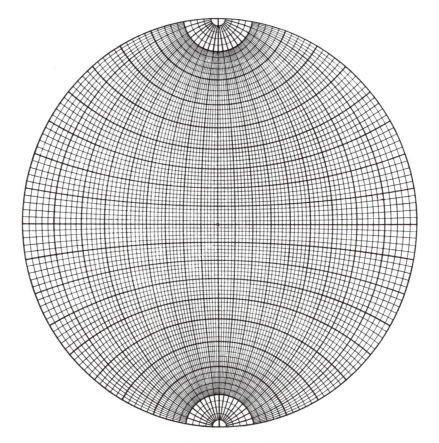

Figure 9.2. *Stereographic or Wulff net.*

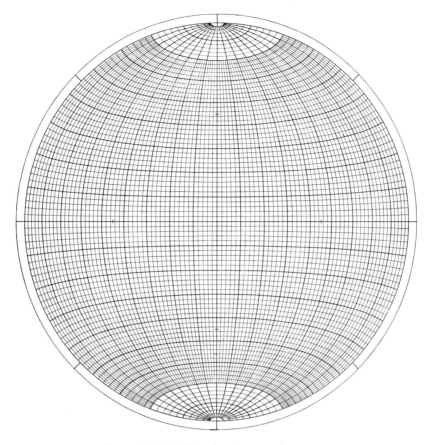

Figure 9.3 Schmidt or Lambert equal area net.

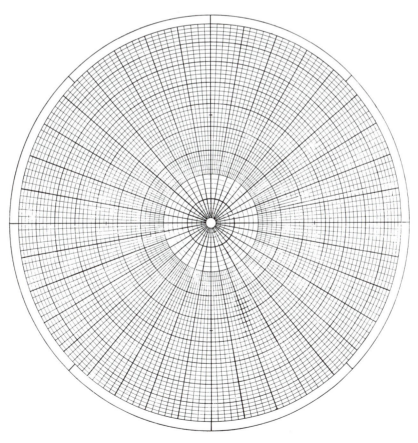

Figure 9.4. Polar or Billings equal area net.

3. It is possible to incorporate very large amounts of data into a single diagram. Significant or insignificant geometric relationships are quickly spotted. The large amount of data means that interpretations can often be made with a statistically high degree of precision. We will see later in Book 2 of this series that projections offer particularly good methods for analysing fold- and fracture-orientation data when it is often possible to incorporate hundreds or even thousands of observed data points to arrive at an answer to a given problem.

4. Although the projection method often offers a good method of primary data analysis, it is not always the best method for publishing data or effectively communicating information. The significance of geometric features set out as pages of projections is not easy to appreciate, and written descriptions of projections make for very dull reading. It is often best to abstract a geometric synthesis from the projections and present this in the form of synoptic maps or diagrams.

5. The projection technique analyses only the angular relationships between the component geometric elements of a structure and does not generally take into account the spatial location of the data. It should not be overlooked that many important features of geometric form are characterized by the relative spatial positions of their components, and not only by angular relationships.

Basic plotting techniques

Orientation features can be classified either as lines or as planes. Curving features, such as bedding planes in a fold, can be thought of as consisting of a number of elemental planar tangents to a curving surface. Because there are only two basic types of information, the techniques for recording this information are few.

Information is recorded on to a semi-transparent overlay. Before making any particular constructions on this overlay mark the circumference of the net and indicate the north point azimuth. If much information is to be plotted on the overlay it is convenient to have the circumference of the overlay marked out in 10° units.

1. Lines

Linear features are recorded by determining the direction or *azimuth a* of the line and the *angle of plunge P*, that is the angle measured in a vertical plane through the lineation and the horizontal (Figure 9.1A).

The plotting sequence for a line with azimuth 190° and plunge 40° is (see Figure 9.5A):

1. Rotate overlay 190° in an anticlockwise sense.
2. On a N–S diameter count from the N point down towards the centre of the net, using small circle intercepts, an angle of 40° from the circumference and mark the point.

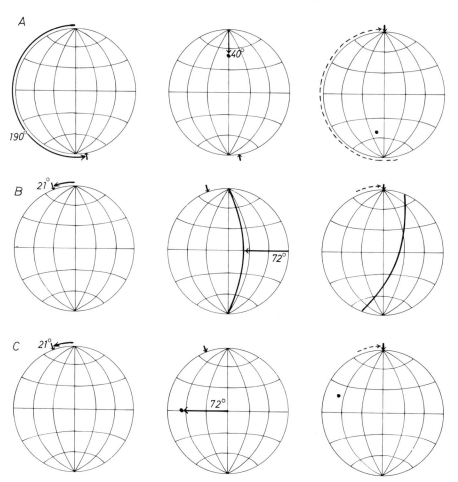

Figure 9.5. *The three basic data plotting techniques:* A, *for lines;* B *for plotting planes as great circles;* C *for plotting plane-normals or π-poles.*

3. Return the overlay to its unrotated position.

2. Planes

Planes are measured by determining the horizontal line direction of strike s, the angle of dip δ and the direction of dip. A plane can be plotted either as a great circle trace or as a pole π, that is the direction of the plane normal. To plot a great circle of a plane with strike 21° and dip 72° SE (see Figure 9.5B):

1. Rotate overlay 21° anticlockwise.
2. Select a great circle on the underlying net which falls in the SE quadrant and which makes an angle of 72° with the circumference of the net. Draw this great circle on the overlay.
3. Return overlay to original azimuth.

To plot this plane as a pole (π) (see Figure 9.5C).

1. Rotate overlay 21° anticlockwise.
2. Find a point on the E–W diameter and in the NW quadrant of the net (i.e. *opposite* to that of the dip direction) making an angle of 72° from the *centre* of the net.
3. Return overlay to original azimuth.

Because the plane and its π-pole are normal, if the plane is steeply inclined, the pole has a low inclination, and vice versa.

Directional sense

The lines or poles to planes that we plot on a projection are normally considered to be directed downwards into the lower hemisphere of the spherical projection. Many of the lines or poles that we measure and plot in structural geology studies have no specific directional sense. If we looked along one direction of a line, observed what we saw, and compared it with what we could see in the other direction, the only difference would be that one view would be the mirror image of the other. There would be no fundamental differences, and such a line has no directional sense. In contrast, consider the pole to a bedding plane. Looking along one direction, all the beds get older, whereas in the other direction all the beds get younger. Clearly this line has a very definite polarity and shows a directional sense. Although lines with directional significance are not common in deformed rocks they do occur. In certain terrains one finds conical-shaped fractures, called shatter cones. Lying along one direction of the cone axis, all the fracture surfaces open away from the observer, whereas looking along the opposite direction of the cone axis, all the fracture surfaces close away from the observer. The axial line of the shatter cone has a well defined polarity. A more detailed discussion of this geometry and its interpretation will be given in Volume 2 of this series, but perhaps we should say a little about why these structures show a definite polarity. The polarity probably arises from the progressive directional sense of propagation of the fracture surfaces, possibly related to the sense of passage of a shock wave producing the fracture.

The reason for introducing here the concept of line polarity is that, for certain features, we might consider the possibility of indicating polarity sense with different sym-

bols in the projection. Most of our data will have no up or down sense and we can plot all lines or poles with the same symbol. However, if we have folded strata with recognizable polarity (normal or inverted bedding, perhaps interpreted from primary sedimentation structures) we are in a position to use differing symbols depending on whether the bedding pole points downwards or upwards towards the older strata (Figure 9.6). We will see later, in our discussion in Volume 2, that such a data separation technique could be especially valuable if we were attempting to construct the fold axis from bedding plane poles.

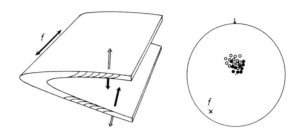

Figure 9.6. Polarity of bedding plane poles for normal and inverted fold limbs are different. In normal limbs the black part of the pole is directed into the lower hemisphere, in inverted limbs the white part is directed downward.

An extension of this technique can be used to define very effectively on a projection the relative displacements of the two sides of shear zone or fault plane. Two sides of a shear surface are displaced relative to each other such that one side has an upward moving polarity relative to the other side with a downward polarity sense (Figure 9.7).

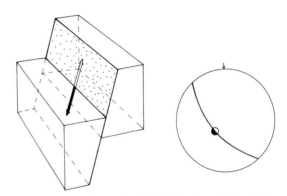

Figure 9.7. Split pole symbol for indicating relative displacement by shearing.

The movement vector can therefore be represented as a split circle pole situated on the great circle representing the shear surface, with the two sides of the circle ornamented in individual ways so as to bring out the relative upward and downward polarity senses. This technique is the only truly correct one for recording displacement sense.

Rotation sense

Occasionally it is necessary to indicate on a projection the effects of a relative rotation of one part of a body about another. Rotations take place about an axis of rotation R through a specific angle of rotation r. Figure 9.8 illustrates the effects of rotation taking place across a fault plane. To convey this effect on a projection we need to indicate the direction of the rotation axis R (in Figure 9.8 it is the

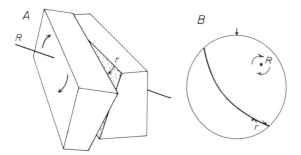

Figure 9.8. *Method of representing rotation about an inclined axis of rotation R.*

perpendicular to or pole of the fault plane) and the sense of rotation. The sense of rotation is given by a short curved line located near to the rotation axis indicating how a line on the upper block oriented a few degrees from this axis moves as a result of the displacement. The angle of rotation *r* as well as the relative movement sense is best indicated on the plane perpendicular to the rotation axis (see Figure 9.8B).

Plotting of planes and lines

Question 9.1

Figure 9.9 illustrates a sharply crested fold developed in alternating limestone and marls in the upper limb of the Morcles nappe, Switzerland. Measurements of the orientations of the two limbs of this fold give the following data:

limb *A*, strike 25°, dip 46° to ESE; limb *B*; strike 53°, dip 79° to SE. Plot these surfaces on to a net as great circles. The two great circles intersect in a point on the net. What is the likely geological significance of the line represented by this point? Determine its orientation, giving its azimuth and angle of plunge.

True dips from apparent dips

Question 9.2

Figure 9.10 shows the appearance of a rock quarry bounded by steep rock faces. On the right-hand side of the quarry these faces show regularly oriented traces of bedding planes, and the faces therefore give the *apparent dip* of the bedding planes, not the true dip of the beds. The main quarry walls (wall *A*) strikes 12°, and is vertical; the other wall (*B*) strikes 87°, and dips 82° to the S. The apparent dips on *A* and *B* are 2° N and 14° E respectively. What is the true dip of the bedding planes? On the left hand side of the quarry the bedding orientation was directly recorded: strike 87°, dip 36° S. Determine the orientation of the fold axis.

Shear zone displacement sense

Question 9.3

Figure 9.11 shows a shear zone with an en-echelon calcite filled vein array of extension fissures. The shear zone strikes 130°, and dips 82° to the SW, and the individual extension veins strike 91°, and dip 76° to the S. What is the orientation of the direction of shear that produced the deformation in the zone? Indicate on the great circle representing the shear plane the direction of the displacement vector, and show, using the split symbol method, the relative sense of displacement of the sides of the shear zone.

Now check the results in the Answers and Comments section and return to the discussion below.

Angles between lines

Two non-parallel lines may or may not be coplanar. If they are coplanar it is geometrically possible to align a single plane so that it contains both lines (Figure 9.12, lines *A* and *B*) whereas if they are not coplanar it is impossible to do this (Figure 9.12, lines *A'* and *B*). The coplanar or non-coplanar property of lines depends on the spatial position of the lines, and not their angular relationships. If lines are coplanar they intersect at a point and it is possible to define a true angle between them ($A \angle B = \alpha$). To measure the angle α between two lines using projection techniques we find the plane on which they lie and measure α in this plane:

1. Plot the poles representing the two lines *A* and *B* (Figure 9.12).

Figure 9.9. *SE–NW section across a synformal fold in the upper limb of the Morcles nappe, Mont à Cavouère, Switzerland. On this mountain side it is relatively easy to measure the limb dips, but not easy to directly measure the fold axis orientation. See Question 9.1.*

Figure 9.10. *Quarry in folded Palaeozoic sandstones, West Yorkshire. The orientation of the bedding surfaces on the left-hand side can be measured directly, the orientations of those on the right-hand side are best determined from apparent dips on steeply inclined joint faces. See Question 9.2.*

Figure 9.11. *En-echelon calcite filled vein array in a deformed limestone from the Morcles nappe. In this shear zone the vein geometry allows us to determine the directions of relative displacement across the zone. See Question 9.3.*

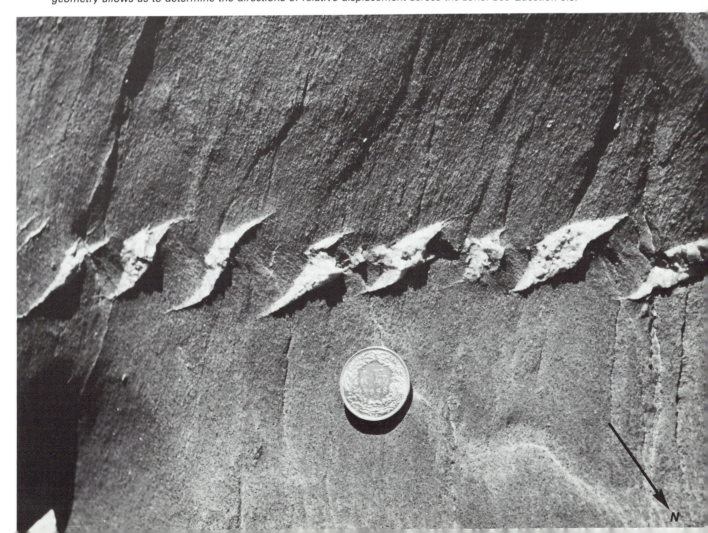

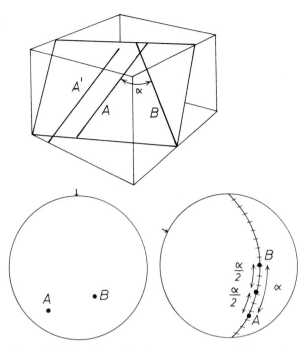

Figure 9.12. *Technique for the measurement of the angle α between two lines A and B.*

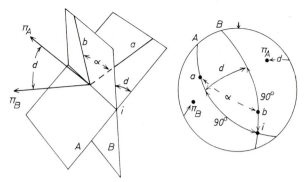

Figure 9.13. *Geometric features arising from the intersection of two planes A and B and their representation on a projection.*

(Figure 9.13, a and b on plane A and B respectively) can vary from 0° to 180° depending on the choice of lines. The *dihedral angle* (Figure 9.13, d) between two planes is unique and is defined as the acute angle made between lines in the two planes which are perpendicular to the line of intersection i. The constructions required to obtain the dihedral angle are as follows:

1. Construct the great circles representing the planes A and B, and determine the intersection i.
2. Find the position of a line in great circle A at 90° to i, and another in great circle B also at 90° to i. The angle between these is the dihedral angle d.

The dihedral angle can also be found directly from the poles of planes A and B (Figure 9.13, the angle between π_A and π_B).

2. Rotate the overlay paper until the two points A and B come to lie on a single great circle of the underlying net.
3. Along the sector of the great circle lying between the two points count the angular intercept made by the small circles which intersect the great circle. This gives a value for α.

Bisector of lines

To bisect an angle α into two equal parts of $\alpha/2°$ we find the point on the great circle containing A and B which separates the sector into equal angular intercepts (Figure 9.12).

Pitch

In our introduction to projections we mentioned that the orientation of a line could be specified either by azimuth and plunge, or by pitch in a plane. The angle of pitch is the angle measured in a plane between a line and the horizontal in that plane. It is measured on a projection by using the small circle intercepts on the great circle representing the pitch measuring plane. The angle of pitch of a line clearly varies with orientation of the reference plane in which it is measured, whereas the plunge of a line element is unique.

Angles between planes

Two non-parallel planes always intersect in a line (Figure 9.13, i), and the orientation of this line can be determined on a projection. Angles between planes are measured between coplanar lines, one of which lies in one plane, one in the other. The angle between any two such lines

Bisectors of planes

Two intersecting planes always have an *acute- and an obtuse bisector* (Figure 9.14). The acute bisector is found by making the appropriate bisection of the dihedral angle, and the obtuse bisector is situated 90° to this in the dihedral plane. The bisector of the angle between the poles to the two planes is the obtuse bisector of the planes. These bisector constructions are extremely important for determining the principal stresses from systems of conjugate shear fractures (see Question 9.5 below).

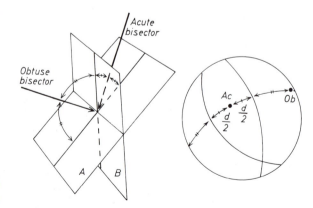

Figure 9.14. *Construction of the acute- and obtuse-bisectors of two non-parallel planes.*

Curved planes and lines

Many, if not most of the circumstances we meet in an investigation of tectonic structures involve the progressive change in the orientations of planar and linear structures from locality to locality. The way these data are handled depends on the local geometric conditions and the purpose of the analysis. In general, however, smoothly geometric variability of the components of the larger structures is analysed by envisaging the structure to be made up of an infinite number of truly planar or truly rectilinear elements in much the same way one begins the study of differential calculus by analysing the slopes of sectors of continuous curves. At each locality we measure a plane surfaces as a tangent to the structure at the locality point, or we choose a measurement of a rectilinear element to represent an infinitesimally small part of a line curved overall. When these data are plotted as to a projection they become disposed in certain characteristic patterns (Figure 9.15). The analysis of these patterns and the geometric form to which they relate is one of the key techniques for the analysis of field data, and we will develop these methods much further in Book 2.

Angles between planes

Question 9.4

Return to the data for the fold of Question 9.1. Determine the *fold interlimb angle*, that is the dihedral angle between the fold limbs. In this fold the axial surface (or axial plane—the plane joining the hinge lines in successive folded strata) bisects the fold limbs. Determine the orientation of the axial surface. What is the *pitch* of the fold axis in the axial surface?

Bisectors of planes

Question 9.5

Figure 9.16 shows a conjugate fracture system cutting through the vertically banded gneiss of the Monte Rosa nappe of southern Switzerland. On one of these faults the displacement is left-handed, on the other it is right-handed. The fault planes have the following orientations: right-hand fault, strike 87°, dip 63° to the North; left-hand fault, strike 120°, dip 83° to NE. The principal axes of stress which led to fracture are related to the faults in the following way: the direction of maximum compression is the acute bisector of the fault planes, the direction of minimum compression is the obtuse bisector of the planes, and the intermediate stress axis follows the intersection of the faults. Determine the orientations of the principal stress directions in this example.

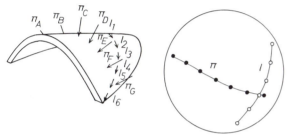

Figure 9.15. *Method of analysing the angular features of curving surfaces and lines with projection techniques.*

Figure 9.16. *Conjugate fracture pattern in granite gneiss—see Question 9.5.*

Now check your work with the Answers and Comments section and either continue with Session 10, or with the starred questions below.

Rotation of data on projections

It is sometimes necessary to compare the angular relationships of lines and planes plotted on one projection with those of another plot, perhaps to compare the structural geometry of one area with that of another. There are two possibilities for systematically modifying the orientations of projection data. The first requires a body rotation of all the data, whereas the second consists of displacing the points according to the rules of some specific strain state with known axial directions, axial ratios and rotations. The second technique requires a knowledge of the effects of three-dimensional strain on planes and lines, and we will leave a discussion of this until we have covered the necessary investigations, but we have enough background now to be able to look into the features of body rotation.

Body rotation can be specified in one of two ways. One is to undertake three rotations successively about mutually perpendicular axes, to rotate first through an angle α about N–S horizontal axis, through angle β about an E–W horizontal axis, and finally through an angle γ about a vertical axis. By choosing α, β and γ appropriately, we can reorient the body in any way we wish. A second way of producing a general rotation is by making a specific angular rotation about an axis which is not necessarily horizontal. By appropriately choosing the orientation of this rotation axis and the angle of rotation it is again possible to reorient the body in any way we wish.

During rotation about an axis all lines (and π poles to planes), describe cones of rotation. Those lines that lie close to the rotation axis show a relatively small deflection, whereas those making a high angle (near 90°) with the rotation axis make large deflections. On a projection we have seen that it is the small circle shapes which relate to cones and, during a body rotation, lines which are not parallel to the rotation axes plot on a projection as points which deflect along small circles.

Rotation about a horizontal axis

All poles representing lines in space move along small circles about the appropriate horizontal rotation axis R. The small circles are those which are printed on the net. The angular distance, equivalent to the rotation r, moved by each pole is measured by counting the appropriate great circle intercepts along the small circle. In Figure 9.17 see, for example, the movement of point A to A' and D to D'. If any point meets the circumference of the net before having completed its rotation, the remainder of the rotation is taken up along a small circle in the opposite quadrant of the net (Figure 9.17, points B to B' and C to C').

Rotations about an inclined axis

The movement of the line elements again takes place on cones around the cone axis R. Because the cone axis is inclined, the small circles indicating the movement paths

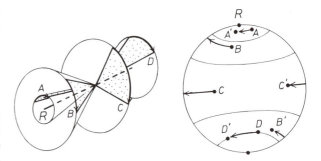

Figure 9.17. Rotation about a horizontal axis. Lines move on constant angle cones around the rotational axis R. On a net the lines move along small circles (e.g. A to A' etc.).

must be constructed; they are not represented by printed lines on the net.

If a few points are to be rotated the construction proceeds as follows (Figure 9.18):

1. Plot the rotation pole R and the pole (A) to be rotated. Measure the angle $R \angle A = \alpha$.
2. Construct the great circle with pole R.
3. Construct a great circle through R and A to meet the great circle with pole R at point a.
4. Find a point a' situated r degrees from a on the great circle with pole R (NB rotation sense, r is the angle of rotation).
5. Draw a great circle through a' and R.
6. Locate A' so that the angle $A' \angle R$ is α.

Repeat for all other points B, C etc.

If there are many points to rotate, this construction is speeded up by constructing a special rotation grid around the rotation axis R (Figure 9.18). Using such a grid it is easier to rotate lines of any orientation through the appropriate angle.

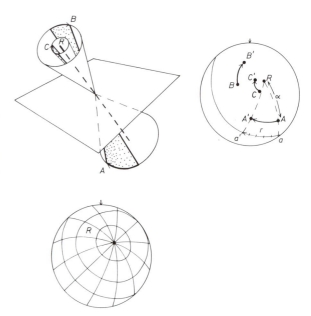

Figure 9.18. Rotation about an inclined axis R. The projection is a specially constructed rotation grid for rotating large numbers of data.

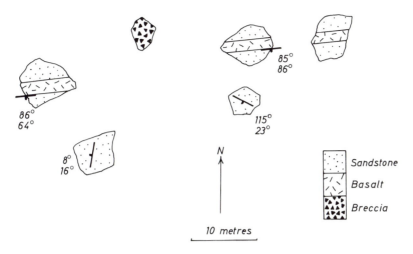

Figure 9.19. Sketch map. See Question 9.6★.

Relative movements across a fault

Question 9.6★

Figure 9.19 is a sketch map showing outcrops of a basaltic dyke which cross cuts inclined sandstone beds. At one locality the rocks are brecciated, suggesting that faulting has taken place. Determine the type of displacement on the fault, and indicate on the map the most likely strike of the fault plane, and its angle and direction of dip.

Drill hole data

Question 9.7★

Cores were obtained from three differently oriented drill holes made at the same locality but at different inclinations. No surface exposures were found but evidence from other localities suggests that the terrain is underlain by folded sediments. What information can be obtained from the drill hole data regarding the position, form and orientation of these folds?

Table 9.1

Hole	Core orientation (azimuth/plunge)	Core length (m)	Bedding inter-section angle with drill core axis
1	Vertical	9·0–14·0	84°
		14·0–36·0	53°
2	90°/60°	10·5–19·5	57°
		19·5–40	30°
3	0°/70°	9·5–14·0	75°
		14·0–36·5	64°
		36.5–40	75°

Rotation conventions

Question 9.8★

A rotation sequence about three orthogonal axes is imposed on projection data as follows:

Rotation 1. 30° about N–S horizontal axis, clockwise looking north
Rotation 2. 40° about E–W horizontal axis, clockwise looking east
Rotation 3. 80° about a vertical axis, clockwise looking down.

Define the total body rotation in terms of a single rotation about an inclined rotation axis. Give the azimuth and plunge of the rotation axis, the angle and sense of the rotation.

ANSWERS AND COMMENTS

Plotting of planes and lines

Answer 9.1

This example is a good one to illustrate how it is possible to determine the orientation of a fold axis using only measurements of the orientations of bedding surfaces in the fold limbs. The mountain side shown in Figure 9.9 is about 1400 metres high and the exposures of narrow hinge zone in terrain of difficult access makes direct measurement of the fold hinge line difficult. In contrast, accessibility of the two main fold limbs is comparatively easy. The normal limb (A) and overturned limb (B) are represented as great circles on an equal area projection in Figure 9.20. The two great circles intersect in a point which defines the line of the fold axis. The fold plunges at 31° with azimuth 60°. The fold plunge is towards the observer of Figure 9.9, and the apparent shape of the syncline gives a false impression of the profile tightness.

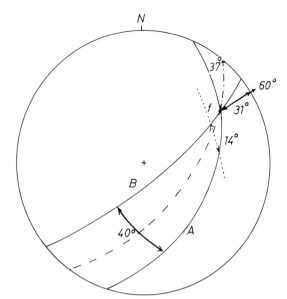

Figure 9.20. *Answer 9.1. The two fold limbs A and B intersect to give the orientation of the fold axis f.*

True dips from apparent dips

Answer 9.2

The axis of the antiformal fold exposed in this quarry is not easy to measure directly in the field because the hinge zone region is comparatively broad. Furthermore, although the fold limb exposed in the left-hand side of the quarry is easy to measure directly, the fold limb on the right-hand side is only exposed on steeply inclined joint surfaces. However, using apparent dips on two such joints it is possible to compute the orientation of the fold limb. The apparent dips on joint faces A and B are recorded in Figure 9.21 and the great circle passing through the two apparent dip lines (points on the net) gives the orientation of the bedding planes as strike 5° and dip 14° E. Intersection of

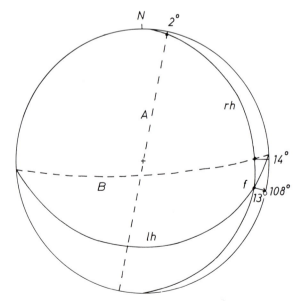

Figure 9.21. *Answer 9.2. The two apparent dips of the bedding (rh) are shown on joint faces A and B. f is the fold axis.*

the two fold limbs gives the orientation of the fold axis as plunge 13°, azimuth 108°.

Shear zone displacement sense

Answer 9.3

The en-echelon nature of the vein geometry indicates a general sinestral or left-handed differential movement across the zone. A plot of the shear zone and extension veins in the zone is shown in Figure 9.22. The intersection

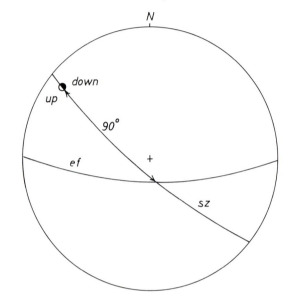

Figure 9.22. *Answer 9.3. Plots of the shear zone sz and extension fissures ef are indicated as great circles. A line at 90° to the intersection gives the shear movement direction x in the shear zone, with split symbol to indicate relative movement sense.*

between the two planar features gives a line which must be perpendicular to the shear movements in the zone, and by finding a point situated on the great circle representing the shear zone at 90° to this intersection we can establish the shear direction in the zone. This has an orientation plunge 14°, azimuth 307°. On the projection the relative displacements across the shear zone are indicated with the split symbol technique.

Angles between planes

Answer 9.4

Figure 9.20 shows the two fold limbs and fold axis. The dihedral angle between the fold limbs is 40°, and the axial plane has a strike of 41° and a dip of 62° towards the SE. The pitch of the fold axes in the axial plane is 37° NE. Note that the pitch angle alone is insufficient to describe the location of the fold axis: it is necessary to specify whether the line lies in the NE or SW quadrants of the projection.

The average slope of the mountainside shown in Figure 9.9 is 40° towards 75°. This means that, on a map, the two fold limbs have limb traces which make an angle of about 14° to each other.

Bisectors of planes

Answer 9.5

The data from the conjugate shear planes of Figure 9.16 are indicated in Figure 9.23. The three directions of maximum, intermediate and least *compression* are designated σ_3, σ_2 and σ_1 respectively and have orientations: σ_3 plunge 30°, azimuth 95°; σ_2 plunge 56°, azimuth 310°; σ_1 plunge 16°, azimuth 195°. The movement directions on the conjugate shears are shown by split circle symbols indicating the relative motions on the two surfaces.

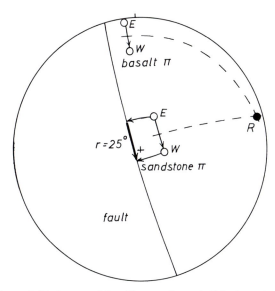

Figure 9.24. Answer 9.6★. The rotation axis R is the pole to the fault plane. Its position is found by using the construction of Figure 19.25 as the two pairs of rotated π poles.

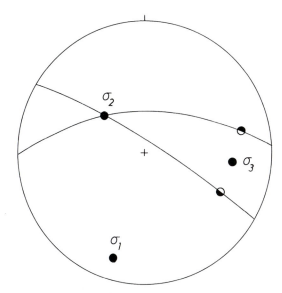

Figure 9.23. Answer 9.5. The axes σ_1, σ_2 and σ_3 represent the directions of principal stresses producing the conjugate shears. The differential shear movements along the two faults are indicated with split circles.

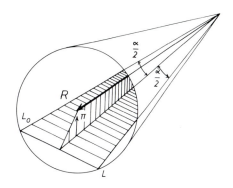

Figure 9.25. Rotational cone formed by the rotation of line L_0 to a new orientation L about a rotation axis R. The rotation axis lies on a plane perpendicular to the plane passing through $L_0 L$ and the right bisector of the angle $L_0 \angle L$.

Rotational movement on a fault plane

Answer 9.6★

The plot of the data is shown in Figure 9.24. The poles of the bedding planes of the sandstone and of the basalt dyke are indicated as open circles. The movement across the fault plane has a rotational sense because no combination of constant displacement with horizontal or vertical component could set up the differences in dip of the various surfaces. Using the construction shown in Figure 9.25 for the two pairs of poles, the rotation pole R is established. This pole must be the pole to the fault surface which, therefore, has a strike of 165° and a dip of 86° to the west. On the map the fault plane will be found on a line with this strike passing through the outcrop of brecciated rock. The rotation angle r is 25°. The west side of the fault has been rotated anticlockwise through this angle relative to the east side of the fault.

Drill hole data

Answer 9.7★

When a drill core cuts through a bedded sequence of rocks we obtain information on the angle between the drill core axis and the bedding surface. When the drill core is withdrawn from the drill hole there is inevitably a relative rotation between the core and the walls, so the bedding plane data can never give a complete definition of the actual orientation with respect to geographical coordinates. One way we can resolve this problem is to interrelate data from differently directed drill holes, and projections offer a very neat way of obtaining the exact orientation from such data. If we know the orientation of the core axis and the angle between the pole to bedding and this axis, we know that the bedding pole must lie somewhere on a cone surface of apical angle δ around the drill core axis (Figure 9.26) and we can represent these possibilities as a small

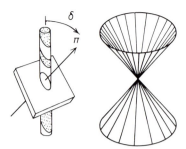

Figure 9.26. *Possible positions of the bedding plane pole π in a drill core. The total range of possible orientations of π trace out a circular cone with apical angle δ.*

circle on a projection. If we obtain data from a second, differently inclined drill hole we obtain another small circle for possible orientations of the bedding pole. This second small circle could touch the first, giving a unique solution to the problem, but generally the two small circles intersect in two points to give two possible solutions. It may be possible to select one of these in preference to the other if it accords better with some known features of the geology. If such a selection cannot be made we must have a third drill hole. A third small circle is constructed from this data and the intersections it makes with the two previous small circle data groups gives a unique solution. Clearly all this analysis depends upon the bedding surfaces being parallel over the area investigated, and that each drill hole has no curvature.

In Question 9.7★ (Figure 9.27) the uppermost layers in all three holes give a triple small circle intersection related to a unique solution, strike 60°, dip 6° SE. The underlying beds also give a triple point intersection related to surfaces of strike 41° and dip 35° SE. In this drill hole the data are similarly oriented to those in the uppermost layers.

This analysis suggests that the rocks are strongly folded by chevron-style structures (planar limbs and very sharp angular hinge zones) with fold axes oriented with plunge of 2° towards 218°. The interlimb angles of the folds are about 30°. The drill hole data suggest that all three holes penetrate first a normal limb of a northwesterly closing antiform, then a more steeply dipping overturned limb of this antiform to finally penetrate an underlying normal limb.

Rotation conventions

Answer 9.8★

The rotation sequence has been applied to three initially mutually perpendicular directions L_0, M_0 and N_0 so that they take up new positions after rotation of L, M and N (Figure 9.28A). The three rotations may also be analysed in terms of a unique rotation about a unique axis. This implies that the join of L_0L, M_0M and N_0N must each lie on small circles (cones of rotation) about this axis. For each data pair like L_0L, the unique axis of rotation R must lie on a plane perpendicular to a plane passing through the two lines and containing the bisector of the angle α between the lines. Figures 9.25 and 9.28B show the constructions necessary for finding one of the planes passing through R using data from L_0 and L. The construction is repeated for data points M_0M and N_0N and all three planes pass through the rotation axis R. The unique angle of rotation related to this axis is found by determining the great circle with axis R as normal, and finding the angle r on this plane between the projections of planes L_0R, and LR (Figure 9.28B). The unique rotation axis R has an azimuth of 69°, and plunge 21°, and the rotation angle r is 53° clockwise looking down this rotation axis.

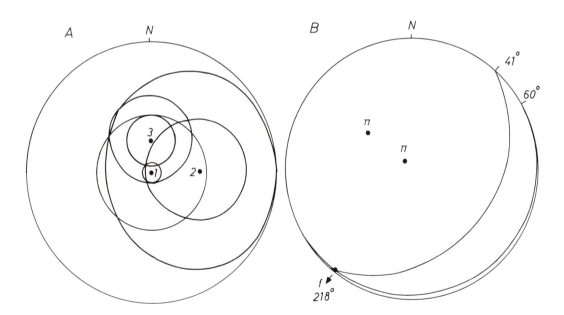

Figure 9.27. A, solutions to the bedding orientation data from the three drill holes of Question 9.7★. B shows the two unique orientations of bedding poles π and the direction of the fold axis f.

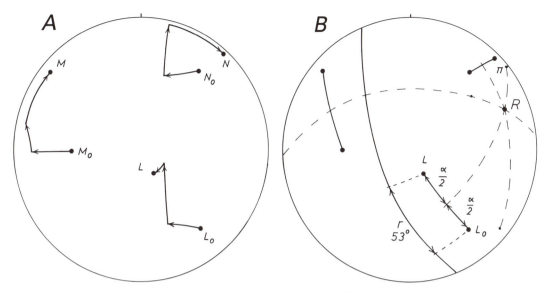

Figure 9.28. A shows the positions of three initially perpendicular lines $L_0 M_0 N_0$ before the three phase rotation sequences described in Question 9.8★ and their positions LMN after rotation. B gives the solution for the unique axis of rotation R, and angle of rotation r.

KEY REFERENCES

Bucher, W. H. (1944). The stereographic projection, a handy tool for the practical geologist. *J. Geol.* **52**, 191–212.

A paper especially written with student needs in mind setting out the main types of construction useful for the analysis of geometric problems arising in structural geology.

Ragan, D. M. (1973). "Structural Geology", 2nd edn. Wiley, New York.

A clearly written and well illustrated book describing many of the geometrical techniques useful in structural geology. Projection techniques are set out in pp. 91–102.

Ramsay, J. G. (1967). "Folding and Fracturing of Rocks," 568 pp. McGraw-Hill, New York.

Chapter 1 of this book sets out a general description of different projection nets and their application to solving various problems in structural geology.

Phillips, F. C. (1954). "The Use of Stereographic Projection in Structural Geology." Edward Arnold, London.

An excellent small book setting out the principal methods of orientation analysis using projections.

Vialon, P., Ruhland, M. and Grolier, J. (1976). "Eléments de Tectonique Analytique." Masson, Paris.

For those who read French this paperback can be recommended as giving a clear, up-to-date description of projection techniques; see pp. 63–98.

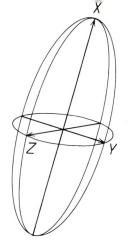

SESSION 10

Strain in Three Dimensions

1. Planar and Linear Fabrics

Various methods of describing the spatial orientations of lines and planes are described. Three-dimensional displacement and strain are analysed, and this analysis leads to the concept of the strain ellipsoid as a method of analysing distortions and rotations. Different types of strain ellipsoids are established and the practical ways of establishing the features of the strain ellipsoid from two-dimensional data described. The geometry of deformed planar and linear features is related to the possible behaviour of crystal particles in deformed rocks, to the development of preferred orientations and to the relationship between cleavage and rock deformation. Different types of penetrative rock cleavage seen in nature are related to deformation sequences and different strain ellipsoid shapes.

INTRODUCTION

Until now we have directed our attention to the geometric effects of strain in a surface. In the next two Sessions we will extend our studies to an analysis of three-dimensional strain. The geometric features of three-dimensional strain will clearly be more complex than those associated with two-dimensional analysis, but it is important to realize that the principles governing the geometry are basically the same as those developed in earlier sessions.

In this Session we first need to extend our methods to describe spatial position and orientation in a convenient mathematical notation. Then we will investigate the most important geometric features of the strain ellipsoid, describe methods for determination of the ellipsoid, and explore some of the geological implications of strain in terms of rock fabric.

Description of the position of a point and the directions of lines

To describe three-dimensional geometry we set up an orthogonal system of reference axes x, y and z. Then any point in space can be described by giving its coordinate position (x, y, z).

The directions of lines can be specified in several ways. The first method uses **direction cosines**. Consider the direction of a line joining the origin of the coordinate system with the point (x, y, z) (Figure 10.1). The orientation of this line can be described by means of the angles between the line and the three coordinate axes x, y, and z. These angles are α_x, α_y and α_z respectively. The direction cosines l, m and n are the cosine values of these angles ($l = \cos \alpha_x$ etc.). From Pythagoras the length of the line joining $(0, 0, 0)$ and (x, y, z) is $(x^2 + y^2 + z^2)^{1/2}$, and it is

therefore possible to relate the direction cosines to the coordinate positions of the ends of the line:

$$l = x/(x^2 + y^2 + z^2)^{1/2}$$
$$m = y/(x^2 + y^2 + z^2)^{1/2} \qquad (10.1)$$
$$n = z/(x^2 + y^2 + z^2)^{1/2}$$

The sign convention for direction cosines conforms to the positive or negative signs of x, y and z. Not all three

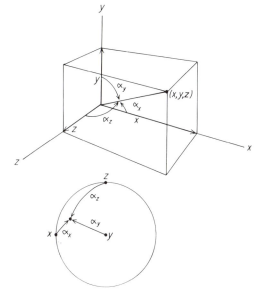

Figure 10.1. *Method of describing the direction of a line in space using the angles between the line and the coordinate direction α_x, α_y, α_z. The direction cosines (l, m, n) are the cosines of these angles ($l = \cos \alpha_x$ etc.).*

167

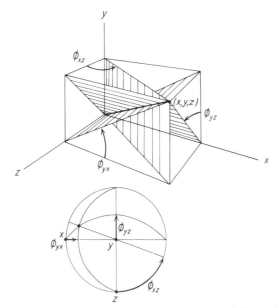

Figure 10.2. *Method of describing the direction of a line using the coordinate plane projection angles ϕ_{xy}, ϕ_{yz} and ϕ_{xz}.*

direction consines are necessary to define an orientation because from Equations 10.1 it follows that

$$l^2 + m^2 + n^2 = 1 \qquad (10.2)$$

The second method of defining direction gives the **coordinate plane projection angles**. These angles are defined as the angles made by the intersection of a plane containing the line and one of the coordinate axes with the plane containing the other two coordinate axes (Figure 10.2). These are ϕ_{xy} (measured in the xy plane between the y direction and the intersection of the plane containing the line and the z-axis with the xy plane), ϕ_{yz} and ϕ_{zx}. The sign convention for these angles is based on the rotation sense around the coordinate axis. It follows the so-called corkscrew rule: if a corkscrew is directed along the positive sense of the coordinate axis, then positive rotations follow the rotation sense of the corkscrew's being driven in that direction (Figure 10.3). The rotation convention we used in our discussion of two-dimensional strain geometry is in accord with this rule. In two dimensions we consider that the positive third axis emerges from the surface towards the observer. The corkscrew rule then gives positive rotations as anticlockwise, negative as clockwise. As with direction cosines, the three coordinate plane projection angles are related. The angles are given by

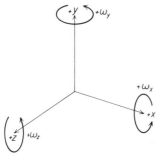

Figure 10.3. *Sign convention according to the corkscrew rule for defining the sign of rotation around the three coordinate axes.*

$$\tan \phi_{xy} = x/y$$
$$\tan \phi_{yz} = y/z \qquad (10.3)$$
$$\tan \phi_{zx} = z/x$$

and it follows that

$$\tan \phi_{xy} \tan \phi_{yz} \tan \phi_{zx} = 1 \qquad (10.4)$$

The relationships between the reference angles of these two methods are given directly from Equations 10.1 and 10.3.

$$\tan \phi_{xy} = l/m \qquad (10.5)$$

A third common method of recording directions is a system of combining the two basic methods described above. This method which uses **Euler angles** should be well known to the reader; it is the latitude–longitude method used to record positions on a global map, and is the basic coordinate reference system of the projection technique we discussed in Session 9. An angle α_y is measured from one of the reference coordinate axes directly to the line being specified. The second angle ϕ_{xz} is that between the plane containing the reference coordinate axis and the line, and another plane containing two coordinate axes, one of which must be the coordinate axis used to define the first angle α. In words this is rather complicated: it should be more easily understood in terms of the Schmidt or Billings net projections shown in Figure 10.4.

Because the different systems may confuse a beginner we have discussed in detail methods for measuring line orientation. The different applications of these systems can be summarized:

1. *direction cosines* offer the best method of formulating angles when variations of elongation, shear strain etc. must be expressed in mathematical form;
2. *coordinate plane projection angles* offer the best method for manipulating angles and planes when the geologist is required to deform or undeform planar and linear structures;
3. *Euler angles* offer the most practical and rapid method for plotting directions on projections.

Displacement in three dimensions

If a point with coordinates (x, y, z) is displaced, it takes up some other position (x', y', z'). The **displacement vector** is defined as the straight line joining these two coordinates. It has three components u, v, and w parallel to the x, y, and z coordinate axes respectively (Figure 10.5) which are

$$u = x' - x$$
$$v = y' - y \qquad (10.6)$$
$$w = z' - z$$

These *displacement vector components* can also be expressed in terms of the length of the displacement vector $(d = (u^2 + v^2 + w^2)^{1/2})$ and two direction cosines $l = \cos d/u$, $m = \cos d/v$.

As in the analysis of two dimensional strain, the displacement vectors encountered during a homogeneous strain vary from point to point in the body, both in length

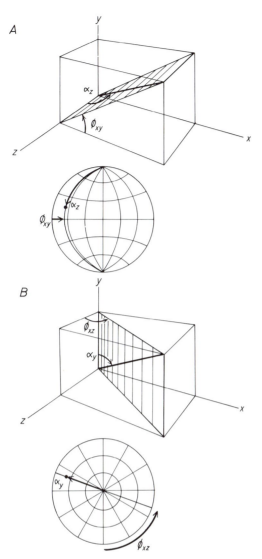

Figure 10.4. *Definition of the two Euler angles* α *and* φ *used to record the direction of a line in space. A illustrates the method with respect to a horizontal reference axis (Schmidt net) and B shows the same system related to a vertical reference axis (Polar net).*

and in orientation. The coordinates of final and initial position are related linearly, and, by analogy with our two-dimensional analysis, we can express this relationship in linear equations

$$x' = ax + by + cz$$
$$y' = dx + ey + fz \qquad (10.7)$$
$$z' = gx + hy + iz$$

where the nine coefficients a–i are constant terms. You will see that on any of the three coordinate planes xy, yz, or zx (where $z = 0$, $x = 0$ and $y = 0$ respectively) these equations reduce to forms identical to those we have already met in our two-dimensional analysis

$$x' = ax + by$$
$$y' = dx + ey$$
$$y' = ey + fz \qquad (10.8)$$
$$z' = hy + iz$$

$$z' = gx + iz$$
$$x' = ax + cz$$

It therefore follows that on each of the coordinate planes an initial circle would be transformed into a strain ellipse, and that, in general, each of these three strain ellipses had a different shape and orientation.

The nine coefficients of Equation 10.7 define the three-dimensional **strain matrix**:

$$\begin{bmatrix} a & b & c \\ d & e & f \\ g & h & i \end{bmatrix} \qquad (10.9)$$

Equation 10.7 is an equation expressing final position in terms of initial coordinates, a form we have previously termed Lagrangian. It is possible to express the **coordinate transformations** in a Eulerian equation (relating initial positions to final coordinates), from which we obtain the **reciprocal strain matrix**:

$$\begin{aligned} x &= Ax' + By' + Cz' \\ y &= Dx' + Ey' + Fz' \\ z &= Gx' + Hy' + Iz' \end{aligned} \qquad \begin{bmatrix} A & B & C \\ D & E & F \\ G & H & I \end{bmatrix} \quad (10.10)$$

where each of the coordinates, like A, can be expressed in the nine terms of the strain matrix, for example:

$$A = \frac{ei - fh}{aei + bfg + cdh - afh - bdi - ceg} \quad (10.11)$$

The importance of the reciprocal strain matrix in our development of three-dimensional strain is that it enables us to effect the transformation of a sphere as simply as we transformed a circle in two dimensions. If we take points (x, y, z) on a sphere of unit radius centred at the origin and given by the equation

$$x^2 + y^2 + z^2 = 1 \qquad (10.12)$$

and transform them to new positions (x', y', z') according to Equation 10.8, we obtain the shape of the deformed sphere. It is given by

$$Px^2 + Qy^2 + Rz^2 + Sxy + Tyz + Uzx = 1 \quad (10.13)$$

where the six coefficients P–U are constants which can be expressed in terms of A–I, and in terms of a–i.

Equation 10.13 is the equation of an ellipsoid centred at the origin, with axes generally inclined to the x, y and

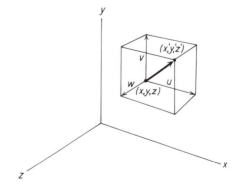

Figure 10.5. *Definition of the displacement vector joining (x, y, z) and (x', y', z'). u, v and w are the three components of the displacement vector.*

z coordinate directions. It is known as the **strain ellipsoid**, and is one of the key concepts for visualizing the properties of homogeneous strain in three dimensions (Figure 10.6).

The strain ellipsoid has three orthogonal principal axes. These are the **three principal axes of finite strain**, of semi-axis lengths $1 + e_1 \geqslant 1 + e_2 \geqslant 1 + e_3$. The extensions e_1, e_2 and e_3 are known as the **principal longitudinal strains**. In describing the geometric properties of the strain ellipsoid it is often most convenient to reorient our reference coordinate frame to coincide with the principal strain axes (Figure 10.6). These new directions are X, Y and Z, parallel to the principal strains of values e_1, e_2 and

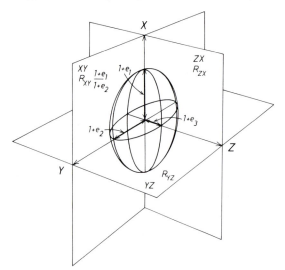

Figure 10.6. *The strain ellipsoid. Notation for principal strains (e_1, e_2 and e_3), principal strain directions (X, Y, and Z), and the principal planes XY, YZ, ZX. R_{xy}, R_{yz} and R_{zx} are the principal plane strain ratios.*

e_3 respectively. The three planes XY, YZ and ZX are known as the **principal planes of finite strain**. The strain ellipses on them are known as the **principal strain ellipses** with strain ratios $R_{xy} = (1 + e_1)/(1 + e_2)$, R_{yz} and R_{xz}. The three principal strain ratios are related, for

$$R_{xz} = R_{xy} \cdot R_{yz} \qquad (10.14)$$

The **volumetric dilatation** is described by the Greek letter "delta" (Δ_V). The volume of the strain ellipsoid is $4\pi(1 + e_1)(1 + e_2)(1 + e_3)/3$ derived from the sphere of volume $4\pi/3$. The proportional dilation is given by

$$1 + \Delta_V = (1 + e_1)(1 + e_2)(1 + e_3) \qquad (10.15)$$

In two-dimensional strain we proved that the orthogonal axes of the strain ellipse were derived from two other orthogonal directions (see Appendix B, Equations B.14, B.15), and we used the change in orientation of these lines to define the rotational component of the deformation (ω). It can be proved that the three principal axes of strain are also derived from three other orthogonal directions with axes X_0, Y_0 and Z_0. If the directions of the principal strain axes X, Y and Z coincide with the directions X_0, Y_0 and Z_0, the strain is termed an *irrotational strain*. Generally they do not coincide and the deformation is a *rotational strain*. Rotation in three dimensions is a more complex concept than that in two dimensions presented in Appendix B.10. The set of axes XYZ and the set $X_0Y_0Z_0$ both require three orientation terms to define their positions in space. For example the set XYZ requires two direction cosines to fix the direction of X. The axis Y, being perpendicular

to X, does not have complete freedom of orientation, and its direction can therefore be specified with one further direction cosine measurement. The third axis Z is fixed absolutely when the first two are located because it is perpendicular to both. To orient the strain ellipsoid and the lines in the initial sphere which have become the ellipsoid axes we require six measurements. To define **rotation** in three dimensions we need to specify how three orthogonal axes before deformations transform into three orthogonal axes after deformation. To do this we clearly need three terms, and several sets of three terms can be chosen to define uniquely the body rotation transformation.

1. The angles between corresponding axes can be specified (Figure 10.7A; $\alpha_x = X \angle X_0$; $\alpha_y = Y \angle Y_0$; $\alpha_z = Z \angle Z_0$). These angles will generally differ and they will not really represent a rotation in the sense of requiring an angle of rotation to transform the three original directions into the three final directions.
2. We can specify three successive body rotations: ω_x about the x coordinate direction, ω_y and ω_z about the y and z axes respectively (Figure 10.7B). The order of the rotation operations must be given (x-rotation, followed by y-rotation followed by z-rotation).
3. It is possible to define the transformation by specifying the orientation of a unique axis of rotation R (this requires two direction cosine measurements) and a single angle of rotation ω about this axis (Figure 10.7C). The positive or negative rotational sense must be given (in accordance with the corkscrew rule).

Of these three methods, the first is of extreme simplicity requiring only the direct measurement of three angles. The second requires the most complex mathematical analysis, but has the advantage that it allows direct comparison of rotation components from locality to locality. The third is probably the most satisfactory from a practical viewpoint in that it reduces the total body movement to a single rotation around a single axis, a concept that is rather easy to visualize geometrically.

Our basic introduction to three-dimensional strain can be summarized thus, to describe how all points in the material change their positions, to develop a general homogeneous strain a set of three coordinate transformation equations are needed. These equations (Equation 10.7) require nine terms for their definition, the *nine displacement components*. These displacements transform an initial unit sphere into the strain ellipsoid. The geometry of the strain ellipse requires six terms for its complete definition (three axial lengths plus three orientations), and the rotations in space made by the ellipsoid axes (the body rotation components) require a further three orientation (or rotation) terms for their specification. The total distortional and rotational effect, therefore, also require nine components for a complete geometric definition. In summary: the *displacement components* relate to *strain and rotation components*

abc		$e_1e_2e_3$	$e_1e_2e_3$	$e_1e_2e_3$	$R_{xy}R_{yz}\Delta_V$
def	$\longrightarrow$	XYZ	or XYZ	or XYZ	or XYZ
ghi		$X_0Y_0Z_0$	$\omega_x\omega_y\omega_z$	$lm\omega$	$X_0Y_0Z_0$

where $e_1e_2e_3$ principal strains
$R_{xy}R_{yz}$ principal plane strain ratios
Δ_V volume dilatation
XYZ orientation of ellipsoid axes

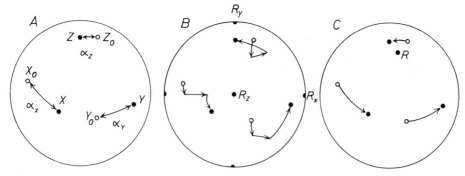

Figure 10.7. Orthogonal directions X_0, Y_0 and Z_0 which become the strain ellipsoid axes X, Y and Z. Three methods for describing the relationships, see text for details.

$X_0 Y_0 Z_0$	initial orientation of lines becoming ellipsoid axes
$\omega_x \omega_y \omega_z$	rotation components about x, y and z coordinate directions
lm	direction cosines of unique rotation axis
ω	rotation around unique rotation axis

As with two-dimensional strain, not all strain and rotation components are equally accessible in naturally deformed rocks. The body rotation components are especially difficult to measure, since we rarely have information from which we can specify the initial orientation of the material. Although the principal strain axes orientations X, Y and Z are relatively easily established, the actual values of the three principal strains are not generally forthcoming. Usually we have to make do with the principal strain ratios $R_{xy} R_{yz}$, the dilatational component Δ_V usually being unobtainable.

Classification of strain ellipsoids

In our discussion of two-dimensional strain we concluded that a strain ellipse can be classified into one of three types (or Fields) depending upon the values of the two principal strains. We can classify strain ellipsoids using similar methods. Two main classification schemes are possible: the first dependent on the absolute values of the principal strains; the second dependent on the ratios of the principal strains.

Table 10.1 sets out the range of possible values for the three principal extensions e_1, e_2 and e_3 and the type of strain.

This table gives the complete spectrum of all the mathematical possibilities, but not all of these are likely to be found as a result of geological deformations. In particular the two end members at the top and bottom of the list are peculiar in that the strain ellipsoid lies either completely outside or completely inside the unit sphere. It is difficult to envisage tectonic processes which would produce such types of strain, although such effects might come about with great volume increases or decreases (e.g. the hydration of anhydrite to gypsum, or the types of volume reduction phase changes which seem likely to occur in certain levels of the earth's mantle). The other five varieties (labelled Types 1 to 5) are all known to occur in naturally deformed rocks. Broadly, they can be classified into ellipsoids of *true flattening*. (e_1 and e_2 + ve) and *true constriction* (e_2 and e_3 − ve). The ellipsoids occupying the boundary zone between these ($e_2 = 0$) are the *plane strain* ellipsoids. Volumetric dilatation in *all* these types of ellipsoids can be either positive or negative.

The second method of ellipsoid classification uses the two independent values of ellipticity of the principal plane strain ellipses R_{xy} and R_{yz}. The scheme is based on a comparison of the two ellipticities and is essentially a description of the ellipsoid shape irrespective of the absolute values of the strain and of volumetric dilatation. It is an extremely practical method of classifying ellipsoid shapes and has wide use in analysis of strain in naturally deformed rocks. A special feature of the method is that it enables ellipsoid shapes to be represented on a two-dimensional graphical plot. The method of representing ellipsoidal shapes in graphs was first suggested by Zingg (1935) as a way of analysing the shapes of pebbles in conglomeratic sediments. Its full potential as a tool for structural analysis was realized by Derek Flinn, and it has quite rightly become known as the Flinn method. The

Table 10.1

Value of e			
+ve	0	−ve	Description of strain
$e_1 e_2 e_3$			Gross +ve dilatation, ellipsoid outside sphere
$e_1 = e_2$		e_3	Uniaxial flattening, Type 1
$e_1 e_2$		e_3	General flattening, Type 2
e_1	e_2	e_3	Plane strain, Type 3
e_1		$e_2 e_3$	General constriction, Type 4
e_1		$e_2 = e_3$	Uniaxial constriction, Type 5
		$e_1 e_2 e_3$	Gross −ve dilatation, ellipsoid inside sphere

Flinn graph plots the value of R_{yz} as abscissa and R_{xy} as ordinate (Figure 10.8). The origin of the graph is not the point $(0, 0)$ but the point $(1, 1)$ because R values of less than unity cannot, by definition, exist. The origin represents a spherical shape. Any ellipsoid is represented by a single point located in the positive sector (Figure 10.8, point p). Flinn suggested the *parameter k* to describe the general position of the ellipsoid plot. The *k-value* is defined as

$$k = \frac{R_{xy} - 1}{R_{yz} - 1} \qquad (10.16)$$

and this represents the tangent of the angle q between the abscissa axis and the line joining p to the point $(1, 1)$. Ellipsoids with k values lying between zero and unity (Figure 10.8, Field A) have more or less pancake forms (**oblate ellipsoids**), and where $k = 0$ the form becomes uniaxial normal to the pancake ($R_{xy} = 1$). Those ellipsoid plots with k-values lying between unity and infinity (Figure 10.8, Field B) have general flattened, cigar-like forms (**prolate ellipsoids**), which, where $k = \infty$, are uniaxial parallel to the cigar axis ($R_{yz} = 1$). Flinn originally described these two fields as flattened-and constricted-ellipsoids respectively, but this terminology is applicable only when the strain ellipsoids show no volume dilatation (flattening and constriction should imply that e_2 is positive or negative respectively). For example, ellipsoids with principal strains given by

	$1 + e_1$	$1 + e_2$	$1 + e_3$	Δ_v
1.	1·2	0·9	0·4	−0.6
2.	3·0	1·1	0·9	+2.0

are constricted-and flattened-ellipsoids respectively, but on the Flinn graph they plot in Fields A and B respectively. Probably the best way of referring in words to the two main fields of the Flinn graph is by the terms *field of apparent flattening* and *field of apparent constriction* (Figure 10.8).

A variant of the Flinn graph, which is especially useful for analyses of deformation sequences, was proposed by Ramsay (1967) and has been used extensively since in discussions of theoretical concepts and for plotting natural strain data. This is the **logarithmic strain graph** in which the abscissa and ordinate are $\log R_{yz}$ and R_{xy} respectively, the logarithmic base being either to an exponential base or to base 10.

Measurement of three-dimensional strain

Question 10.1

It is most unusual to be able to observe the three-dimensional form of a deformed object in sufficient detail to establish directly the form of the strain ellipsoid. We normally construct the form and orientation of the strain ellipsoid by putting together a number of two-dimensional strain ellipses. A strain ellipse represents a section of the strain ellipsoid. It passes through the ellipsoid centre. The orientations and values of the three-dimensional strains directly control the orientation and values of the two-dimensional strain ellipse. It is mathematically possible to determine the orientations and values of the three-dimensional principal strains from *any three non-parallel strain ellipses*, and in Volume 3 we will investigate the techniques required for this construction. In practice features are often available in the strained material from which the three principal axes can be determined with sufficient accuracy to locate the principal strain planes. If we are able to choose our strain ellipse constructions to coincide with *any two of the principal planes*, we can construct the strain ellipsoid. Although measurements made on the third principal plane are not necessary for an answer, they do provide additional data which act as an overall check on the accuracy of the computation.

Figure 10.9 shows the appearance of two principal plane surfaces from an outcrop of Cambrian slate in North Wales.

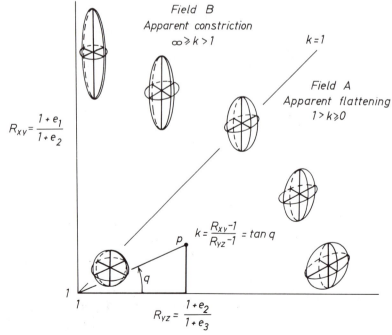

Figure 10.8. The Flinn graph for representing the shape of any strain ellipsoid. An ellipsoid plotting at point p has a k-value of tan q.

Figure 10.9. Slates with deformed spots. See Question 10.1.

The slate is dark purple (on account of the presence of finely crystalline flakes of hematite), and the pale areas are light green as a result of predeformational diagenetic chemical processes which took place in the muddy sediment. In undeformed regions these green spots are almost perfectly spherical in shape. Photograph A is a section perpendicular to the slaty cleavage planes. The surface contains the maximum and minimum finite strains parallel to the X and Z axes of the strain ellipsoid. Photograph B shows the appearance of the spots on a cleavage surface, and the surface contains the maximum and intermediate principal strains (an XY principal plane).

Calculate the shapes of the two-dimensional strain ellipses on the two principal planes. From these data construct the form of the strain ellipsoid in terms of the principal plane strain ratios R_{xy} and R_{yz}. Express the data also in the form $1 + e_1:1 + e_2:1 + e_3$ as $a:b:1$ and $c:1:d$. Plot the ellipsoid on a Flinn graph. Calculate the k-value of the ellipsoid. Describe the form of the ellipsoid in words.

Types of strain ellipsoid

Question 10.2

Finite strains result from the superposition of a sequence of deformation events. The shape of the finite strain ellipsoid depends on the history of this deformation sequence. One way that we can investigate the factors which control the finite strain ellipsoid is to envisage the strains being built up in successive small but finite incremental steps. The next problem will look into the geometry of coaxially superposed small strains.

Before attempting this we should check that we are quite sure what happens when an incremental strain is coaxially superposed on a previously established strain state. If the incremental strain ellipsoid has principal extensions e_{i1}, e_{i2} and e_{i3}, and these are superposed on an ellipsoid with semi-axes of lengths $1 + e_1$, $1 + e_2$ and $1 + e_3$, then the resulting ellipsoid has semi-axes which are the products of the two sets of semi-axes, i.e. $(1 + e_1)(1 + e_{i1})$, $(1 + e_2)((1 + e_{i2})$ and $(1 + e_3)(1 + e_{i3})$.

Figure 10.10 shows in diagrammatic form four different coaxial deformation sequences developed by deforming an initial cube with edges of unit length into rectangular blocks of different shapes. Table 10.2 presents the data from each of these four sequences in detail, and gives the values of the principal strains at successive stages during the deformation, together with features of the volume dilatation and principal plane strain ratios that can be derived from them. Note that all the numerical data for each of the four sequences show identical successions of values in the tectonic $1 + e_3$ data rows, and indicate the same progressive incremental shortenings producing progressive total shortening in the Z direction in 10% stages. Also note that the tabulated data for the shapes of the deformed blocks are presented as principal strain ratios measured *along the tectonic X, Y and Z directions*, and that these directions do not always coincide with the X, Y and Z directions of the finite strain ellipsoid at all stages in the sequence.

From the data tables and diagrams determine what the main features of each deformation sequence are (e.g. plane strain, apparent constriction, true flattening, progressive volume loss etc.).

Table 10.2

1.

Tectonic X	1.00	1.11	1.25	1.43	1.67	2.00	2.50	3.33	5.00
Tectonic Y	1.00	1.00	1.00	1.00	1.00	1.00	1.00	1.00	1.00
Tectonic Z	1.00	0.90	0.80	0.70	0.60	0.50	0.40	0.30	0.20
$1 + \Delta$	1.00	1.00	1.00	1.00	1.00	1.00	1.00	1.00	1.00
X_t/Y_t	1.00	1.11	1.25	1.43	1.67	2.00	2.50	3.33	5.00
Y_t/Z_t	1.00	1.11	1.25	1.43	1.67	2.00	2.50	3.33	5.00
$\log_{10} X_t/Y_t$	0.00	0.05	0.10	0.16	0.22	0.30	0.40	0.52	0.70
$\log_{10} Y_t/Z_t$	0.00	0.05	0.10	0.16	0.22	0.30	0.40	0.52	0.70

2.

Tectonic X	1.00	1.10	1.20	1.32	1.51	1.74	2.09	2.65	3.73
Tectonic Y	1.00	10.2	1.04	1.07	1.10	1.15	1.20	1.26	1.34
Tectonic Z	1.00	0.90	0.80	0.70	0.60	0.50	0.40	0.30	0.20
$1 + \Delta$	1.00	1.00	1.00	1.00	1.00	1.00	1.00	1.00	1.00
X_t/Y_t	1.00	1.08	1.15	1.23	1.37	1.51	1.74	2.10	2.78
Y_t/Z_t	1.00	1.13	1.30	1.53	1.83	2.30	3.00	4.20	6.70
$\log X_t/Y_t$	0.00	0.03	0.06	0.09	0.14	0.18	0.24	0.32	0.44
$\log Y_t/Z_t$	0.00	0.05	0.11	0.18	0.26	0.36	0.48	0.62	0.83

3.

Tectonic X	1.00	1.06	1.13	1.21	1.33	1.50	1.75	2.17	3.00
Tectonic Y	1.00	1.00	1.00	1.00	1.00	1.00	1.00	1.00	1.00
Tectonic Z	1.00	0.90	0.80	0.70	0.60	0.50	0.40	0.30	0.20
$1 + \Delta$	1.00	0.95	0.90	0.85	0.80	0.75	0.70	0.65	0.60
X_t/Y_t	1.00	1.06	1.13	1.21	1.33	1.50	1.75	2.17	3.00
Y_t/Z_t	1.00	1.11	1.25	1.43	1.67	2.00	2.50	3.33	5.00
$\log X_t/Y_t$	0.00	0.03	0.05	0.08	0.12	0.18	0.24	0.34	0.48
$\log Y_t/Z_t$	0.00	0.05	0.10	0.16	0.22	0.30	0.40	0.52	0.70

4.

Tectonic X	0.50	0.56	0.63	0.71	0.83	1.00	1.25	1.67	2.50
Tectonic Y	1.00	1.00	1.00	1.00	1.00	1.00	1.00	1.00	1.00
Tectonic Z	1.00	0.90	0.80	0.70	0.60	0.50	0.40	0.30	0.20
$1 + \Delta$	0.50	0.50	0.50	0.50	0.50	0.50	0.50	0.50	0.50
X_t/Y_t	1.00	1.11	1.25	1.41	1.20	1.00	1.25	1.67	2.50
Y_t/Z_t	2.00	1.61	1.26	1.01	1.38	2.00	2.50	3.33	5.00
$\log X_t/Y_t$	0.00	0.05	0.10	0.16	0.08	0.00	0.10	0.22	0.40
$\log Y_t/Z_t$	0.30	0.21	0.10	0.00	0.14	0.30	0.40	0.52	0.70

Plot the ellipsoids from each of the four sequences, first on a Flinn diagram (R_{yz} abscissa, R_{xy} ordinate, $(1, 1)$ as origin), and second on a logarithmic graph ($\log_{10} R_{yz}$ abscissa, $\log_{10} R_{xy}$ ordinate). Describe and account for the geometric features of the lines joining the succession of points, noting especially whether lines are straight or curved and where they change direction. What are the main differences between the two graphical plots?

Return to the plot of the measured strain ellipsoid from the slates of North Wales (Question 10.1). Suggest a number of interpretations of the strain history of this rock that might account for the ellipsoid shape.

Check the results of these questions in the Answer and Comments section and return to the questions below.

Change of orientation of planes as a result of strain

As a result of deformation the orientation of a plane and its pole are altered. In Session 9 we investigated the rules for determining how lines and planes were reoriented as a result of a body rotation, and now we will extend this investigation to see what modifications take place as a result of a homogeneous finite strain.

Figure 10.11A shows a body containing an inclined surface s with pole π_s. The reference axes $X_0 Y_0 Z_0$ are the three perpendicular directions that will become the principal axes of the strain ellipse. The surface s intersects the three planes $X_0 Y_0$, $Y_0 Z_0$ and $X_0 Z_0$ and the intersections

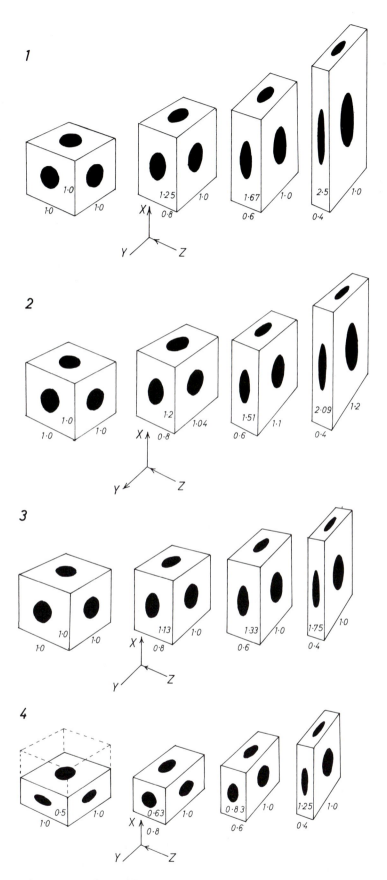

Figure 10.10. Diagrammatic representations of four co-axial deformation sequences. Further data given in Table 10.1. See Question 10.2.

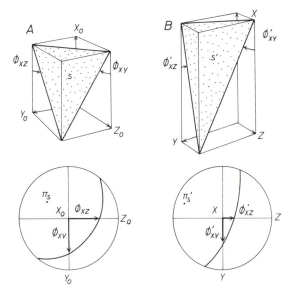

	Strike	Dip
1.	110°	70° to SW
2.	150°	60° to SW
3.	120°	30° to SW
4.	0°	45° to W
5.	135°	vertical
6.	90°	45° to S
7.	90°	vertical

Figure 10.11. *Change of orientation of a plane s as a result of a homogeneous finite strain.*

make angles of ϕ_{xy}, ϕ_{yz} and ϕ_{xz} with the X_0, Y_0 and Z_0 axes respectively. The surface s changes its orientation to s' as a result of straining (Figure 10.11B). X, Y and Z are the three principal strain directions, and ϕ'_{xy}, ϕ'_{yz} and ϕ'_{xz} are the angles made on the three principal planes of strain with X, Y and X respectively. The relationships between the angles $\phi_{xy}\phi_{yz}\phi_{xz}$ and their deformed counterparts ϕ'_{xy}, ϕ'_{yz} and ϕ'_{xz} can be computed from the equation introduced in Session 8 (Equation 8.3, see also proof in Appendix D, Equation D.13). We can express the relationships thus:

$$\tan \phi'_{xy} = \tan \phi_{xy}/R_{xy} \qquad (10.17)$$

and derive two similar equations for the two other principal planes, where R_{xy} is the XY principal plane strain ratio. This function is plotted in Figure 10.12. If we know two of the principal plane strain ratios, we can use these relations to compute the new orientations of the plane (Figure 10.11B).

Question 10.3

On a Schmidt net plot a plane (plane 1 below) with orientation: strike 110°, dip 70° to SW as a great circle and as a π-pole. Consider that the initial and final orientations of the strain ellipsoid axes $X_0Y_0Z_0$ and XYZ are vertical, horizontal N–S and horizontal E–W respectively. Determine the angles ϕ_{xy} and ϕ_{xz}. Deform the plane by a homogeneous strain with principal plane strain ratios $R_{xy} = 2\cdot0$, $R_{xz} = 4\cdot0$. Using the graphs of Figure 10.12, determine the new angles ϕ'_{xy} and ϕ'_{xz}. Find the new position of the plane and its pole. Repeat for the planes 2–7.
Observe the changes in position of the poles to each plane as a result of strain and determine the general rules governing the rearrangement of variably oriented planes in a body as a result of a homogeneous strain.

Change of orientation of lines as a result of strain

Figure 10.13A shows the same body as that in Figure 10.11A, but here it contains a line element, l. In geometric terms this line can be envisaged as formed at the intersection of two planes containing the Y_0 and Z_0 axes of the strain system, the two planes being fixed by the coordinate plane projection angles ϕ_{xy} and ϕ_{xz}. The deformed body is shown in Figure 10.11B. The angular modifications of ϕ_{xy} and ϕ_{xz} to ϕ'_{xy} and ϕ'_{xz} are identical to those of Equation 10.17, and we can therefore use this geometry to reconstruct the planes containing the deformed lineation and the Y and Z axes of the system. Hence we can construct the new position l' of the line.

Question 10.4

On a Schmidt net plot a pole representing a line with azimuth 20°, and plunge 20° (line 1). Using the same deformation scheme as that used in Question 10.3, construct the deformed position of the line. Repeat the construction for the lines 2–7.

	Azimuth	Plunge
1.	20°	20°
2.	60°	30°
3.	30°	60°
4.	90°	45°
5.	45°	0°
6.	0°	45°
7.	0°	0°

Comment on the overall scheme of change in orientation of the line elements as a result of finite strain.
Check the answers to Questions 10.3 and 10.4, then continue to Session 11 or to the starred questions below.

STARRED (★) QUESTIONS

Computation of three-dimensional strain from two sections which are not principal planes

Question 10.5★

The ratios of the principal two-dimensional strains R_{PQ} and R_{RS} in two planes which are not principal planes of the strain ellipsoid are $R_{PQ} = 4\cdot00$, $R_{RS} = 3\cdot14$. Plane PQ has

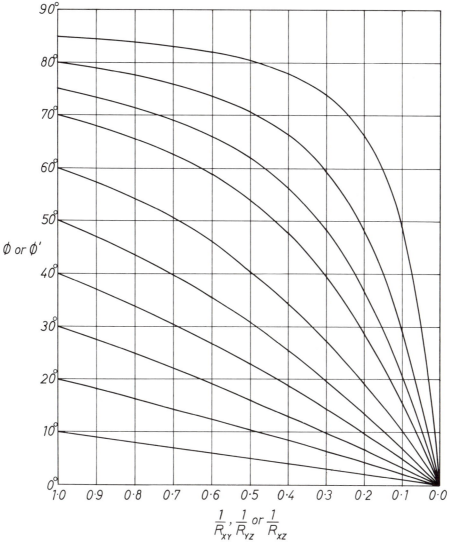

Figure 10.12. *Graphical representation of the function* $\phi' = tan^{-1}(tan\,\phi/R)$.

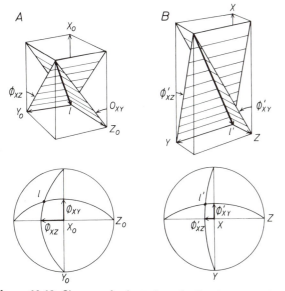

Figure 10.13. *Change of orientation of a line l as a result of a homogeneous finite strain.*

a strike of 152°, and dips 46° NE: plane RS has a strike of 49° and dips 83° NW. The maximum extension direction P on plane PQ has an azimuth 18° and plunge of 38°; the maximum extension direction R on plane RS has an azimuth 246° and plunge 68°.

The rock is cut by a slaty cleavage with strike 30° and dip 75° NW. On this cleavage is a stretching lineation (X-axis of the principal finite strain with azimuth 300° and plunge 75°). Assuming that the slaty cleavage is the XY principal plane of the strain ellipsoid, determine the principal strain ratios R_{xy} and R_{yz}. *Hint*, you will need to cross-correlate the strains on the two planes PQ and RS using their common line of intersection.

Change of orientations of planes and lines as a result of strain

Question 10.6★

In Question 10.3 we used a graphical method with a Schmidt net to determine how the orientation of a plane was modified as a result of strain. Determine the pair of equations which relate the change of orientation of a plane with initial strike ϕ and angle of dip δ to new orientation strike ϕ' and dip δ' in terms of the principal plane strain ratios R_{xy} and R_{yz}.

If you have access to a computer, run a program to deform a randomly oriented series of initial planes. Determine the concentration effects of a strain with $R_{xy} = 2\cdot0$ and $R_{xz} = 4\cdot0$.

Question 10.7★

Determine, in a similar way to the previous question, a pair of equations describing the change of orientation of a line of initial azimuth ϕ and plunge angle δ to new position azimuth ϕ' and plunge δ' in terms of the principal plane strain ratios R_{xy} and R_{yz}.

Run a program on a computer to study how a randomly oriented group of lines is modified by a strain with $R_{xy} = 2\cdot0$, $R_{yz} = 4\cdot0$.

ANSWERS AND COMMENTS

Measurement of three-dimensional strain

Answer 10.1

Figure 10.9A shows the principal XZ plane, and the strain ratio R_{xz} is $5\cdot52$, B is the principal XY plane with $R_{xy} = 1\cdot77$. $R_{yz} = R_{xz}/R_{xy} = 3\cdot11$. The ellipsoid has the ratios of principal strains $5\cdot52:1\cdot77:1\cdot00$ or $3\cdot11:1\cdot00:0\cdot56$. The k-value of the ellipsoid is $0\cdot36$. The ellipsoid is oblate, and lies in the field of apparent flattening. These observations about strain in slates corroborate those of Haughton in 1856.

Types of strain ellipsoid

Answer 10.2

The strain increments are always coaxial with the principal strains of any previously established ellipsoid; therefore the total strain is always irrotational. The tectonic increments of shortening have been arranged so that the total tectonic shortening along the tectonic Z-direction leads to a finite shortening of 10%, 20% ..., 80%, in 10% increments.

Sequence 1: This is a plane strain ($e_2 = 0$ for all stages of deformation) at constant volume.

Sequence 2: This is a progressive, true flattening deformation (e_2 is positive and shows a progressive increase with shortening along the Z-direction). It is a constant volume deformation, so at any stage in the deformation sequence the amount of the extension along the X-direction is less than that for sequence 1 at a similar stage.

Sequence 3: This sequence is a plane strain ($e_2 = 0$) and shows a progressive loss of volume with decrease in length along the Z-axis. For a shortening of 10% along Z, the volume loss is 5%, for 20% the volume loss is 10% etc. As a result of this progressive volume reduction, the elongation along the X-direction is much less than that of the equivalent stage of sequence 1. The best everyday analogy to this sequence is the squeezing of

a sponge between two parallel plates. In geology such deformations appear to be common, particularly in porous sediments. The ellipsoids produced by such a deformation all have an oblate shape and lie in the field of apparent flattening.

Sequence 4: This sequence consists of a plane strain superimposed on a block which had an original compaction (parallel to the later imposed tectonic X-direction) of 50%. No further volume reduction took place during the tectonic deformation. In geology such a sequence might occur when a tectonic deformation was superimposed on a sediment compacted by diagenetic processes. The resulting ellipsoids combining primary compaction and tectonic deformation are complex, passing from a field of apparent flattening, into a field of apparent constriction, and returning to the field of apparent flattening. In the first two deformation fields the X-direction of the total strain ellipsoid is parallel to the Y-direction of the tectonic strain component. Only in the last field do the X-directions for total and tectonic strain coincide.

The various ellipsoids of these four sequences are plotted in Figure 10.14. The principal features of the plots are as follows:

Flinn Graph, *Figure 10.14A*

1. Straight line, slope 1 through the origin $(1, 1)$, the ellipsoid plots all having k-values of unity.
2. Curved line in apparent flattening field originating at $(1, 1)$, and with the ellipsoids having progressively decreasing k-values.
3. Straight line, in apparent flattening field, with slope of $0\cdot5$ passing through the origin $(1, 1)$. The ellipsoids all have the same k-value ($k = 0\cdot5$).
4. Curved line (concave upwards) originating at point $(2, 1)$ (uniaxial flattening) with a negative slope and running to an apparent uniaxial constriction $(1, 1\cdot43)$. The path then returns along the same line to $(2, 1)$ and then moves into the apparent flattening field with a positive slope of $0\cdot5$, and with ellipsoids of progressively increasing k-values.

Logarithmic graph, *Figure 10.14B*

1. Straight line of unit slope passing through the origin $(0, 0)$. Ellipsoids have a logarithmic K-value (defined as $\log R_{xy}/\log R_{yz}$) of 1.
2. Straight line in apparent flattening field through $(0, 0)$ with a slope of $0\cdot5$. All the ellipsoids have $K = 0\cdot5$.
3. Curved line (concave upwards) in apparent flattening field originating at $(0, 0)$, with ellipsoids having progressively increasing K-values ($1 > K > 0\cdot5$).
4. Straight line from $(0\cdot3, 0)$ to $(0, 0\cdot16)$, then returning along same path, then a straight line of slope l in apparent flattening field, the ellipsoids showing progressively increasing K-values from $0\cdot0$ towards $1\cdot0$.

The main difference between the two plots is that in the Flinn graph the initial stages of deformation plot close to the origin whereas the points representing higher strains are spread widely from the origin. The logarithmic plots give a more even distribution of points with increase in deformation. Where the incremental strains are constant (e.q. sequence 2) curved lines on the Flinn graph occur as straight lines on the logarithmic graph.

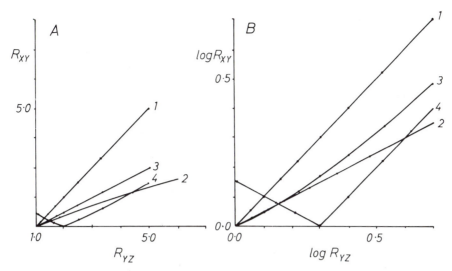

Figure 10.14. *Plots of the shapes of the strain ellipsoids from the four deformation sequences of Question 10.2: A, a Flinn graph; B, a logarithmic graph.*

The strain ellipsoid we plotted in Question 10.1 falls into the field of apparent flattening, and the possible modes of deformation can be listed as follows:

1. True flattening deformation with positive extension along Y as well as along X.
2. Deformation with volume loss in plane strain.
3. Flattening deformation with volume loss.
4. Deformation of a previously compacted sediment under plane strain conditions without further volume loss.
5. Deformation of a previously compacted sediment under plane strain with further volume loss.
6. Deformation of a previously compacted sediment by a flattening deformation with or without volume loss.

It is clear that from the *shape* of a strain ellipsoid we have insufficient data to specify what the real features of the strain are. To resolve the problem it is necessary to seek other information. If we can obtain one measurement of absolute extension in any direction in the ellipsoid three more pieces of information become available. It is then possible to compute the absolute values of the principal extensions, to determine whether the deformation is a real flattening or not, and to discover whether or not there has been a volume change (e.g. using the techniques we explored in Questions 6.1, 8.3).

The relationship between the fabric of deformed rocks and the finite strain state

As a result of the overall change of shape of a rock mass, the individual mineral components generally undergo some mechanical rearrangement, and new mineral components sometimes crystallize in preferred orientations controlled by the strain increment directions. Deformed rocks therefore frequently show preferred orientations of the minerals which comprise them: this geometric organization of the various components we term **fabric**.

Rock fabrics are of two main types, planar and linear. A rock may possess only a planar fabric, only a linear fabric, or may show various combinations of planar and linear elements.

Planar fabrics may be primary or secondary features of the rock. A primary fabric can be caused by the deposition or segregation of mineral components into layers such as primary bedding in sediments, gravitational banding in igneous rocks, or lithological layering in metamorphic rocks. Primary planar fabrics can also develop from the preferred shape orientation of certain mineral species. For example, muscovite crystals sedimented in a shale or silt-stone usually show a preferred orientation with the crystal plates lying sub-parallel to the bedding. In metamorphic rocks certain crystals (e.g. staurolite, amphibole) grow in layers with a shape orientation controlled by the easy availability of the chemical components necessary for their development. Secondary fabrics are created either by the mechanical reorientation of pre-existing minerals or by the result of stress- or strain-controlled new mineral growth.

In tectonically deformed rocks the most common planar fabrics are **cleavage** and **schistosity**. Cleavage is a planar fabric found in low grade metamorphic rocks (such as slates, where the cleavage is termed *slaty cleavage*), and schistosity is the equivalent fabric found in rocks above the greenschist facies. In many regions there is a gradual transition between the cleavage exhibited by the rocks of the low grade metamorphic areas into the schistosity found in the higher grade metamorphic terrains. **Slates** are those rocks where it is only possible to see the individual mineral grains with a microscope, whereas in **schists** it is possible to see the minerals with the naked eye. **Phyllite** is the name given to a rock with grain size characteristics intermediate between those of slates and schists. In phyllites one can usually make out the individual mineral grains with a hand lens. Slates, phyllites and schists are all characterized by showing a preferred planar alignment of many (if not all) of the component minerals (Figure 10.15). The phyllosilicates and amphiboles generally show this alignment most strongly, but other components (such a quartz, carbonates and feldspars) may also possess a similar preferred orientation. Minerals with a platy habit generally lie with their planes sub-parallel to the cleavage and minerals with a prismatic or an acicular habit generally have

Figure 10.15. Thin section of the slate shown in Figure 10.9 illustrating the preferred alignment of phyllosilicates and quartz grains known as slaty cleavage, × 200.

Figure 10.16. Cleavage–bedding intersection in a slate, Engelberg, Central Switzerland. The top of the photograph is a bedding surface with linear features produced by the intersection of the slaty cleavage. The inclined surface on the right is a cleavage plane and shows intersections of the lithological layering.

their longest axes in the cleavage surface. The preferred orientation of the minerals is not perfect, but is of a statistically significant degree of certainty. In a true cleavage or schistosity, however, all the minerals of a particular species show the same degree of preferred alignment, and the fabric, which pervades the whole rock, is termed a **penetrative fabric**.

Cleavage and schistosity practically always cross-cut the primary lithological layers, Figure 10.16 (bedding or metamorphic banding), and if the layering is folded the cleavage usually shows some well marked geometric relationship to the fold geometry. Cleavage is sometimes sub-parallel to the axial surfaces of folds and is termed **axial plane cleavage** (Figure 10.17). Cleavage usually shows systematically converging or diverging forms known as **cleavage fans** (Figures 10.18), 10.21). Cleavage fans are termed **convergent- or divergent-cleavage fans** depending upon whether adjacent cleavages converge or diverge *when traced from the outer to an inner arc of a fold*. Convergent cleavage fans are diagnostic of the more competent layers in a folded sequence, whereas divergent fans are found in the less competent layers. Cleavage exactly parallel to bedding planes is very rare indeed, and close observation of many relationships termed *bedding plane cleavage* reveal a small but detectable angle between the two planar structures. Where cleavage passes from one lithological layer to another, the orientation of the planar fabric changes (Figures 10.19, 10.21). This change of orientation is known as **cleavage refraction**, by analogy with the deflection that is made by a beam of light as it passes through media of differing optical refractive index. The angle between cleavage and lithological surfaces is always highest in the most competent layers and lowest in the least competent layers. The change in angle that the cleavage makes as it

passes from layer to layer is a function of the competent contrast. Cleavage is most strongly developed where the cleavage–bedding angle is small (Figure 10.19). Where lithological changes are gradational, as for example where one finds a gradational upward increase in clay content towards the top of a graded turbidite bed, then the cleavage shows a gradational change in angle through the graded unit (Figure 10.22). Sorby (1853) correctly related the phenomenon of cleavage refraction to changes in deformability of differing lithologies. He suggested that the difference in orientation of cleavage between a sandstone and a shale was caused by the greater deformation by volume loss suffered by the shale. An exceptionally clear discussion of this idea is to be found in Harker's masterly review of slaty cleavage (1885, pp. 17–21).

Where competent objects, such as conglomerate pebbles or hard diagenetic concretions, exist in cleaved rocks, the cleavage is deflected around the competent object (Figure 10.20).

There is a vast literature on the significance of cleavage and schistosity. As a starting point we strongly recommend three excellent reviews, by Harker (1885), Siddans (1972) and Wood (1974). Although some of the fine details of the mechanisms controlling the development of cleavage are still under debate, there is a generally accepted view that cleavage and schistosity are related to the finite strain state. The fabric forms perpendicular to the shortest axis (*Z*-direction) of the finite strain ellipsoid and increases in intensity with the strain ratio R_{xz}. These conclusions were first put forward in the mid-nineteenth century as a result of studies of cleaved rocks containing objects suited to deformation measurement (Sharpe, 1847, 1849; Sorby, 1853, 1856, Haughton, 1856; Phillips, 1857; Harker, 1885). All subsequent work on cleavage *which has employed strain*

Figure 10.17. Axial plane cleavage in folded Ordovician siltstones, Iglesias, Sardinia.

Figure 10.18. Cleavage fans in folded Ordovician sediments, Rhosneigr, North Wales.

Figure 10.19. Cleavage refraction in Carboniferous sandstones and shales (slates). Note the change of intensity of cleavage and the changes in lithology.

Figure 10.20. Deflection of cleavage around a competent fragment of quartzite in an argillaceous pelite, Srinagar, India.

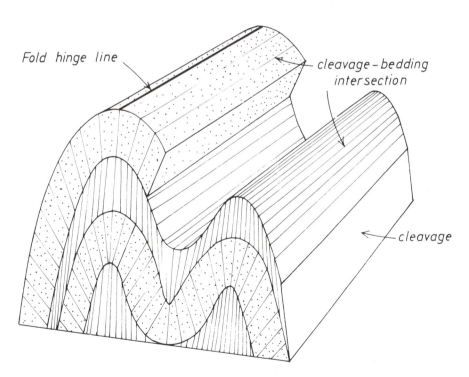

Figure 10.21. Geometric form of cleavage fans and cleavage–bedding intersection lineations in folded rocks. The most competent layers are stippled, the least competent layers unstippled. In simply folded rocks the cleavage–bedding intersection lineation is parallel to the fold hinge lines.

Figure 10.22. *Cleavage refraction in a lithologically graded bed from Flysch sediments in the Wildhorn nappe of the Central Helvetic Alps. The cleavage orientation changes abruptly at the base of the graded unit, and shows a progressive change when traced upwards through the graded bed. Note how the intensity of development of the cleavage is a function of orientation (and rock lithology).*

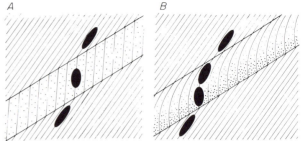

Figure 10.23. *Cleavage refraction and strain variations in layers of differing lithology. The more competent layers are stippled. A shows a competent layer embedded in a less competent matrix; B illustrates the curved cleavage found in graded beds (cf. Figure 10.22).*

measurement techniques has, without exception, supported these results (Heim, 1878; Cloos, 1947; Breddin, 1957; Siddans, 1977; Wood, 1974). It follows that cleavage planes represent the traces of the XY planes of adjacent strain ellipsoids and that cleavage must obey the geometric rules of finite strain trajectories. In our investigations of two-dimensional, heterogeneous strain we concluded that the convergence or divergence of X-finite strain trajectories went together with increase and decrease in finite strain values respectively. The same holds true for three-dimensional XY-finite strain trajectories and cleavage planes. Where cleavage planes converge, the finite strain XZ ratios increase, and this convergence goes together with an increase in the intensity of the cleavage. The phenomenon of cleavage refraction can be easily interpreted from these geometric rules. A change of orientation of cleavage passing through rocks of differing lithology implies a spacing change between cleavages, and this can be directly related to the change in strain state (Figure 10.23). The geometric boundary conditions of layers with refracted cleavage are precisely those we investigated in Session 3 in our discussion of shear zones (pp. 46–47). We concluded there that the only compatible strain variations were: (1) heteroge-

neous simple shear; (2) volume change perpendicular to the layer boundaries; (3) a homogeneous strain. In cleavage refraction probably all three are operational. A study of deformed fossils and other strain markers generally indicates that layered strata undergo a layer-parallel shortening before folding (3, above). During the folding process differential shear often develops parallel to the layers (1, above) and differential volume loss (2, above) caused by decrease of pore space, pressure solution and mineral transformations (e.g. clays to phylosilicates) also occurs frequently. We will return to a more detailed analysis of cleavage variation in Volume 2 when we come to discuss the geometric features of folded rocks.

Once the geometric relationship of cleavage and finite strain state was established in the mid-nineteenth century, there followed an investigation into the reasons for the dependence. Sorby suggested that mechanical rotation of the original flaky and acicular mineral particles comprising the rock was a prime factor in inducing a preferred orientation parallel to the XY plane of the strain ellipsoid. He used the equations relating line rotation and strain ratio to develop a mathematical basis for this view (Sorby, 1853, 1856). Later, however, he noted that the mineralogy of slates and schists was quite different from that of the original clay-shales from which they were derived. Slates are composed of various micas and chlorites which could only have been formed by chemical changes during metamorphic processes (Sorby, 1858). He also noted that quartz veins were often abundant in slates yet did not generally occur in shaly rocks. This suggested that there were chemical links between the mineral changes taking place during cleavage formation and the release of silica together with mobilization of this silica probably in high temperature solutions. His invention of the petrographic microscope enabled him to confirm the presence of marked mineralogical changes in clays and schists. It provided him with further evidence of the chemical redistribution of material and showed how mineral particles could change shape by preferential solution on certain faces and redeposition of this solved material on other faces. This pressure solution process enabled initially equidimensionally shaped clastic quartz sand grains and carbonate fossil fragments to take on a markedly elongate form parallel to the cleavage planes (Sorby, 1879). Microscopic investigations of cleaved limestones also showed how individual calcite crystals could be deformed and elongated in the cleavage direction by processes of plastic flow. The processes leading to cleavage formation suggested by Sorby can be listed as follows:

1. Mechanical reorientation of initial platy and acicular minerals.
2. Mechanical reorientation of minerals crystallizing during metamorphism.
3. Preferred growth of new minerals in directions controlled by stresses at time of growth or by deformational anisotropy already present at time of growth.
4. Plastic flow of individual crystals inducing shape anisotropy related to the principal strain directions.
5. Change of shape of crystals by pressure solution and redeposition.

Most geologists today would probably agree that Sorby's list contains the principal factors leading to rock cleavage. Discussion would probably focus on their relative importance. Probably the variations we observe in cleavage development and cleavage type can be attributed to changing importance of these various factors in differing geological environments.

Our high opinion of the work on cleavage most of which was carried out over 100 years ago, must be apparent. The appreciation of the field facts, the realization of the critical nature of strain measurements and the logical way that the data were analysed seem to us to provide models of scientific methodology.

Before the next section we must comment briefly on some new suggestions on the origin of cleavage which emerged from Maxwell's study of slates in the Appalachian mountains (1962). Maxwell observed that the sedimentary or clastic dykes he found in cleaved sediments were consistently parallel to the slaty cleavage planes. He therefore proposed that cleavage formed at a very early stage in the tectonic history, when the water content of the sediments was very high and when they were probably in a soft, unlithified state. He suggested that the expulsion of pore water led to an alignment of the clay particles and hence to the initiation of cleavage. Subsequent to Maxwell's original work, a number of others have supported this *dewatering hypothesis* for the origin of cleavage. This hypothesis is not acceptable to us for several reasons. Where the strain state can be measured, cleavage is always seen to be parallel to the *XY* principal plane. Had the cleavage been initiated very early in the deformation, there would be no reason why the initial orientation should always guide the later strain, nor is there any reason why increase in finite strain should be directly correlated with cleavage intensity. We will see later that many folded rocks containing cleavage have local neutral strain points related to specific geometrical features of these folds. However, no embryonic cleavage is evident in these regions. There is much evidence that many cleaved sediments were in a lithified state before cleavage formed. For example, cleavage sometimes passes through large fossils or conglomerate pebbles with little or no deflection. Many igneous rocks show a rock cleavage, yet it is difficult to imagine how a dewatering phenomenon might account for this. Finally, close investigation in the field has shown that cleavage cross-cuts the clastic dykes which must therefore have existed before the cleavage developed. There is no field evidence to support the hypothesis of grain orientation by dewatering.

Sequence of fabric development arising from the deformation of sediments

We saw from Question 10.2 that during the deformation of a previously compacted sediment strain ellipsoids with a considerable range of shapes and orientations could develop. A progressive change in rock fabric generally accompanies the progressive change in the strain ellipsoids and we will now investigate the likely geological effect of tectonic strain imposed on a previously compacted argillaceous sediment. Six main stages of fabric development can be determined with transitions between these stages. Figure 10.24 sets out the stages A to F.

A. Undeformed condition

Most tectonically undeformed shales and marls possess a fabric as a result of sedimentation and diagenetic processes. This fabric is parallel to the bedding planes and develops first from the stable hydrodynamic orientation of large platy minerals (e.g. muscovite) parallel to bedding and second (and probably more important) from the effect of diagenetic compaction on the clay minerals present in the initial mud by overburden pressure and expulsion of pore water. This is a primary planar fabric known as bedding plane *fissility*.

B. Earliest deformation stage

The earliest stages of tectonic deformation lead to a contraction sub-parallel to the bedding surface. This deformation is usually accompanied by considerable volume loss as a result of mechanical closure of pore spaces left after diagenetic compaction and the expulsion of pore water contained in the initial shale. The effects of deformation are analogous to the effect of compressing a sponge: the elongation in the tectonic *X*-direction taking place normal to the contraction direction *Z* is small compared with the amount of shortening taking place. As a result of volume loss, this elongation in no way compensates for the degree of the contraction. The initial fissility of the shale therefore becomes slightly modified: the bedding plane fabric becomes rather less well marked as a result of rotation of platy elements around the tectonic *Y* axis, and a very weak linear orientation of any acicular minerals begins to form as they rotate towards the *Y*-tectonic direction. The rock now has a weak bedding plane fabric and a very weak (but probably undetectable) linear fabric.

C. Pencil structure

With increasing tectonic strain, perhaps accompanied by further volume loss by pore reduction, the strain ellipsoid takes on a more prolate form, eventually producing a uniaxial prolate or cigar-shaped ellipsoid. The longest axis of this ellipsoid is parallel to the tectonic *Y*-direction. At this stage the tectonic shortening is slight, but sufficient to compensate in the *XZ* principal plane for the initial diagenetic compaction. Ellipsoids with shapes of 2:1:1 are typ-

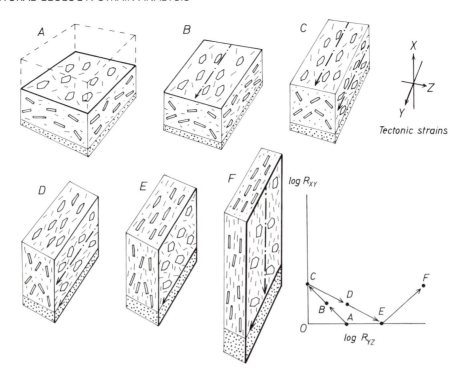

Figure 10.24. *Progressive stages in fabric development arising from the tectonic deformation of a shale. A, initial compacted shale; B, earliest deformation stage; C, pencil structure stage; D, embryonic cleavage stage; E, cleavage stage; F, cleavage with stretching lineation.*

ical, and tectonic shortenings ranging from 10 to 25% are general. At this stage the mineral rotations that have taken place are usually sufficient to produce a fabric strong enough to give a mechanical weakness to the rock. The rock tends to break into elongated fragments giving a characteristic **pencil structure** (Figures 10.24C, 10.25A and 10.26). This structure has sometimes been called **pencil cleavage**, but this terminology is not recommended because a genuine pencil structure has no truely planar cleavage elements. Pencil structure is best formed in perfectly homogeneous argillaceous rock. The cross sectional shapes of the pencils are usually bounded irregular polygonal or curvelinear walls (Figures 10.25A and 10.26). If folds are present in the rock, the pencil lineation is usually sub-parallel to the fold axes.

D. Embryonic cleavage stage

The next stage in increasing tectonic strain leads to a weakening of the prolate nature of the strain ellipsoid and the progressive movement of the ellipsoid into the apparent flattening field of the logarithmic strain plot (Figure 10.24, stage D). These changes are caused by progressive shortening in the tectonic Z-direction and elongation in the X-direction. The total ellipsoid XY' plane (which coincided with the bedding plane between stages A and C) now cross cuts the bedding plane to coincide with the tectonic XY plane, but the total ellipsoid X and Y axes coincide with the tectonic Y and X axes respectively. At this stage mineral rotations (and perhaps the growth of new phyllosilicates) leads to the production of a weak or imperfectly formed cleavage crossing the bedding, generally associated with a weak pencil structure linear fabric parallel to the tectonic Y direction. The weak cleavage may intersect the bedding

lithology to produce a crudely developed **intersection pencil structure** (Figures 10.24D, 10.25B) parallel to the hinge lines of nearby folds. In its early stages of formation this intersection pencil structure may be parallel to the true pencil structure, but genetically the two are unrelated. In fact, one sometimes observes that the true pencil structure is slightly oblique to the intersection pencil structure.

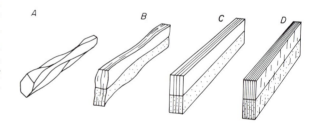

Figure 10.25. *Types of linear fabric produced during progressive deformation of a shale. A, stage C, pencil structure; B, stage D, embryonic cleavage with irregularly oriented intersection pencil structure; C, stage E, cleavage with regular intersection pencil structure; D, stage F, strong cleavage with stretch direction lineation and regular intersection pencil structure.*

At this stage of cleavage formation the fabric is rarely sufficiently strongly formed for the rock to be exploited to form good workable slates.

E. Cleavage stage

At this stage in the deformation sequence the tectonic strain imprint is sufficiently strong to form the dominant rock fabric, a planar cleavage cross-cutting the bedding planes. Platy and acicular minerals are oriented in the cleavage plane. However, they are generally without any

Figure 10.26. Pencil structure in Cretaceous marl from the Wildhorn nappe of the Central Helvetic Alps, Valais.

Figure 10.27. Intersection pencil structure.

linear alignment in the cleavage. This is because the total strain ellipsoid now has a form close to that of a uniaxial pancake. Linear fabrics do arise, but they are intersection pencil structures (Figures 10.24E, 10.25C, 10.27) and, in contrast to the true pencil structure, the linear pencils have very regularly oriented cross sectional shapes controlled by the orientation of the two dominant planar surfaces (Figures 10.25C, 10.27). The intersection lineations are generally parallel to the hinge lines of folds developed during the cleavage-forming deformation (Figure 10.20).

At this stage, the fabric is generally sufficiently well formed and regular in orientation for good roofing slates to be quarried. Sometimes this possibility is precluded if the bedding plane intersection structure is well developed, or if lithological variations set up marked refraction of the cleavage planes.

F. Strong cleavage with stretching-lineation

The final tectonic stage leads to an increasing perfection of the planar cleavage together with a progressively intensifying linear structure in the cleavage surface parallel to the tectonic X direction (Figures 10.24F, 10.25D). This linear structure, sometimes termed grain (longrain in the Ardennes) by quarrymen, produces differing breaking properties in different directions in the slaty cleavage surface. Very intensely linear slates may be unexploitable for roofing material. Most slates at this stage of development also show a well developed intersection lineation between the cleavage and bedding planes. It is quite common to find this intersection lineation at an angle to the Y-direction of the strain. The intersection lineation can, in fact, lie quite close to the X-direction stretch linear fabric, and often undulates wildly on regularly oriented cleavage surfaces. The reason for this considerable variation in orientation is that, because of the generally high ratio of R_{xy} in the cleavage (cf. stage E where this is not so), slight variations in bedding orientation become amplified in proportion to the increasing ellipticity of the XY strain ellipse on the cleavage surface. It is not uncommon to find cleavage–bedding lineations sub-parallel to the maximum extension direction over quite large areas where the regional strains are great, and where shortenings in the Z-direction exceed 60%.

Before we leave this introductory discussion on the significance of fabrics we should stress that those we have discussed here are essentially all-pervasive or penetrative through the whole rock mass. Other types of discontinuous or *non-penetrative fabrics* are also termed cleavage (e.g. *crenulation cleavage* and *close-joints cleavage*), but their origin and mechanical significance are different from the slaty cleavage and schistosity discussed above. We will discuss these types of fabric in Volume 2.

Changes in orientation of planes and lines as a result of three-dimensional strain

Answer 10.3

The original and new positions of the poles to the planes 1 to 7 are shown on an equal area projection in Figure 10.29A. The numerical data for the new orientations are as follows:

	Strike	Dip
1.	100°	85° to SW
2.	131°	78° to SW
3.	106°	64° to SW
4.	0°	63° to W
5.	117°	vertical
6.	90°	76° to S
7.	90°	vertical

The strain ellipsoid has a k-value of 1·0. The poles of planes tend to move towards the axis of maximum shortening Z, but not along the shortest angular route, since there is a component of movement towards the Y axis. Poles which originated on the principal strain planes remain on those planes and move towards the minimum strain axis in that plane. Planes with poles parallel to a principal finite strain axis do not alter their position.

If there are many planes with initially randomly oriented poles, these poles show a decrease in concentration around the X-axis, a concentration around the Y-axis and a very strong concentration around the Z-axis (see Answer 10.7★).

Answer 10.4

The new positions of the deformed lines are shown in the projection of figure 10.29B, and the numerical data are as follows:

	Azimuth	Plunge
1.	36°	51°
2.	74°	52°
3.	49°	79°
4.	90°	63°
5.	63°	0°
6.	0°	76°
7.	0°	0°

Line elements tend to move towards the maximum extension axis X, but there is a component of movement towards the Y-axis. Line poles plotting on a principal plane move towards the axis of greatest extension in that plane, and lines parallel to a principal strain direction show no change in orientation.

An assemblage of randomly oriented lines shows a decrease in concentration around the Z-axis, a concentration towards the Y-axis and a strong concentration towards the X-axis (see Answer 10.7★).

Answer 10.5★

The solution to this problem is based on finding values of extensions in the three principal planes of the strain ellipsoid by converting the known strains from the two given planes into principal plane data. If we knew the absolute values of the principal strains in the two planes PQ and RS we could derive these principal plane strains directly, but because we know only strain ratios in the PQ and RS planes, we have to relate the strain ellipses of PQ and RS to each other via the intersection i. This intersection has the same extension in both planes, and therefore forms a

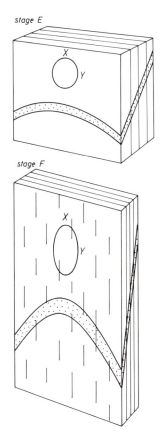

Figure 10.28. *Modification of the orientation of cleavage–bedding intersection lineation in passing from cleavage stage E to strong cleavage with stretching lineation stage F.*

link relating the principal strains in PQ to those in RS. The most logical line of analysis proceeds as follows:

1. In PQ choose a standard value (k) for the length of the major semi-axis $(1 + e_p)$ of the strain ellipse along direction P.
2. The choice of k sets a value of the semi-axis length of the minor axis of the strain ellipsoid $(1 + e_Q)$.
3. From these values calculate the value $1 + e_i$ along the intersection line, knowing the angle $P \angle i$.
4. In plane RS, knowing the elongation along direction i and the angle between i and the greatest extension R, we calculate a value for the length of the major semi-axis $1 + e_R$.
5. Knowing the principal strain ratio in the RS plane, calculate the value of $1 + e_S$. We now have the two strain ellipses standardized one with respect to the other and we have all the principal elongations in terms of the same units.
6. Calculate the values of the strains made by the intersections of the planes PQ and RS with the three principal strain ellipsoid planes XY, YZ, ZX.
7. On each of the three principal planes we have two known strains and the orientation of the principal strain axes, we can therefore calculate the values of the principal strains.
8. All the principle strain values will have a multiplying factor k of unknown value. However, this factor will disappear when we put the data in the form of the two principal plane strain ratios R_{xy}, R_{yz}. Two values of each principal strain will be obtained, and this forms a good check on the accuracy of the work.

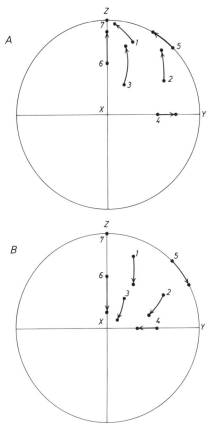

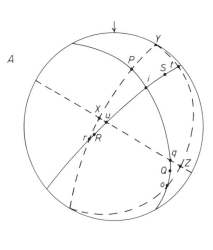

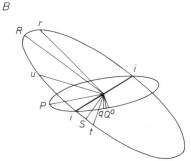

Figure 10.29. *Changes in the orientations of (A) the poles to planes and (B) linear elements as a result of a homogeneous strain with $R_{xy} = R_{yz} = 2.0$.*

Figure 10.30. *Data plot for Question 10.5★. A is a projection of the planar and linear data and B illustrates the geometric cross-correlation between the two strain ellipses and their common intersection line i.*

The data from Question 10.5★ are plotted on a projection in Figure 10.30A. The points P, Q and R, S represent the two-dimensional principal strain directions in the two surfaces. The greatest extension P in plane PQ lies on the XY principal plane of the strain ellipsoid, but the greatest extension R in plane RS does not lie on the XY plane. It is important to note that it is *not valid to assume that the maximum two-dimensional extention in any general surface cutting a strain ellipsoid coincides with the cleavage trace (XY plane trace) on that surface*. In certain sections of some types of ellipsoid the difference in position may be quite large.

The XY, YZ and ZX principal planes of the strain ellipsoid were constructed directly from the cleavage (XY), cleavage plane pole (Z) and given X-direction in the cleavage. The two planes PQ and RS intersect these three principal planes at positions P, o, q and r, t, u respectively, and intersect each other at i.

In plane PQ let the extension along P be e_p, such that $1 + e_p = k$. Because $R_{PQ} = 4.0$, the minor semi-axis of the strain ellipsoid along Q has a length $1 + e_Q = 0.25 k$. We now calculate in PQ the $1 + e$ values of the intersections with the principal planes of the strain ellipsoid and along i using Equation 6.5 (see also Equation D.7). This equation gives the lengths of lines in terms of the principal strains and the angle θ' that the line makes with the greatest extension:

$$1 + e = \left[\frac{\cos^2 \theta'}{(1 + e_p)^2} + \frac{\sin^2 \theta'}{(1 + e_Q)^2} \right]^{-1/2} \quad (10.18)$$

The data are as follows:

along P, $\theta' = 0$, $\quad 1 + e_p = k$

o, $\theta' = o \angle P = 78°$, $\quad 1 + e_o = 0.26 k$

q, $\theta' = q \angle P = 80°$, $\quad 1 + e_q = 0.25 k$

i, $\theta' = i \angle P = 20°$, $\quad 1 + e_i = 0.60 k$

We now use the strain along i to standardize the strains in RS to those in PQ (see in diagrammatic form in Figure 10.30B the link of the two ellipsoids along i). Knowing that $R_{RS}^2 = \lambda_S'/\lambda_R' = (3.14)^2$ and that $i \angle R = 66°$, we can write Equation 10.18 in the form

$$\frac{1}{(1 + e_i)^2} = \lambda_R' \cos^2 66 + \lambda_R' R_{RS}^2 \sin^2 66$$

From the previously calculated value of $1 + e_i$ we obtain

$$1 + e_R = 1.75 k$$

and

$$1 + e_S = 0.56 k$$

We now calculate the values of the lengths $1 + e$ of the intersection of plane RS with the three principal planes using Equation 10.18:

RS/XY intersection r, $\theta' = r \angle R = 8°$, $1 + e_r = 1.62 k$

RS/YZ intersection t, $\theta' = t \angle R = 74°$, $1 + e_t = 0.58 k$

RS/ZX intersection u, $\theta' = u \angle R = 20°$, $1 + e_u = 1.23 k$

On each principal plane we now have two known strains in known directions with respect to the principal three-dimensional strain axes. We can therefore obtain values

of the lengths $1 + e_1$, $1 + e_2$, $1 + e_3$ using the technique developed in Session 6 (Question 6.1) to calculate the principal strains from two stretched belemnites. Solving the two simultaneous equations we obtain:

in XY, $\quad 1 + e_r = 1.62 k$, $\quad \theta' = r \angle X = 26°$

$\quad 1 + e_p = k$ $\quad \theta' = P \angle X = 50°$

$\quad \therefore 1 + e_1 = 2.65 k$, $\quad 1 + e_2 = 0.75 k$

in YZ $\quad 1 + e_t = 0.58 k$, $\quad \theta' = t \angle Y = 17°$

$\quad 1 + e_o = 0.26 k$, $\quad \theta' = o \angle Y = 70°$

$\quad \therefore 1 + e_2 = 0.77 k$, $\quad 1 + e_3 = 0.25 k$

in ZX $\quad 1 + e_u = 1.23 k$, $\quad \theta' = u \angle X = 10°$

$\quad 1 + e_q = 0.25 k$, $\quad \theta' = q \angle X = 75°$

$\quad \therefore 1 + e_1 = 2.59 k$, $\quad 1 + e_3 = 0.24 k$

The numerical cross checks are consistently good. The principal strain ratios are:

$$R_{xy} = (1 + e_1)/(1 + e_2) = 3.32$$
$$R_{yz} = (1 + e_2)/(1 + e_3) = 3.16$$

Answer 10.6★

Figure 10.31 shows an inclined plane cutting the three principal planes of the strain ellipsoid. The principal axes of strain are arranged so that X is vertical, Y is E–W horizontal, and Z is N–S horizontal. The intersection of the inclined plane with the YZ principal plane (strike) makes an angle of ϕ_{yz} with Y, and the dihedral angle between the inclined plane and YZ (dip) is δ.

If AC is x, then

BC is $x \sin \delta$

AB is $x \cos \delta$

BD is $\dfrac{x \cos \delta}{\sin \phi_{yz}}$ and BE is $\dfrac{x \cos \delta}{\cos \phi_{yz}}$

then

$$\cot \phi_{xz} = \tan \delta \cos \phi_{yz}$$

and

$$\cot \phi_{xy} = \tan \delta \sin \phi_{yz}$$

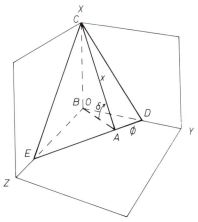

Figure 10.31. Relationship between a plane CDE and the axes X, Y and Z of the finite strain ellipsoid. See Answer 10.6★.

If, after deformation, all angles are modified (ϕ_{xy} to ϕ'_{xy}, etc.), then

$$\cot \phi'_{xy} = R_{xy} \cot \phi_{xy} = R_{xy} \tan \delta \sin \phi_{yz} = \tan \delta' \sin \phi'_{yz}$$

$$\cot \phi'_{xz} = R_{xz} \cot \phi_{xz} = R_{xz} \tan \delta \cos \phi_{yz} = \tan \delta' \cos \phi'_{yz}$$

and it follows that

$$\tan \delta' = \tan \delta (R_{xy}^2 \sin^2 \phi_{yz} + R_{xz}^2 \cos^2 \phi_{yz}) \quad (10.19)$$

also

$$\tan \phi'_{yz} = (\tan \phi_{yz})/R_{yz} \quad (10.20)$$

Answer 10.7★

The trigonometric relationships follow closely those of the previous example. If the line has an azimuth relative to the Y strain axis of θ_{yz}, and a plunge of δ, before deformation, then after deformation

$$\tan \delta' = \tan \delta (R_{xy}^2 \cos^2 \theta_{yz} + R_{xz}^2 \sin^2 \theta_{yz}) \quad (10.21)$$

$$\tan \theta'_{yz} = (\tan \theta_{yz})/R_{yz} \quad (10.22)$$

The March model for deformed planes and lines

In the discussion of the significance of cleavage and schistosity we commented briefly on the ideas put forward by Sorby to explain the preferred orientation of particles by rotation, and we showed how he used a two-dimensional mathematical model to indicate how platy and acicular crystals might rotate towards the maximum extension direction. The equations we have established above extend these calculations to cover a three-dimensional array of planes or of lines deformed by a three-dimensional strain. Such calculations were first made by March in 1932, and the results of his work have had a strong influence on structural geologists interested in explaining rock fabrics.

Figure 10.32 illustrates on projections the effects of superimposing different types of three-dimensional finite strain on initially random aggregates of 500 planar and 500 linear elements. It plots the redistributed poles to planes and the linear directions. The patterns of redistribution which result are functions only of the two principal plane strain ratios, and do not depend on the absolute values of the principal strains. The reorientations arising in three different types of strain are shown. In all examples X is vertical, Y is E–W horizontal, Z is N–S horizontal.

A. Apparent constriction, $R_{xy} = 3\cdot6$, $R_{yz} = 1\cdot2$, $k = 13\cdot0$, Figure 10.32A

The poles of planes are distributed around the periphery of the net in a Y–Z girdle, and the lines are concentrated in a maxim concentrated along the X-axis. The fabric produced is a strongly linear one known as an **l-fabric**.

B. Ellipsoid on the boundary between apparent constriction and apparent flattening, $R_{xy} = 2\cdot0$, $R_{yz} = 2\cdot0$, $k = 1\cdot0$

The poles to planes are also concentrated in a Y–Z girdle arranged like that of the previous example A. However, they are not uniformly distributed as in A, but show a

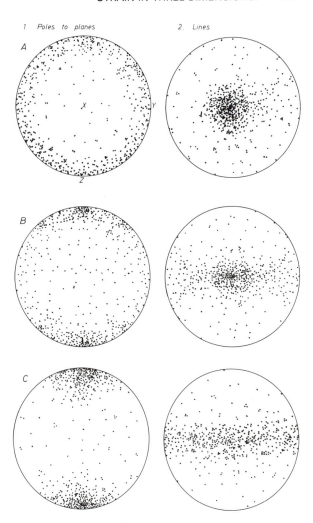

Figure 10.32. Changes in distribution of 500 randomly oriented poles to planes (left) and 500 lines (right) as a result of deformation by strain ellipsoids with k-values of (A) k > 1, (B) k = 1 and (C) k < 1.

concentration in the girdle around the Z-axis. The line elements are also distributed in a non-uniform girdle in the XY plane, with a concentration parallel to the X-axis. This fabric is a composite one, with planar and linear elements, planar in XY and linear in X. It is a mixed type known as an **l–s fabric**.

C. Apparent flattening, $R_{xy} = 1\cdot2$, $R_{yz} = 3\cdot6$, $k = 0\cdot08$

This type of strain shows a completely reversed distribution of planar poles and lines compared with that produced by an apparent constriction. The poles to planes form a concentration about the Z-axis, whereas the linear elements uniformly spread around a complete girdle aligned parallel to the XY principal plane. This arrangement produces a strongly planar arrangement, and is generally known as an **s-fabric**. The "s" in this term comes from the German *Schichtfläche* or *Schieferung* (bedding and cleavage respectively).

In naturally deformed rocks there is often a relationship between linear-, linear-planar-, and planar-fabrics and the general form of the strain ellipsoid that accords quite well with the predictions of the March model (Figure 10.33).

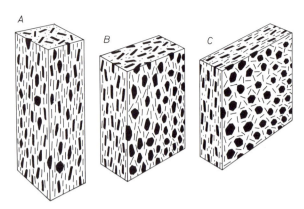

Figure 10.33. *Representation of rock fabrics developed in a mineral aggregate of platy and acicular crystals as a result of deformation by different types of strain ellipsoids: A, apparent constriction, B, k = 1; C, apparent flattening.*

As an example, Figure 10.34 shows an outcrop of strongly deformed gneiss derived by the Alpine deformation of a Hercynian basement granite. Measurement of the shapes of xenoliths in this rock show that the average values of the strain ellipse axes are in the proportion 20:1:1. The strain ellipsoid is an almost uniaxial prolate type, and the overall fabric of the rock is in accord with this symmetry. The rock shows a strongly developed *l-fabric* with almost perfect parallelism of prismatic crystal components parallel to the *X*-axis, and with a well developed girdle arrangement of the micas with the nomals to the basal plates of the crystals arranged in a girdle perpendicular to *X*.

Although the March model predicts preferred orientations of crystals as a function of their shape and the shape of the strain ellipsoid relates reasonably well to the fabrics seen in naturally deformed rocks, it cannot explain all the features of fabric development. The model is one where the planar and linear elements move in a completely passive way, and the model assumes that the planar and linear elements are able to change their shapes in accordance with the elongation required for the particular strain state. Clearly, mineral components do not have this degree of freedom to change shape. The theories put forward to explain the behaviour of rigid particles in a deforming system show that the behaviour is very complex. It depends on the strain sequence, the grain size and the degree of packing of large and small components. When we look more carefully at the fabrics of naturally deformed rocks, we see evidence for a more heterogeneous behaviour of the particles than the March model predicts: rotations not in accord with the theory, crystals which interfere with one another as a result of collision and blockage during rotation, crystals which kink or become folded (indicating that they have varied mechanical properties), particles which have undergone pressure solution and evince nucleation with growth under metamorphic conditions. Like many theories, the March model explains some features of fabric formation, but it does not provide a completely satisfactory explanation of the complex behaviour of naturally deforming crystalline rocks.

Figure 10.34. *Example of a linear- or l-fabric in granite gneiss. The rock face on the left contains the X-direction of the strain ellipsoid, and the rock face on the right is perpendicular to this X-axis. Cristallina, Ticino, Switzerland.*

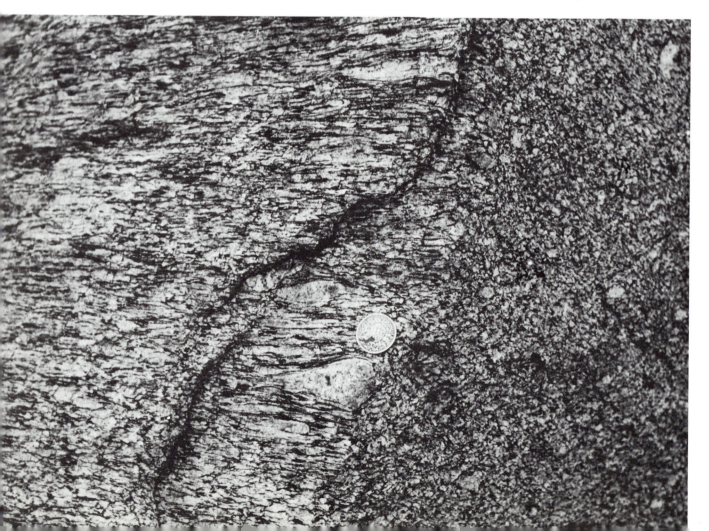

DEFINITIONS AND KEYWORDS

Cleavage (Schistosity)   A dominantly planar or *s*-fabric produced by the preferred alignment of platy minerals (generally phyllosilicates). This alignment is not perfect, but is of a statistical nature: it imparts a special property to the rock whereby the rock preferentially splits in a direction parallel to the cleavage planes. **Slaty cleavage** is found in **slates**: tectonically deformed rocks (often strongly strained) found in regions of low grade regional metamorphism (greenschist or lower). **Schistosity** is found in **schists**: deformed rocks with platy or acicular minerals easily visible to the eye and crystallized above greenschist facies metamorphic conditions.

Cleavage fan   Non parallel cleavage planes arranged into a fan-shaped group. Cleavage fans are generally geometrically related to folds and the relations of the fan to the fold curvature enable **convergent** and **divergent cleavage fans** to be distinguished (see Figure 10.20). In convergent fans the cleavage planes come closer when traced from the outer to inner arc of a folded layer, in divergent fans the cleavage planes diverge in this sense.

Cleavage refraction   A change of orientation of cleavage planes when they are followed through layers of differing lithology (Figure 10.23).

Fabric   The geometric and spatial relationships between the crystal components making up a rock. The fabric can relate to the preferred orientations of grain shapes, to grain sizes, and to crystallographic orientations of the components. Preferred orientations of platy- and needle-shaped crystals can give rise to **planar (*s*-fabrics), linear (*l*-fabrics)** or to **mixed linear–planar (*l–s* fabrics)**.

Flinn graph   A graphical representation of the shape of an ellipsoid made by plotting R_{yz} as abscissa and R_{xy} as ordinate.

Logarithmic graph   A graphical representation of the shape of an ellipsoid made by plotting $\log R_{yz}$ as abscissa and $\log R_{xy}$ as ordinate.

Pencil structure   A planar fabric which enables a rock to be split into long narrow pencil-like fragments (Figure 10.26). It may be formed as the result of the development of strain ellipsoids of prolate type, or it may result from the intersection of crossing planar fabrics where it is known as **intersection pencil structure** (Figure 10.27).

Reciprocal strain ellipsoid   The original ellipsoid which, after deformation, is transformed into a sphere of unit radius.

Reciprocal strain matrix   The matrix representing the nine terms of the linear Eulerian coordinate transformation equations (Equation 10.10) describing a reciprocal homogeneous strain.

Strain ellipsoid   The ellipsoid derived by the homogeneous strain of an original sphere with unit radius. It has three orthogonal axes, the **principal axes of strain** with directions X, Y and Z, with semi-axis lengths $1 + e_1$, $1 + e_2$ and $1 + e_3$. The parameters $e_1 \geq e_2 \geq e_3$ are the **three principal longitudinal strains**. The planes containing two of the principal axes are known as the **principal planes**. The strain ratios on these principal planes are the **principal plane strain ratios** R_{xy}, R_{yz}, R_{zx}.

Strain matrix The three by three matrix formed by the nine terms of the linear Lagrangian coordinate transformation equations (Equation 10.9) describing a homogeneous strain.

Volumetric dilatation Δ_V The proportional change in volume resulting from a homogeneous strain (Equation 10.15).

KEY REFERENCES

Flinn, D. (1962). On folding during three dimensional progressive deformation. *Q. Jl Geol. Soc. Lond.* **118**, 385–428.

This paper has exerted a great influence on the development of modern structural geology, in particular the application of mathematical methods to account for geometric features of naturally deformed rocks. It was the first to relate in detail the geometric features of different types of strain ellipsoids with fold formation and boudinage. The mathematical nomenclature used is not that currently favoured, and rather complex looking equations arise because of the notation.

Means, W. D. (1976). "Stress and Strain", 339 pp. Springer-Verlag, New York.

A discussion of the concept and properties of the strain ellipsoid is presented on pp. 139–150.

Ramsay, J. G. (1967). "Folding and Fracturing of Rocks", 568 pp. McGraw-Hill, New York.

Pages 121–184 set out many of the basic geometric features of strain ellipsoids, and relate them to the geometric features of naturally deformed rocks.

Ramsay, J. G. and Wood, D. S. (1973). The geometric effects of volume change during deformation processes. *Tectonophysics* **16**, 263–277.

This describes, from a theoretical viewpoint, the effects of initial (diagenetically formed) volume dilatations, and tectonically induced volumetric dilatations on the shapes of successively formed finite strain ellipsoids. It would form appropriate background reading to the discussion of Question 10.2.

Cleavage

There is an immense literature on cleavage and schistosity in rocks. We recommend below only particularly informative articles and reviews containing extensive reference lists for further reading.

Graham, R. H. (1978). Quantitative deformation studies in the Permian rocks of the Alpes Maritimes. *Proc. Goguel Symp.* (*Bur. Rech. Géol. Mines, France*), 220–238.

This paper describes changes in the types of cleavage in a cross section of the Western Alps and relates different types of fabric to changing strain states.

Harker, M. A. (1885). On slaty cleavage and allied rock structures with special reference to the mechanical theories of their origin. *Rep. Br. Ass.* 55th meeting, 1–40.

We have to admit that this paper may be difficult to find in many libraries. We think it one of the classic works of structural geology and we recommend most strongly that the student tries to obtain a copy. It is a model of how excellent and pertinent field observations can be integrated with sound mathematical analysis. Many of the conclusions in this paper have been subsequently rediscovered by other workers in the field and even today publications appear with results which were set out clearly in this work. Do not miss this paper.

Maxwell, J. C. (1962). Origin of slaty and fracture cleavage in the Delaware Water Gap Area, New Jersey and Pennsylvania. *Geol. Soc. Am. Buddington volume* 281–311.

This paper develops the dewatering theory for the origin of slaty cleavage suggesting that cleavage forms during the mass transport of water along the axial planes of folds as the result of stressing porous sediments.

Siddans, A. W. B. (1972). Slaty cleavage, a review of research since 1815. *Earth Sci. Rev.* **8**, 205–232.

An excellent, well illustrated review account with a very good list of references.

Siddans, A. W. B. (1977). The development of slaty cleavage in a part of the French Alps. *Tectonophysics* **39**, 533–557.

This describes changes in cleavage morphology across a 45 km long cross section, and links these with variations in finite strain, metamorphic grade, rock texture and the associated minor structures.

Wood, D. S. (1974). Current views of the development of slaty cleavage. *A. Rev. Earth Planet. Sci.* **2**, 369–401.

Wright, T. O. and Platt, L. B. (1982). Pressure dissolution and cleavage in the Martinsburg Shale. *Am. J. Sci.* **282**, 122–135.

A review article which emphasizes the importance of observations of finite strain as a technique to solve the problem of the origin of cleavage.

The strains developed during cleavage development were studied using the shapes of deformed graptolites. Because the undeformed dimensions of thecal spacing is well known, strains and volume change could be computed. These data indicate a volume loss of 50% in a direction perpendicular to the cleavage.

Anisotropy

Rock deformation leads to the development of orientation anisotropy in the crystalline components. This anisotropy is sometimes recorded in other types of physical anisotropy. In the future it seems likely that measurement of such parameters as seismic, magnetic and thermal anisotropy will, with the help of suitable conversion factors, offer a method of determining finite rock strain. The following papers can be recommended as a lead to these anisotropy correlations:

Kligfield, R., Owens, W. H. and Lowrie, W. (1981). Magnetic susceptibility anisotropy, strain and progressive deformation in Permian sediments from the Maritime Alps (France). *Earth Planet. Sci. Lett.* **55**, 181–189.

Kneen, S. (1976). The relationship between the magnetic and strain fabrics of some haematite bearing Welsh slates. *Earth Planet. Sci. Lett.* **31**, 413–416.

Owens, W. H. and Bamford, D. (1976). Magnetic, seismic, and other anisotropic properties of rocks. *Phil. Trans. R. Soc. Lond.* A **283**, 55–68.

Tuck, G. J. and Stacey, F. D. (1978). Dielectric anisotropy as a petrofabric indicator. *Tectonophysics* **50**, 1–11.

Wood, D. S., Oertel, G., Singh, J. and Bennett, H. F. (1976). Strain and anisotropy in rocks. *Phil. Trans. R. Soc. Lond.* A **283**, 27–42.

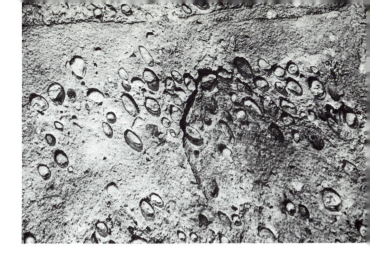

SESSION 11

Strain in Three Dimensions

2. Review of Methods and Representation of Strain State

This Session gives a review of the types of features found in deformed rocks which are useful for the determination of two- and three-dimensional finite strain and summarizes the principal methods whereby knowledge of two-dimensional strains can be used to establish the strain ellipsoid. It discusses the methods whereby different valued strain states may be recorded and compared one with another, and it describes cartographic methods for representation of strain parameters. Strain ellipse fields are related to different strain ellipsoid types. The geometry of surfaces of no finite longitudinal strain and of circular sections of the ellipsoid is established and methods of strain determination using the directions of no finite longitudinal strain evaluated.

INTRODUCTION

Strain measurement: a summary

The measurement of finite strain in naturally deformed rocks uses, for its basis, the shapes of marker objects of which some predeformational geometric features are known. We will begin this Session by bringing together some of the ideas we have already established and making some general comments on the practical methods for determining finite strain.

The key to strain determination lies in finding objects with some known characteristic initial shape (e.g. ooids, conglomerate pebbles), initial packing arrangement (e.g. clastic particles in sediments), or some features which enable final lengths or angles to be computed (e.g. fossils, folded or boudinaged competent layers).

The main objects useful for strain determinations are as follows:

1. Spherical, sub-spherical and elliptical objects: Ooids, pisolites, pebbles and other clastic (or volcano-clastic) fragments, diagenetic alteration spots in sediments (so-called reduction spots), diagenetic concretions, microfossils (especially certain species of Foraminifera and Radiolaria), amygdales, vesicles and spherulitic concretions in volcanic or intrusive igneous rocks, spotted rocks in thermal aureoles of igneous intrusions.

2. Circular objects, disks and tubes: *Scolithus-* and other worm borings, certain fossils—e.g. crinoid ossicles, pipe amygdales in lavas.

3. Line elements: Boudinaged and ptygmatically folded competent layers, certain fossils (belemnites, crinoid columns).

4. Angular elements: Fossils—with or without bilateral symmetry, polygonal desiccation cracks in continental sediments, polygonal cooling fractures in lavas and small igneous intrusive bodies, trace fossils, cross-bedding in sediments, ripple marks.

It is clearly most important to attempt to determine completely the orientation and values of the three axes of the strain ellipsoid. Many of the features in the above list are two dimensional, and we have to build the form of the strain ellipsoid by combining two-dimensional data taken from differently oriented surfaces. The different strain ellipses must all come from an environment where the strain is effectively homogeneous and sometimes it is difficult to assess the degree of homogeneity. Folded strata must always have heterogeneous strain states through the structure. At any locality where the bedding has a constant orientation the strain may be homogeneous, but it is to be anticipated that differing lithologies are likely to have different strains. The shapes of deformed concretions and conglomerate pebbles are not always simple indications of the shape of the strain ellipsoid, partly because of initially non-spherical shapes and preferred alignments of the initial axes of the particles, and partly because of competence contrasts between the particle and its matrix. The presence of competence contrasts can often be assessed from the

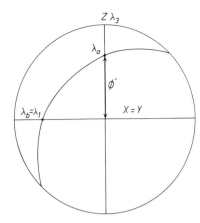

Figure 11.1 *Projection of a plane with known strain state in a uniaxial strain ellipsoid ($\lambda_1 = \lambda_2 > \lambda_3$).*

geometric form of rock cleavage. If the cleavage planes vary in orientation as they are traced through layers of differing lithology (cleavage refraction), or if they are deflected around objects (Figure 10.20) it is certain that the strain is heterogeneous, whereas if the cleavage planes are constantly oriented strain homogeneity is indicated. It is most important to be aware of the scale of strain heterogeneity in an area if the results of strain measurement are to be correctly interpreted. In some terrains the scale of heterogeneity is such that regional maps of strain variation may be prepared, whereas in others such a technique is quite impossible.

Two-dimensional strain ellipse data can be used to determine various features of the strain ellipsoid under the following conditions:

1. One strain ellipse, plus knowledge of the orientations of the three principal strains: In general, it is very difficult to find practical solutions to this problem. If the strain ellipse lies on a plane cutting all three principal planes it is possible to make a mathematical solution (Ramsay, 1967, pp. 148–149) to obtain the principal longitudinal strains, or their ratios. The accuracy of this solution is very sensitive indeed to slight variations and imperfections in the input measurements and so it is not always of practical interest. If the strain ellipsoid can be assumed to be uniaxial (either $e_1 = e_2$, or $e_2 = e_3$), then it is possible to compute the strain along the other two principal axes. From the relationships shown in Figure 11.1 it can be seen that, knowing the quadratic extensions λ_a (on a principal plane) and λ_b on any surface containing the rotation axis of the ellipsoid, in the plane of the strain ellipse (AB) we can write:

$$\lambda_a' = \lambda_1' \cos^2 \phi' + \lambda_3' \sin^2 \phi'$$

$$\lambda_b' = \lambda_1'$$

which gives values for $\lambda_1 = \lambda_2$ and λ_3 ($\lambda = 1/\lambda'$).

2. Two strain ellipses parallel to two principal planes: This is probably the most practical of all the methods of strain analysis, especially where initially spherical or ellipsoidal particles are available for shape analysis. The principal planes can be determined from specific tectonic structures in the rock (cleavage parallel to the XY plane, stretching lineation in the cleavage parallel to X), but it may not always be easy to recognize immediately the principal planes. For example, in weakly deformed rocks possessing poorly developed tectonic fabrics it may be very difficult to identify the principal strain axes by inspection, and other methods requiring three strain ellipses may be necessary (see 4, below).

The computations to determine the principal strains (or strain ratios) are very simple (Question 10.1). It is sometimes useful to check the results from two principal planes by making measurements on the third principal plane.

3. Two strain ellipses, plus knowledge of the orientations of the principal planes (Question 10.5★): This is often a useful method for strain analysis when some of the strain ellipse information comes from a bedding surface which is not a principal plane of the strain ellipsoid. In general, the principal two-dimensional strain axes will not have any simple relationship to the positions of the three-dimensional principal planes. It is unwise to assume, for example, that the trace of a cleavage plane on a bedding surface is parallel to the maximum extension in the bedding (see pp. 190 and 209 for further discussion of this point). The computations necessary to determine the values of the three principal strains (or their ratios) are not difficult in principle, but rather time consuming in practice. Computer programming of the data is advisable if this technique needs to be applied frequently.

4. Three strain ellipses, no knowledge of the orientations of the principal planes (Ramsay, 1967, p. 142): When two-dimensional strain ellipses can be constructed on any three non-parallel planes it is possible to determine both the orientations and values of the three principal strains. The techniques for making these calculations are somewhat complicated, and we will discuss them in Volume 3 of this series. Computer methods are almost essential if practical use is to be made of this method. This method provides objective results and is especially good for low as well as for high strain ratios (see 2 above). From each ellipse one obtains three data (two strain values, one orientation of the principal two-dimensional strain axes), and from three differently oriented strain ellipses we therefore have a total of nine data. Because the finite strain ellipsoid requires only six terms to define its orientation and principal strain values it follows that the measured data exceed the minimum required for the solution, and that we therefore have internal checks on the accuracy and consistency of the measurements.

Strain values and their ratios

In general we find that the material we employ for calculation of strain provides us only with the ratios of the principal strains on the principal planes, i.e. R_{xy}, R_{yz} and R_{xz} ($R_{xz} = R_{xy} \cdot R_{yz}$). From these data we can assess the shape of the strain ellipsoid (whether oblate, prolate, uniaxial etc.), but we cannot determine the volume dilatation, or relate the ellipsoid shape to a true flattening or true constriction.

If, however, an *absolute length change in any direction* in the strain ellipsoid can be determined, we are able to transform the ellipse strain ratio data into absolute values of principal extension, and as a result we can compute the volumetric dilatation. It cannot be stressed too highly how important such information is for furthering our understanding of rock deformation. We still know surprisingly little about volume changes taking place during deformation and how volume change relates to mechanical opening or closing of pore spaces, tectonic cementation, and chemical and mineralogical transformations taking place as a result of metamorphic action during deformation. When a strain ellipsoid form has been established it is recommended that the single extra datum is sought to transform this information into absolute strain values. Perhaps with a single observation of a stretched crinoid column, or a single ptygmatic fold we can do this. The relationships of length changes to orientation of the line in the strain ellipsoid expressed in terms of the direction cosines l', m' and n' and the reciprocal quadratic extensions (λ'_1, λ'_2 and λ'_3) is:

$$\lambda' = \lambda'_1 l'^2 + \lambda'_2 m'^2 + \lambda'_3 n'^2 \tag{11.1}$$

The proof of this equation can be found in Ramsay (1967, p. 126). It might be useful to point out that the equation for extension in a two-dimensional situation is contained in Equation 11.1, where $n' = 0$, and

$$\cos^{-1} l' = 90 - \cos^{-1} m' \tag{11.2}$$

Replacing these special relationships in Equation 11.1 we obtain Equation 6.5. Using Equation 11.1 together with that of 10.2, and the definitions of the principal plane strain ratios given by $\lambda'_2 = \lambda'_1 R_{xy}^2$, $\lambda'_3 = \lambda'_1 R_{xz}^2$ we obtain:

$$\lambda_1 = \lambda(l'^2(1 - R_{xz}^2) + m'^2(R_{xy}^2 - R_{xz}^2) + R_{xz}^2) \tag{11.3}$$

From this formula it can be seen that, knowing the quadratic extension λ along a line with directions given by direction cosines l' and m', the value of the principal quadratic extension λ_1 can be computed. Values of λ_2 and λ_3 follow directly from a knowledge of the principal plane strain ratio, and volume dilatation Δ_V from

$$(1 + \Delta_V)^2 = \lambda_1 \lambda_2 \lambda_3 \tag{11.4}$$

Recording and comparing strain states

One of the main aims of strain analysis is to relate differing states of deformation in a region to each other, and to assess the geological significance of these variations. It is therefore necessary to develop methods whereby we can set down our data in a form suited for this analysis so that

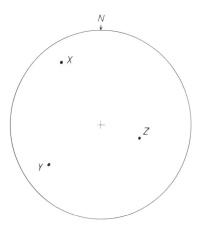

Figure 11.2. Method of using a projection to record the orientation components X, Y and Z-axes of the strain ellipsoid.

we can objectively determine the differences in strain state.

We have already seen that the strain ellipsoid requires nine terms to define the complete relationships of the rock mass before and after deformation, and that the three of these terms which relate to body rotation are generally impossible to establish. This leaves us with six terms: three related to ellipsoid orientation and three to principal strain values. Although the physical nature of these two groups of three seems very different (angles versus extensions) it should be emphasized that mathematically they all have the same characteristics being numbers or scalars. This is appreciated very quickly from a consideration of the nine numbers defining the coordinate transformation equations and the deformation matrix, and the fact that each of our ellipsoid parameters can be expressed as functions of these nine numbers. In a heterogeneous strain field there is an intimate connection between orientations and values of the principal strains which can be expressed mathematically. Remember, therefore, that it is basically *only for our convenience that we separate our strain ellipsoid data into angles and extensions*, and that *both are equally important aspects of a strain field*.

The *angular relationships* can be expressed on a projection, and a plot of the X, Y and Z axes on a net gives a diagrammatic method of recording certain aspects of the *principal strain axis orientation field* (Figure 11.2). If the region has relatively constantly oriented principal strain axes this method of data presentation is useful. Such a region will generally have rather constant principal strain values. If the principal strain values vary greatly, then it is usual for the principal strain orientations to also show considerable variations. The projection method is not suited to bringing out these variations as it gives no separation of data according to spatial position. Changes in the direction field are probably best recorded by plotting the directions directly on to a map, perhaps using symbols with a scaling factor to bring out the variations in plunge of the X-lines and dips of the XY surfaces. This is the method which was used very effectively by Ernst Cloos

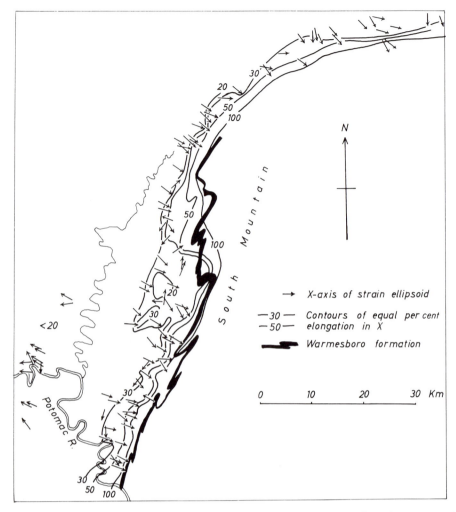

Figure 11.3. *Map showing variations in orientation of strain ellipsoid X-axes and values of maximum extension in the region west of South Mountain, Maryland, USA. After Cloos (1947).*

(1947) to demonstrate the angular relationships of the strain ellipsoids in the Appalachian fold belt (Figure 11.3).

The *deformational aspects* of the strain ellipsoid can be formulated and graphically represented in several ways. The three terms required to define distortion can be expressed in four ways:

1. In terms of the values of the principal strains:

$$e_1 \quad e_2 \quad e_3$$

or lengths of the ellipsoid semi-axes:

$$1 + e_1 \quad 1 + e_2 \quad 1 + e_3$$

2. In terms of the principal plane strain ratios R and volumetric dilatation Δ_V:

$$R_{xy} \quad R_{yz} \quad 1 + \Delta_V$$

3. In terms of values of the principal natural-, or logarithmic strain parameters $\varepsilon_1 = \ln(1 + e_1)$

$$\varepsilon_1 \quad \varepsilon_2 \quad \varepsilon_3$$

4. In terms of the principal plane strain ratios expressed as logarithmic strain parameters and logarithmic volume dilatation:

$$\varepsilon_1 - \varepsilon_2 \quad \varepsilon_2 - \varepsilon_3 \quad \ln(1 + \Delta_V)$$

Note that

$$\ln R_{xy} = \ln(1 + e_1) - \ln(1 + e_2) = \varepsilon_1 - \varepsilon_2$$

The advantages of systems 2 and 4 are that they separate the deformational features of the strain ellipsoid into two parts, the first two terms related to the differential distortions, the third term describing an overall uniform distortion related to the volume change. We have already seen that the two terms of the differential distortion can be represented either on the Flinn graph or Ramsay logarithmic plot, and that an ellipsoid shape can be shown on these plots as a single point.

The major problem facing us is how can we best formulate the data relating to differential distortion in a way that is fundamental, or which aids our appreciation of geometric variation. We have to answer two questions: first, what is the ellipsoid type, and second, how much strain?

We have already met some answers to the first question, for it is possible to describe an ellipsoid form in terms of its ***k*-value** on a Flinn plot, and that this parameter expresses the degree of oblateness or prolateness of the ellipsoid (Figure 11.4A).

$$k = \frac{R_{xy} - 1}{R_{yz} - 1} \qquad (11.5)$$

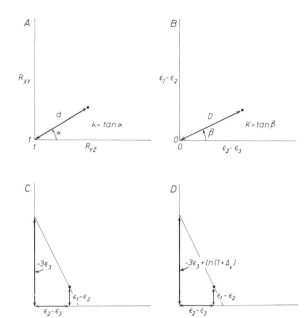

Figure 11.4. *Methods for representing strain ellipsoid graphs. A, the Flinn graph using principal plane strain ratios; B, the Ramsay logarithmic plot using logarithmic principal plane strain ratios; C and D show the relationships between the logarithmic maximum shortening ε_2 and the ellipsoid shape in cases of no volume change or a volume dilatation Δ_V respectively.*

Table 11.1

Ellipsoid type	k	K	ν
Uniaxial prolate	∞	∞	-1
General prolate	$\infty > k > 1$	$\infty > k > 1$	$0 > \nu > -1$
Boundary prolate–oblate	1	1	0
General oblate	$1 > k > 0$	$1 > K > 0$	$+1 > \nu > 0$
Uniaxial oblate	0	0	$+1$

Ramsay (1967) used a similar parameter for the logarithmic plot (Figure 11.4B) where the ellipsoid **K-value** is

$$K = \frac{\varepsilon_1 - \varepsilon_2}{\varepsilon_2 - \varepsilon_3} = \frac{\ln(R_{xy} - 1)}{\ln(R_{yz} - 1)} \qquad (11.6)$$

There is no simple mathematical relationship between k and K, because points on the Flinn graph with constant k-value fall on a curved line in the logarithmic plot and therefore have variable K-values. There are, however, some special coinciding values (see Table 11.1).

Another parameter that describes ellipsoid shape was introduced into the structural geology literature by Jake Hossack in 1968. This is known as **Lode's parameter ν** after its introduction into the study of deformed metals by Lode (1926). This is expressed in terms of logarithmic principal strains

$$\nu = \frac{2\varepsilon_2 - \varepsilon_1 - \varepsilon_3}{\varepsilon_1 - \varepsilon_3} \qquad (11.7)$$

This parameter can be written in terms of the abscissa and ordinate values of the logarithmic graph:

$$\nu = \frac{(\varepsilon_2 - \varepsilon_3) - (\varepsilon_1 - \varepsilon_2)}{(\varepsilon_2 - \varepsilon_3) + (\varepsilon_1 - \varepsilon_2)}$$

Dividing numerator and denominator by $\varepsilon_2 - \varepsilon_3$ we can express the relationship between the K-value of an ellipsoid and Lode's parameter:

$$\nu = \frac{1 - K}{1 + K} \text{ or } K = \frac{\nu + 1}{\nu - 1} \qquad (11.8)$$

The relationships between the three commonly used parameters to describe strain ellipse shape are indicated in Table 11.1.

Any one of these shape parameters can be usefully represented in contoured form on a map. A good example of the way such maps bring out very clearly the variations in ellipsoid form is shown in Figure 11.5. This is taken from Hossack's work on strains determined from pebble shape analysis of the Bygdin conglomerate in the Jotun nappe, Norway, and illustrates how great the variations can be in a comparatively small area. Rodded conglomerates are found in the areas of prolate strain ($\nu-$ve) and lie in channel-like regions surrounded by regions of oblate strain. The deformation pattern is clearly very complex and cannot be explained by simple shear displacement along the thrusts. One explanation for the strain pattern is that the contacts between the overriding nappe and underlying sole were not completely planar and perhaps groove-like features localized the areas of prolate deformation. Another explanation is that the channels of prolate ellipsoids might be related to regions which had an initial predeformational pebble fabric of depositional origin.

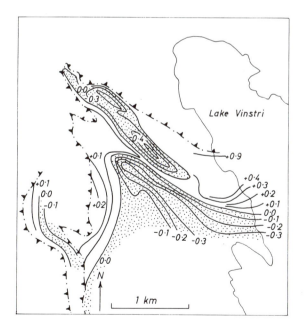

Figure 11.5. *Variations in Lode's parameter in the Jotun nappe, Norway. After Hossack (1968). The regions of prolate ellipsoids (negative values of ν) are stippled. Thrust planes are indicated with dash–dot lines.*

We will now move to a discussion of how we might represent the amount of deformation and, at the start, we point out that this is a problem with no easy solution. If two ellipsoids have the same k-, K- or ν-value then it is clearly possible to find a method to say if one ellipsoid is farther from the origin of the Flinn graph or logarithmic graph and hence to say that it is more deformed than the other. However, when the shape parameters are different, then it is not so obvious how to go about making a deformation comparison, and how we can compare the amount of deformation in a cigar-shaped strain ellipsoid with that in a pancake-shaped one.

Two methods have been used: the first attempts to evaluate the degree of shortening along the Z-direction (or elongation along the X-direction), the second tries to define how much three-dimensional distortion is implied by the two principal plane strain ratios.

Shortening or extension measurement

If no volumetric dilatation occurs, it is possible to compare the amount of shortening in the Z-direction of different ellipsoids. In the logarithmic plot we find ellipsoids of equal X-shortening by reorganizing the equation relating the logarithmic principal extensions to dilatation given by:

$$\varepsilon_1 + \varepsilon_2 + \varepsilon_3 = \ln(1 + \Delta_V) = 0 \qquad (11.9)$$

into a form expressed in units of the ordinate ($\varepsilon_1 - \varepsilon_2$) and abscissa ($\varepsilon_2 - \varepsilon_3$) of the logarithmic plot:

$$(\varepsilon_1 - \varepsilon_2) = -2(\varepsilon_2 - \varepsilon_3) - 3\varepsilon_3 \qquad (11.10)$$

This is a straight line function with a uniform negative slope of minus two which intersects the ordinate axis at a position depending upon the value of the shortening ($\varepsilon_1 - \varepsilon_2 = -3\varepsilon_3$). If there is no volume change it is possible to determine a value for ε_3 for a particular ellipsoid and use this as a basic parameter to record on a map (Figure 11.4C). However, if there is a volumetric dilatation, then Equation 11.9 becomes:

$$\varepsilon_1 - \varepsilon_2 = -2(\varepsilon_2 - \varepsilon_3) - 3\varepsilon_3 + \ln(1 + \Delta_V) \quad (11.11)$$

This is a line with the same slope of minus two, but for a given shortening it shifts upwards or downwards depending upon whether Δ_V is positive or negative respectively (Figure 11.4D).

The measurement of elongation along the principal X-axis of the strain ellipsoid can also be used as a deformation parameter. This was the basic method used by Ernst Cloos in his Appalachian studies (1947). Lines of equal extension along X are located on a logarithmic plot where

$$\varepsilon_1 - \varepsilon_2 = -\tfrac{1}{2}(\varepsilon_2 - \varepsilon_3) + \tfrac{1}{2}(3\varepsilon_1 - \ln(1 + \Delta_V)) \quad (11.12)$$

An example of a plot of the extension values obtained by Cloos over part of an area of some 4000 square kilometres is illustrated in Figure 11.3. This map shows the remarkably systematic increase in deformation from NW to SE across this fold belt into the core of the fold belt in the Blue Ridge and South Mountain areas.

Another system of describing deformation was introduced by Hans Breddin and was suggested as a method of presentation of data from deformed fossils. Fossils can give excellent two-dimensional strain values for bedding surfaces, but they are generally less useful for determining strain on other surfaces in the rock mass. We have seen that from a single strain ellipse it is only practical to determine the ellipsoid shape if the ellipsoid is uniaxial. Breddin's method of strain computation assumes that the ellipsoid in the slates he investigated was of a uniaxial type, with maximum and intermediate strains of equal value ($e_1 = e_2$), and with the XY plane of the ellipsoid coinciding with the slaty cleavage plane. With this assumption it is possible to calculate a single principal plane strain ratio ($R_{xz} = R_{yz}$, $R_{xy} = 1\cdot0$) diagnostic of the ellipsoid. An example of a deformation map presenting this type of data is shown in Figure 11.6. The characteristic R_{xz} ratio of each locality is presented in a form standardized to an ellipsoid with a volume of 10^6 units. Zero deformation would be represented by 100/100, and a strain ellipsoid with a principal plane ratio $R_{xz} = 2\cdot2$ would be represented by the number 59/130 (implying a uniaxial ellipsoid with dimensions $130:130:59$). The map shows clearly that there is a systematic change in the deformation from NW to SE, but the assumptions built into the data make the significance of the results open to question. Although many slates do have strain ellipsoids with oblate shapes, the k-values of the slate deformation field generally lie in the range $1 > k > 0$ (see Ramsay and Wood, 1973). The method is therefore not completely valid for slaty rocks, and cannot be generally recommended.

The biggest problem arising in all the methods of assessing the amount of deformation using shortening and elongation parameters concerns the amount of volume change, and how this affects the actual total shortening values calculated from an ellipsoid shape. We have seen that a strain ellipsoid shape can be defined by two principal plane strain ratios. We can transform these values into a shape type using the k-, K-, or ν-values. Is it possible to combine this with a fundamental parameter describing degree of deformation? In some way the degree of deformation is linked to the distance of the point representing a strain ellipsoid on a Flinn- or logarithmic plot from the plot origin. It is perfectly feasible to define parameters d and D on the two types of plot and call these parameters measures of the amount of deformation (Figure 11.4). The two parameters can be specified exactly in terms of the principal plane strain ratios:

$$d = [(R_{xy} - 1)^2 + (R_{yz} - 1)^2]^{1/2} \qquad (11.13)$$

$$D = [(\varepsilon_1 - \varepsilon_2)^2 + (\varepsilon_2 - \varepsilon_3)^2]^{1/2} \qquad (11.14)$$

Thus an ellipsoid can be exactly specified by k- and d-values, or by K and D values. There is no simply expressed mathematical formula relating d and D, because the relationships depend upon the actual values taken by the principal plane strain ratios. It therefore follows that neither of these parameters expressing amount of deformation has an *absolute* significance. Is there an absolute parameter describing how much distortional deformation is present in a particular ellipsoid? The problem is one of finding an effective method for comparing, for example, a prolate with an oblate ellipsoid.

It has been suggested by Nadai (1963) that the amount of shear strain on a plane making equal intersections with all three principal axes of strain, and known as the *octahedral plane* might provide an absolute measure of amount

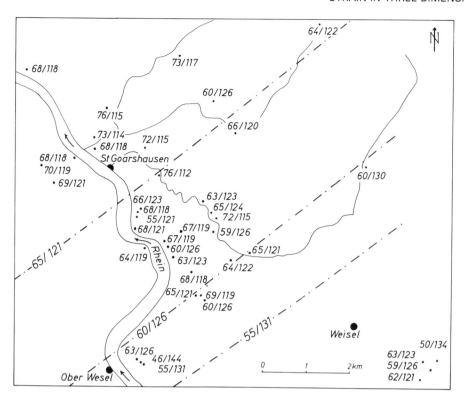

Figure 11.6. *Strain map of part of the Rheinisches Schiefergebirge (West Germany) determined from deformed fossils. For the significance of the numerical method of representing strains see text. After Breddin (1957).*

of the distortion in a strain ellipsoid. He arrived at this conclusion by determining the work done in deforming a block of material with linear stress–strain properties in an irrotational deformation increment sequence. He found that a unit, which he termed the natural octahedral unit shear $\bar{\gamma}_{oct}$, was directly proportional to the amount of work done in the system. This unit has a value related to the logarithmic expressions of the principal strains:

$$\bar{\gamma}_{oct} = \frac{2}{3} \left[(\varepsilon_1 - \varepsilon_2)^2 + (\varepsilon_2 - \varepsilon_3)^2 + (\varepsilon_3 - \varepsilon_1)^2 \right]^{1/2} \quad (11.15)$$

and Nadai defined the *amount of strain* by another unit $\bar{\varepsilon}_s$ related to the octahedral unit shear according to:

$$\bar{\varepsilon}_s = \frac{\sqrt{3}}{2} \bar{\gamma}_{oct} \quad (11.16)$$

It therefore follows that the amount of strain may be expressed as a function of the three principal logarithmic strains

$$\bar{\varepsilon}_s = \frac{1}{\sqrt{3}} \left[(\varepsilon_1 - \varepsilon_2)^2 + (\varepsilon_2 - \varepsilon_3)^2 + (\varepsilon_3 - \varepsilon_1)^2 \right]^{1/2} \quad (11.17)$$

Hossack used this parameter together with Lode's shape parameter to plot strain ellipsoids derived from the deformed conglomerates of Bygdin. The graphical method he used followed suggestions of Nadai (1963) and Hsu (1966), and used a plot based on a 60° sector of a circle. The $\bar{\varepsilon}_s$ parameter was plotted radially, and the value of Lode's parameter indicated by changing the direction of the radial line. Figure 11.7 reproduces Hossack's data, and should clarify the plotting technique.

An important point to discuss is to decide if the $\bar{\varepsilon}_s$ parameter really is a more fundamental measure of the

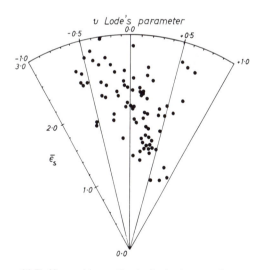

Figure 11.7. *Hossack's method of plotting strain ellipsoid shapes using Lode's parameter and the parameter $\bar{\varepsilon}_s$ (Hossack, 1968).*

amount of distortion than any other. It was derived by considering minimum work criteria under the special conditions of irrotational strain. It cannot therefore be a fundamental parameter applicable to the strain ellipse shape in general because, in geology, practically all strain ellipsoids are built up during non-coaxial rotational strain sequences. To define the amount of work in such sequences we would have to specify the nature of the sequence, implying that the same shaped end product could be produced from very different inputs of work. We are uncon-

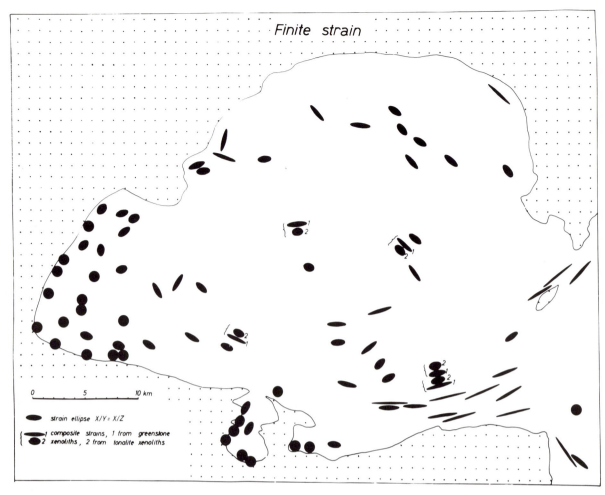

Figure 11.8. *Representation of strain state in granitic rocks of the Chindamora batholith, Zimbabwe. The ellipsoids are of uniaxial oblate type ($e_1 = e_2 > e_3$), and all the strain data can be represented in the form of a single ellipse. The strains were developed as a result of balloon-like magmatic inflation of the mass. At some localities composite strains could be recorded for from foreign and from cognate xenoliths.*

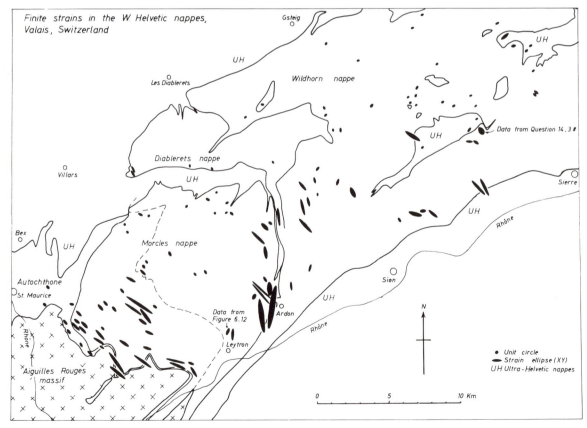

Figure 11.9. *Map showing the shapes of XY principal plane strain ellipses from the Helvetic nappes of the Valais, W. Switzerland. The nappes form a succession of tectonically superposed fold and thrust sheets plunging N 60° E, overlain by the higher, earlier emplaced Ultrahelvetic nappes.*

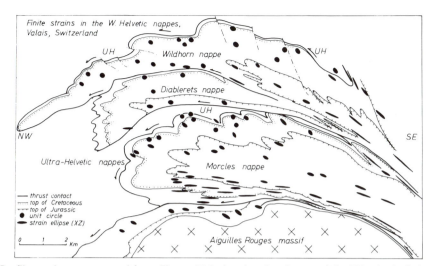

Figure 11.10. Profile section constructed from Figure 11.9 showing the shapes of the XZ principal plane strain ellipses.

vinced that the $\bar{\varepsilon}_s$ parameter has any specially meritorious properties that would justify its use instead of the more simply expressed D-value suggested above. In fact, a comparison of Equations 11.14 and 11.17 will show that the differences in mathematical formulations of amount of deformation expressed by D and $\bar{\varepsilon}_s$ are not very significant.

In concluding this discussion on methods of recording strain we might recommend some especially effective methods for the presentation of data. For publication purposes, although it is probably necessary to present numerical data in graphical or list form, it is often a good idea to produce maps or sections with the appropriate strain ellipses drawn at correct scale and with the correct spatial orientation at the location point of the observation point. This technique enables the main features of the data to be rapidly assimilated, and probably offers the best method for putting the data in a form where their implications can be most easily appreciated (Figures 11.8, 11.9, 11.10).

Three-dimensional strain computed from two-dimensional data

Question 11.1

Four rock specimens A, B, C and D were collected from a cleaved calcareous slate in a quarry. The spatial relationships of these four specimens are indicated in the block diagram of Figure 11.11. The orientation marks on surfaces are as follows:

- A. strike 30°, dip 40° SE
- B. strike 104°, vertical
- C. strike 104°, vertical
- D. strike 20°, dip 86° E.

The calcareous sediments contain ossicles of *Pentacrinus*. In one layer the bedding is penetrated by worm borings. In undeformed terrain the axes of the tubular borings are perpendicular to the bedding planes, whereas in specimen C they are deflected from the perpendicular and make an angle of 52° with the bedding. On this section the angle between bedding and cleavage trace is 9°. The steeply dipping strata in the quarry are cut by a dyke of vesicular

basalt which also shows features showing that it has been deformed. The shapes of the deformed vesicles are indicated in A and B. The dyke also contains a number of curved calcite-filled extension veins. The calcite in these veins shows a fibrous texture, and the fibres are curved in the manner shown in B.

1. Calculate the true orientation of the bedding planes and of the dyke.
2. Determine the states of finite strain on the surfaces of the four specimens A, B, C and D and compute the shape of the strain ellipsoid in the bedded calcareous slate and in the deformed dyke. Calculate the d- and k-values of these ellipsoids.
3. What is the deformational significance of the small scale structures seen in these rocks?

Relationships of strain ellipse fields and strain ellipsoid types

In Session 4 we saw that any strain ellipse can be classified into one of three fields depending upon the values taken by the two principal longitudinal extensions. In Session 10 we defined a number of types of strain ellipsoid and we will now discuss the inter-relationship of different strain ellipse sections in the strain ellipsoid. Before we do this it might be a good idea to check over these ideas (pp. 65–66, 171–172).

In any strain ellipsoid it is possible to determine the orientations of lines having a given longitudinal strain. Equation 11.1 can be combined with Equation 10.2 and put into a form expressing the relationships of direction cosines l' and m' of lines having a particular reciprocal quadratic extension λ':

$$m' = \left[\frac{l'^2 (\lambda'_1 - \lambda'_3) + \lambda'_3 - \lambda'}{\lambda'_3 - \lambda'_2} \right]^{1/2} \tag{11.18}$$

Any chosen values of λ'_1, λ'_2, λ'_3, λ' and l' will give either no solution for m' (if the expression in the brackets is negative) or two values of m' equal in value and numerically opposite in sign. Solving this equation for different λ' values with a computer it is possible to plot on a projection a diagram showing how longitudinal strains vary with

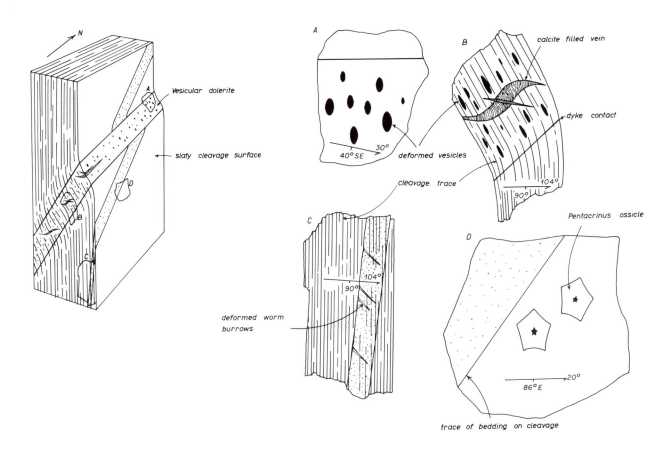

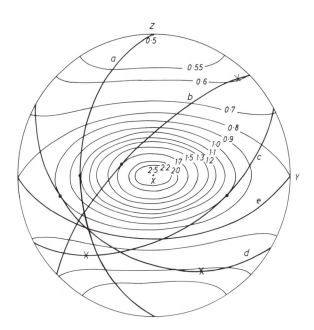

Figure 11.11. Data for Question 11.1.

Figure 11.12. Projection of contours giving varying longitudinal strains in a strain ellipsoid. The five great circles (a)–(e) represent sections through the ellipsoid. See Question 11.2.

orientation in the strain ellipsoid (Figure 11.12). This figure relates to an ellipsoid with vertical X-axis $(1 + e_1 = 2 \cdot 5)$, E–W horizontal Y-axis $(1 + e_2 = +0 \cdot 8)$, N–S horizontal Z-axis $(1 + e_3 = +0 \cdot 5)$, and the curving lines represent joins of all lines having the same values of $1 + e$.

Shapes of strain ellipses in a strain ellipsoid

Question 11.2

In the contoured projection diagram of Figure 11.12 are drawn five great circles a–e. These represent planes passing through the ellipsoid, and the two-dimensional strain on each of these sections can be determined from the values of $1 + e_1$ transected by the plane and the curving contour lines. In each plane the $1 + e$ values generally reach some maximum (indicated by a heavy dot) and a minimum value (indicated by a cross). For each surface, these directions are perpendicular in the surface, and they represent the axes of maximum and minimum length of the strain ellipse. Determine the characteristic features of the two-dimensional strain ellipse fields on the three principal planes of the strain ellipsoid (XY, YZ, ZX) and on the five planes a to e. Plot the data graphically according to the method shown in Figure 4.10, and evaluate the range of possible strain ellipse fields present in this particular three-dimensional finite strain state.

Now pass on to the Answers and Comments section, then proceed to the starred questions below or to Session 12.

Relationship of directions of no finite longitudinal strain to the values of the principal longitudinal strains

In deformed metamorphic rock complexes, especially those of gneissic terrain, it is often very difficult to find the objects, discussed earlier in this Session, which are suited to establishing the strain state. In such complexes there may be an abundance of veins of differing composition from that of their host rock (e.g. quartz veins, aplite and pegmatite dykes) and which, as a result of competence contrasts during deformation, have become folded or boudinaged. There is sometimes a sufficiently wide range of

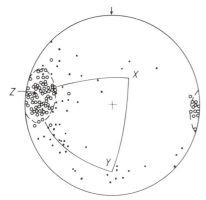

Figure 11.13. *Projection of poles to quartz veins which have undergone stretching (open circles) and shortening (filled circles), from the Easdale slate belt, Scottish Caledonides. The dashed line shows the separation of the data to compute the position of the surface of no finite longitudinal strain in the ellipsoid. From Talbot (1970).*

vein orientation that it is a practical possibility to determine the directions of no finite longitudinal strain demarcating the contraction and stretching fields within the strain ellipsoid. Figure 11.13 illustrates such a plot from the work of Talbot (1970) using data from deformed quartz veins cutting slates in western Scotland. From the symmetry of the data field it is possible to determine the general orientations of the principal planes and axes of the strain ellipsoid. The problem which faces us now is to decide if we can go further and obtain values of the principal longitudinal strains. As a start to seeing the problems here Question 11.3★ considers the equivalent problem in two dimensions.

STARRED (★) QUESTIONS

Determination of principal strains knowing the directions of lines of no finite longitudinal strain

Question 11.3★

In the surface of a Field 2 type finite strain ellipse there are always two directions of no finite longitudinal strain $(e = 0)$. These directions are symmetric about the principal strain axes and the direction of maximum elongation bisects the angle $2\phi'$ between the lines. If we know the value of $2\phi'$ what constraints (if any) can be place on the values of the two principal longitudinal elongations e_1 and e_2?. *Hint*: check the mathematical relationships set out in Equations D.10 and B.21 of the Appendices and obtain an expression for λ_1 in terms of ϕ'.

ANSWERS AND COMMENTS

Three-dimensional strain computed from two-dimensional data

Answer 11.1

The data are plotted on to a projection in Figure 11.14A. On surface A the angle between the strike line and the dyke-cleavage intersection is 12°. This will be a line in the true plane of the dyke. On surface B the dyke has an apparent dip of 42° and this is also a line in the dyke plane. There is only one great circle passing through the two points and its orientation gives the true strike (12°) and dip (42°W) of the dyke. Similarly two apparent dips of the bedding enable a true strike (8°) and dip (86°W) to be computed.

The three-dimensional strains are computed from two two-dimensional ellipses. From the geometric relationships of the deformed vesicles surface A is a principal XY plane, and B is an XZ plane. Measurement of the vesicle shapes gives $R_{xy} = 2 \cdot 15$, and $R_{xz} = 4 \cdot 30$. R_{yz} can be computed from their quotient and is $2 \cdot 00$. Note that the shapes and orientations of the vesicles are fairly constant, implying that they were initially spherical. Using the definitions in Equations 11.5, 11.6, 11.13, 11.14 we obtain the following parameters for the strain ellipsoid $d = 1 \cdot 52$, $k = 1 \cdot 15$, and $D = 1 \cdot 03$, $K = 1 \cdot 1$. The ellipsoid is slightly prolate.

In the calcareous slate the cleavage plane gives the

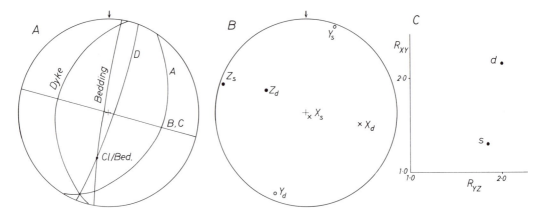

Figure 11.14. Answer 11.1. A shows a projection of orientation data; B illustates the principal strain directions of the ellipsoid in the sediment (X_s, Y_s, Z_s) and dyke (X_d, Y_d, Z_d); C shows the ellipsoid shape data for sediment (s) and dyke (d).

principal XY plane of the ellipsoid. On the XY surface the crinoids can be used in various ways to determine the two-dimensional strain ratio R_{xy}. One crinoid is bilaterally symmetric and can be used to determine the direction of the principal axes, and using the techniques described in Session 8 the value of R_{xy} is 1·30. In section C the deformed worm tubes show an angular shear strain of 38° with respect to the direction of the bedding and, using the cleavage trace as the X-direction a value of R_{xz} of 2·4 is found (Mohr circle construction or Breddin graphs). Combining this data with that on the XY section, $R_{yz} = 1·85$, and we find the following parameters for the strain ellipsoid $d = 0·90$, $k = 0·35$ and $D = 0·67$, $K = 0.43$. The ellipsoid lies in the field of apparent flattening and is oblate.

The orientations of the ellipsoids are shown in Figure 11.14B together with a Flinn plot of the shape components (C).

It should be clear that the slaty cleavage represents the XY trajectory of adjacent strain ellipsoids. The cleavage is refracted close to the dyke–sediment contact but the refraction curve is rather gradual. The smaller angle between cleavage and dyke contact in the sediment is consistent with the higher strains measured in the sediment and with the fact that the dyke was more competent than the sediment (Figure 10.23). The gradual change of refraction angle suggests that the deformational properties of the sediment were modified along the contact zone of the dyke by thermal metamorphism before all or most of the subsequent deformation. The differing k-values of the ellipsoids can be interpreted in several ways. Because of compatibility constraints across the dyke–sediment contact there must be some connection between the $1 + e_2$ values of the two ellipsoids, although if the dyke was intruded during deformation it is possible that the sediment might contain strain components which were not taken up by the dyke. If assumptions are made that the dyke took up all the strain and underwent plane strain with little or no volumetric dilation then the low k-value of strain in the sediment could be explained by volume reduction in the sediment. If this was the case and if complete compatibility occurred along the Y-direction it is possible to compute the values of the principal longitudinal strains in the sediment, for $1 + e_1 = R_{xy}$, and $1 + e_3 = 1/R_{yz}$ giving values of 1·3 and 0·54 respectively. Using Equation 10.15 these

values give a volumetric dilatation Δ_v of minus 30%. Although this value is large, such high values have been described in the structural geology literature where very strong pressure solution has occurred. Other interpretations that are possible could be based on an assumption that the system has suffered a real elongation parallel to the direction of intermediate strain, or that the dyke was intruded during the deformation sequence. A choice between these various possibilities necessitates further data.

The outcrop is probably part of a folded rock sequence. The worm tubes and cleavage–bedding relationships suggest that a synform lies to the west of this exposure and an antiform to the east. The fold axes are likely to be oriented in a direction parallel to the cleavage-bedding intersection line (Figure 10.20) which plunges at 52° towards an azimuth of 192°. The dyke–cleavage intersection will not give the direction of the major fold structures in the bedding, but if the dyke is involved in the same fold structures the fold axis within the dyke should be sub-parallel to this intersection.

The carbonate filled veins show a sigmoidal geometry suggesting a complex sequence of incremental strains. If you return to the discussion of Answers 2.11 and 6.1 on vein and fibre geometry the best explanation of this geometry would be that an extension fissure first opened in the direction indicated by the fibrous crystals at the centre of the largest vein, then the maximum incremental extensions moving in a clockwise sense on section B (Figure 11.11) to give rise, finally, to the thin vein which cross cuts the larger one. Such a sequence would be in agreement with the build-up of deformation leading to the finite strain ellipse that we see on this section.

Shapes of strain ellipses in a strain ellipsoid

Answer 11.2

The strain ellipsoid is of a constrictional type with $k = 3.5$, and $d = 2.2$, and has suffered no volume dilatation ($\Delta_v = 0$).

The shapes of the three principal plane strain ellipses are shown in Figure 11.15A, and their positions in the ellipse field graph are shown in B. The trace of the XY

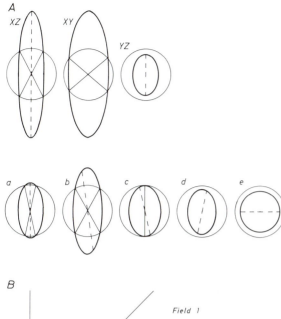

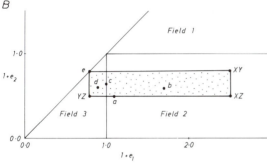

Figure 11.15. A: *Shapes of the three principal plane strain ellipses, and the special strain ellipses on sections (a)–(e) of Figure 11.12. The traces of the principal XY plane on the strain ellipse sections is indicated with a dashed line.* B *indicates the positions of the strain ellipses in a two-dimensional ellipse Field plot. All possible strain ellipses of the ellipsoid fall in the stippled area.*

principal plane on the other two principal plane sections is indicated as a dashed line: this could represent the trace of cleavage on these sections. Of the three ellipses, two lie in Field 2, and one lies completely within the original circle from which it was derived and is in Field 3. The strain ellipses on surfaces that are inclined to the three principal axes of the strain ellipsoid are indicated in (a)–(e): lines of no finite longitudinal strain are found in Field 2 ellipses (a), (b) and (c) and occur wherever these ellipses cut the original unit circle. In (c) the strain ellipse just touches the unit circle, and the two lines of no finite longitudinal strain coincide. The trace of the principal XY plane on each ellipse is shown as a dashed line. Only in one ellipse (a) does this trace coincide with the direction of greatest longitudinal strain of the strain ellipse. This geometric conclusion is very important in geology and implies that the trace of cleavage on a surface should not generally be used to define the directions of maximum and minimum elongation in that surface. Ellipses (d) and (e) lie in Field 3 and, because they lie inside the unit circle, lines of no finite longitudinal strain do not exist in these sections. Section (e) shows a negative area dilatation but no differential distortion—the strain ellipse has an aspect

ratio R of one. This section passes through the intermediate direction of principal strain and all extensions take on a value e_2. All ellipsoids show such surfaces of uniform extension, and these are known as the **circular sections of the strain ellipsoid**. Uniaxial ellipsoids are special in that they have one circular section parallel either to the XY (Type 1) or to the YZ (Type 5) principal plane. All other ellipsoids have a conjugate pair of circular sections intersecting along Y and symmetrically bisected by the X-direction. The orientation of these circular sections depends only on the ratios of the principal plane ratios, and is found as follows: from Equation 11.1 circular sections will occur at positions where

$$\lambda_2' = \lambda_1' l'^2 + \lambda_2' m'^2 + \lambda_3' n'^2$$

rearranging, and substituting the equation for direction cosines (Equation 10.1)

$$0 = l'^2(\lambda_2' - \lambda_1') - n'^2(\lambda_3' - \lambda_2')$$

$$0 = [l'(\lambda_2' - \lambda_1')^{1/2} + n'(\lambda_3' - \lambda_2')^{1/2}]$$

$$\times [l'(\lambda_2' - \lambda_1')^{1/2} - n'(\lambda_3' - \lambda_2')^{1/2}] \quad (11.19)$$

This is the equation of a pair of planes intersecting along the Y axis and, because $\cos^{-1} l$ and $\cos^{-1} n$ lie in the XZ principal plane with $\cos^{-1} l + \cos^{-1} n = 90°$, it follows that the two planes make angles with the X axis given by

$$\pm \cos^{-1}\left(\frac{\lambda_3' - \lambda_2'}{\lambda_3' - \lambda_1'}\right) = \pm\cos^{-1}\left(\frac{R_{xy}^2(R_{yz}^2 - 1)}{R_{xy}^2 R_{yz}^2 - 1}\right) \quad (11.20)$$

Because the longitudinal strains are all equal in these circular sections, the surfaces have features of special geological interest. If the components making up a rock show little or no competence contrast, the relative spatial relationships of these components on a circular ellipsoid section are identical to that of the original rock. Figure 11.16 shows a cut block of Carrara marble breccia from North Italy, a rock that will be familiar to many from its use as a building facing stone. The dark face shows highly deformed fragments, this section being close to an XZ principal plane. In contrast, on the well lit upper part of the block the fragments look almost unstrained, being close to one of the circular sections of the ellipsoid. The appearance of this surface reflects the fabric of this breccia before deformation and its geometric characteristics give a valuable guide to choice of analytical method for determining the two-dimensional strain ellipses on the other deformed faces. For example, it can be seen that the fragments were originally angular—a feature that is not always so apparent on the strongly deformed faces. It will also be seen that there was very little initial preferred orientation of the long axes of the breccia fragments, a feature which is especially important if the R_f/ϕ technique discussed in Session 5 is to be employed as a method of strain analysis.

The various strain ellipses (a)–(e) are plotted in Figure 11.15B. The total range of possible strain ellipses in this particular ellipsoid lie in a rectangular area bounded by corners which plot at ellipse points representing the principal section strains and the ellipse of the circular section. Only Field 2 and Field 3 strain ellipses are possible in this ellipsoid.

Figure 11.16. *Cut block of Carrara marble showing appearance of deformed breccia fragments on the XY plane (dark) and on the circular section (light). In many quarries the two circular sections of the strain ellipsoid are preferentially selected on aesthetic grounds for facing stone.*

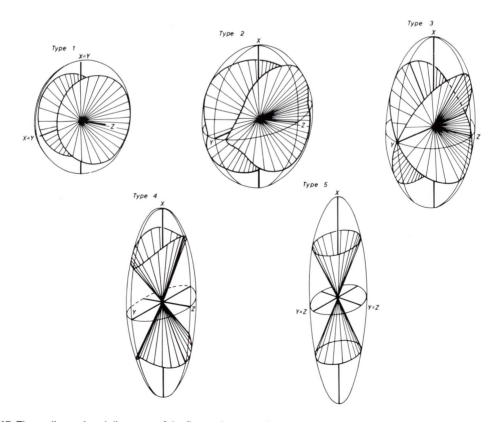

Figure 11.17. *Three-dimensional diagrams of the five main types of strain ellipsoids and the directions of no finite longitudinal strain. After Ramsay (1967).*

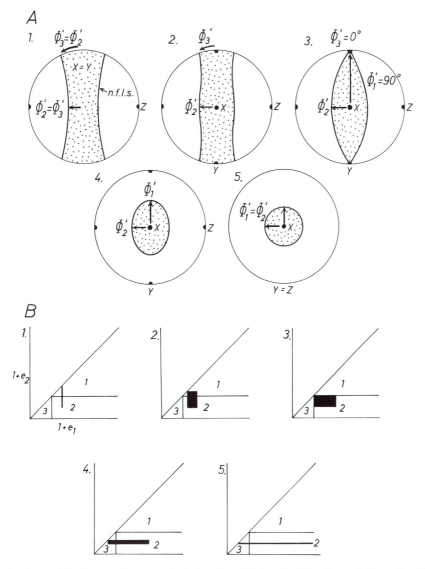

Figure 11.18. A: Projections of the five main types of strain ellipsoid showing directions of lines of no finite longitudinal strains separating stretched (stipped) and contracted (unstippled) regions. B shows the possible Fields of two-dimensional strain ellipses (in black) that are possible in these ellipsoids.

In our discussion of the geological potentials of different strain ellipse fields (p. 65) we noted that certain fields of strain ellipses arise in certain types of ellipsoids. Figure 11.17 illustrates three-dimensional diagrams of the five main types of strain ellipsoid occurring as a result of tectonic processes. It shows the positions of **lines of no finite longitudinal strain (n.f.l.s)** which are arranged on surfaces separating sectors within the ellipsoid where positive and negative longitudinal strains are developed.

The line on the surface of the ellipsoid marking out the directions of no finite longitudinal strain can be thought of as the line of intersection of the original sphere with that of the ellipsoid. The lines of no finite longitudinal strain radiating from the centre of the ellipsoid define a conic surface, the form of which depends upon the strain ellipsoid type (Figure 11.17). In uniaxial ellipsoids (Types 1 and 5) the cones are circular with their axes of symmetry parallel to the Z and X-directions respectively. For the special case of plane strain (Type 3) the conic surface degenerates into a pair of planes, symmetrically disposed

to the X-axis, and intersecting in a line parallel to the Y-axis. In the more general cases of flattening (Type 2) and constriction (Type 4) the cones are elliptical and intersect the XZ and YZ, and XY and XZ planes respectively. These elliptical cones can be completely defined by the angles made by their lines of intersection on the principal planes and one of the principal strain axes (angles Φ_1', Φ_2', Φ_3') (Figure 11.18A and Table 11.2). The values

Table 11.2

	XY section	XZ section	YZ section
1. Uniaxial flattening	—	$\Phi_2' = \Phi_3'$	$\Phi_2' = \Phi_3'$
2. General flattening	—	Φ_2'	Φ_3'
3. Plane strain	90°	Φ_2'	0°
4. General constriction	Φ_1'	Φ_2'	—
5. Uniaxial constriction	$\Phi_1' = \Phi_2'$	$\Phi_1' = \Phi_2'$	—

of these angles depend upon the values of the principal extensions, and from Equation D.10 they can be expressed in a form using the principal quadratic extensions λ_1 etc.

$$\tan^2 \Phi'_1 = \frac{\lambda_2(\lambda_1 - 1)}{\lambda_1(1 - \lambda_2)} \qquad (11.21)$$

$$\tan^2 \Phi'_2 = \frac{\lambda_3(\lambda_1 - 1)}{\lambda_1(1 - \lambda_3)} \qquad (11.22)$$

$$\tan^2 \Phi'_3 = \frac{\lambda_3(\lambda_2 - 1)}{\lambda_2(1 - \lambda_3)} \qquad (11.23)$$

The importance of these surfaces of no finite longitudinal strain is seen when we consider the possible strain ellipse fields which arise in different types of strain ellipsoid. From an inspection of the projections of Figure 11.18A it can be seen that it is always possible to construct a great circle which cuts the surface of lines of no finite longitudinal strain. This implies that in all ellipsoids of Type 1 to Type 5 it is always possible to find some sections with Field 2 strain ellipses. In ellipsoid Types 1, 2 and 3 it is possible to draw great circles so that they lie completely within the positive extension sector, and in ellipsoids of Types 3, 4 and 5 great circles can lie completely in the negative extension sector. These two groups of ellipses clearly lie in Field 1 and Field 3 respectively. The complete range of possible ellipses for each of the five ellipsoids shown in Figure 11.18A is illustrated in Figure 11.18B. It should be clear how all the implications of the various ellipse Fields discussed in Session 4 relate to the three-dimensional form of the strain ellipsoid.

Principal strains from lines of n.f.l.s.

Answer 11.3★

The orientations Φ' of lines of no finite longitudinal strain ($e = 0$, $\lambda = 1$) in a strain ellipse are given from Equation D.10 (the prime indicates that the angle is measured in the deformed state):

$$\tan^2 \Phi' = \frac{\lambda_2(\lambda_1 - 1)}{\lambda_1(1 - \lambda_2)}$$

Combining this equation with that expressing the area dilatation Δ_A and the principal quadratic extensions $(1 + \Delta_A)^2 = \lambda_1 \lambda_2$ we can obtain an equation expressing λ_1 in terms of the angle $2\Phi'$ between the two lines of no finite longitudinal strain and the area dilatation Δ_A:

$$\lambda_1^2 \tan^2 \Phi' - \lambda_1(1 + \Delta_A)^2(\tan^2 \Phi' + 1)$$

$$+ (1 + \Delta_A)^2 = 0 \qquad (11.24)$$

Because this is a quadratic function it follows that, for a given value of Φ' and Δ_A, two possible values exist for the principal quadratic extension λ_1, and two corresponding values for λ_2. The two pairs of solutions for the principal strains refer to the two possibilities where the direction of λ_1 bisects the acute or obtuse angle $2\Phi'$ (Figure 11.19A).

If there is no area change Equation 11.24 simplifies into a factorizable form

$$(\lambda_1 \tan^2 \Phi' - 1)(\lambda_1 - 1) = 0 \qquad (11.25)$$

giving two solutions, $\lambda_1 = 1/\tan^2 \Phi'$ and $\lambda_1 = 1$. In a deformed state the second of these is geologically spurious, and so one unique pair of λ_1, λ_2 values can always be found.

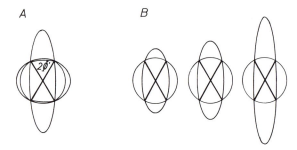

Figure 11.19. *Solutions to Question 11.3★. For a given angle Φ' between the lines of no finite longitudinal strain various strain ellipses can be found depending upon area dilatation Δ_A.*

If it is possible to specify the direction of the maximum elongation relative to the acute and obtuse angles between the lines of no finite longitudinal strain then Equation 11.24 and the values of the principal quadratic extensions are only a function of area dilatation Δ_A. It must be emphasized, however, that only rarely do we have any knowledge of Δ_A, and without this information there are an infinite number of solutions to the problem (see Figure 11.19B). We may therefore summarize the results of this two-dimensional analysis as follows:

1. If nothing is known of the area dilatation there are an infinite number of possible $\lambda_1 \lambda_2$ pairs for fitting differing strain ellipses to the Φ' value. If we can specify the ellipticity of the strain ellipse a unique solution is possible.

2. If we know the area dilatation Δ_A, but cannot distinguish the maximum and minimum strain directions two possible $\lambda_1 \lambda_2$ pairs are possible solutions to the problem. These two solutions often show markedly different strain ellipse aspect ratios R, and it should be possible by inspection of other structural features, to select one as the most likely solution to the problem.

3. If we know the area dilatation and can identify the direction of maximum extension the solution for a $\lambda_1 \lambda_2$ pair is unique.

From the discussion of the analytical solutions of this problem in two dimensions it should not be a surprise to find that an even greater number of possible solutions arise in three-dimensional problems. In three dimensions we have seen that the cone defining the directions of no finite longitudinal strain can be specified by the *two* angles (Φ'_1, Φ'_2 or Φ'_2, Φ'_3) between two of the principal directions and the intersection of the cone on two principal planes. As we hope to discover *three* unknown principal strain values the problem is mathematically insoluble unless further information is available. In two dimensions we saw that knowledge of strain ratio or area dilatation was the key to obtaining solutions; the same holds true in three dimensions—we need to know the shape of the strain ellipsoid (ratios R_{xy}, R_{yz}) or the extent of volumetric dilatation Δ_V before a solution can be obtained.

If we know the shape of the strain ellipsoid then Equation 11.3 enables an immediate solution to the problem. If this information is unavailable there are two main groups of solution: one relating to the pointed cone of lines of no finite longitudinal strain arising in constriction, the other to the open cone produced during flattening deformations.

Constriction deformations

The conical surface of lines of no finite longitudinal strain cuts the principal XY and XZ principal planes and makes angles of Φ_1' and Φ_2' with the X-axis on the XY and XZ principal planes respectively. The relationships defining these angles, the volumetric dilatation Δ_V and the principal quadratic extensions are described in Equations 11.21, 11.22 and 11.23.

Eliminating λ_2 and λ_3 from these three equations it is possible to obtain a single cubic equation for λ_1 expressing values of the maximum quadratic extension in terms of the unknown factors Φ_1', Φ_2', Δ_V.

$$\lambda_1^3 \tan^2 \Phi_1' \tan^2 \Phi_2' - \lambda_1^2 (1 + \Delta_V)(1 + \tan^2 \Phi_1')$$

$$\times (1 + \tan^2 \Phi_2') + \lambda_1 (1 + \Delta_V)^2$$

$$\times (\tan^2 \Phi_1' + \tan^2 \Phi_2' + 2) - (1 + \Delta_V)^2 = 0 \quad (11.26)$$

By selecting different values of Φ_1', Φ_2' and Δ_V it is possible to obtain, as is normal for a cubic equation, three real solutions for λ_1', or one real and two mathematically imaginary solutions (involving the square root of minus one). From the real valued solutions of λ_1 it is possible to use Equations 11.21 and 11.22, to compute λ_2 and λ_3. From possible values of real solutions of λ_1 in Equation 11.26 we may find that λ_2 or λ_3 have negative values, which are clearly geologically impossible (λ must always be positive, as it is the square of a real number).

Examining the real valued solutions to Equations 11.26, 11.21 and 11.22 we may conclude:

1. There is a minimum value of Δ_V below which *no* real geometric solutions are possible.
2. Above this minimum value of Δ_V there are two possible ellipsoids for a given negative volumetric dilatation.
3. Where volume dilatation is zero one of the two possible ellipsoids becomes a unit sphere and the other provides a unique solution to the data.
4. For a specific positive volume dilatation only one ellipsoid is geometrically possible.

Figure 11.20, curve A, is an illustration of the range of possible strain ellipsoids, all having cones of no finite longitudinal strain with $\Phi_1' = 70°$, $\Phi_2' = 30°$.

1. No ellipsoids can be found with volumetric dilatation Δ_V less than -0.155 (i.e. $15 \cdot 5\%$ volume loss). At this critical point a unique ellipsoid solution is found (Figure 11.20; a with $R_{xy} = 1 \cdot 26$ and $R_{yz} = 1 \cdot 40$).
2. For a given volume loss (e.g. $\Delta_V = -0.1$) two ellipsoids are possible (Figure 11.20; b with $R_{xy} = 1 \cdot 08$, $R_{yz} = 1 \cdot 17$, and b' with $R_{xy} = 1 \cdot 60$, $R_{yz} = 1 \cdot 61$).
3. Where the dilatation is zero one of these ellipsoids is a sphere (point c, $R_{xy} = R_{yz} = 1 \cdot 0$), the other has $R_{xy} = 1 \cdot 94$, $R_{yz} = 1 \cdot 69$ (point c').
4. Where the dilatation is positive (e.g. point d) a unique ellipsoid solution can be found.

Flattening deformations

In a flattening deformation the cone of no finite longitudinal strain intersects the XZ and YZ principal planes making

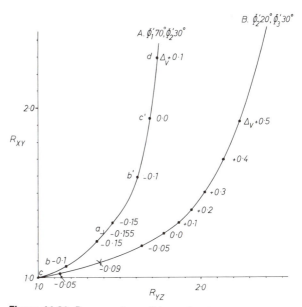

Figure 11.20. Ranges of possible strain ellipsoids with given orientation of the surfaces of no finite longitudinal strain (defined by Φ_1', Φ_2' or Φ_2', Φ_3') and volume dilatation Δ_V.

angles with the X and Y axes of Φ_2' and Φ_3' respectively. The relationships of these angles to the principal quadratic extensions are given in Equations 11.22 and 11.23. Combining these with the dilatation relationships (Equation 11.4) we obtain a cubic equation for λ_3:

$$\lambda_3^3 - \lambda_3^2 (1 + \Delta_V)^2 (\tan^2 \Phi_2' + 1)(\tan^2 \Phi_3' + 1) + \lambda_3 (1 + \Delta_V)^2$$

$$\times [(\tan^2 \Phi_2' + 1)\tan^2 \Phi_3' + (\tan^2 \Phi_3' + 1)\tan^2 \Phi_2']$$

$$- \tan^2 \Phi_2' \tan^2 \Phi_3' (1 + \Delta_V)^2 = 0$$

The solutions to this equation for λ_3 proceed along the same lines as that discussed for constrictive strains, and we find the following possibilities.

1. A negative value of dilatation below which no real solutions are possible.
2. For a negative value of dilatation above the cut off point two ellipsoids are feasible.
3. Where there is no volume change one ellipsoid is spherical (and spurious), a second gives a unique answer to the problem.
4. A deformation with positive dilatation produces a single unique ellipsoid solution. A numerical example of such a range of ellipsoids is shown in curve B, Figure 11.19, with $\Phi_2' = 20°$ and $\Phi_3' = 30°$.

It should be apparent from the discussion above that the determination of the strain ellipsoid using the geometric features of lines of no finite longitudinal strains (Talbot's method, 1970) is a practical one and is likely to be of interest when it is required to determine strains in gneissic terrains containing variably oriented sheets or dykes (Figure 6.14A). The technique must be used with care, however, bearing in mind the different possible solutions that can arise from a given data angle pair, and it may be necessary to use additional criteria to select the appropriate ellipsoid solutions, perhaps by using the overall rock fabric to select an ellipsoid with the most appropriate shape to explain the fabric. Some knowledge of the dilatation is clearly critical, for in Figure 11.20 it will be seen that quite small changes in dilatation may give rise to quite large differences in the computed principal strain values.

KEYWORDS AND DEFINITIONS

Circular sections of an ellipsoid	Sections of the strain ellipsoid passing through the Y-axis on which longitudinal strains are equal in all directions (i.e. the strain ellipse has a circular form). The angle made by these sections and the X-axis of the ellipsoid is a function of the principal plane strain ratios (Equation 11.20).
Cone of directions of no finite longitudinal strain	The surface produced by joining lines of unchanged total length passing through the centre of a strain ellipsoid (Figure 11.17). The surface can be specified by noting the two angles made between the intersection of the surface with the principal planes and the principal strain axes (Φ_1, Φ_2 or Φ_2, Φ_3—see Figure 11.18A and Equations 11.21, 11.22, 11.23).
Ellipsoid *d*-value	A parameter expressing the *intensity of distortion* of a strain ellipsoid as measured by the distance of an ellipsoid plot from the origin in a Flinn diagram (Equation 11.13, Figure 11.4A).
Ellipsoid *D*-value	A parameter expressing the *intensity of distortion* of a strain ellipsoid as measured by the distance of an ellipsoid plot from the origin of a logarithmic ratio plot (Equation 11.14, Figure 11.4B).
Ellipsoid *k*-value	A parameter expressing the *shape* (oblateness or prolateness) of a strain ellipsoid defined in terms of the principal plane ratios (Equation 11.5, Figure 11.4A).
Ellipsoid *K*-value	A parameter expressing the *shape* of a strain ellipsoid, defined in terms of the logarithmic principal plane ratios (Equation 11.6, Figure 11.4B).
Lode's parameter ν	A parameter describing the *shape* of a strain ellipsoid in terms of logarithmic values of the semi-axis lengths of the strain ellipsoid (Equation 11.7).

KEY REFERENCES

Strain

There is now an extensive literature on strain measurements in rocks. Those recommended below are particularly relevant to the problems of representing and evaluating the significance of finite strain states.

Breddin, H. (1957). Tektonische Fossil- und Gesteinsdeformation im Gebiet von St. Goarshausen (Rheinisches Schiefergebirge), *Decheniana* **110**, 289–350.

This gives an account of strains determined from several species of deformed fossils, and presents the results on deformation maps. These maps illustrate systematically changing shortening across the region which is linked with changes in the intensity and orientation of slaty cleavage.

Cloos, E. (1971) "Microtectonics", 234 pp. Johns Hopkins University, Baltimore.

This book gives a beautifully illustrated account of strain measurements carried out over a region of some 4000 square kilometres in extent and is an extension both in area and in appreciation of the significance of the structural features described in his classic paper of 1947.
Here are documented the results of a lifetime study, and as a demonstration of the systematic relationships of small scale tectonic structures, strain state and large scale tectonic phenomena it probably has no equal. The writing is concise and informative, and the two large finite strain maps, excellently drawn diagrams and clear plates showing the appearance of rocks in thin section and hand specimen provide a model text.

Hossack, J. R. (1968). Pebble deformation and thrusting in the Bygdin area (S. Norway). *Tectonophysics* **5**, 315–339.

This paper describes strain measurements determined from the shapes of deformed pebbles in the Bygdin conglomerate and which were deformed during nappe transport in Caledonian times. Useful discussions of methods of representing the amount of deformation and ellipsoid shapes are given.

Ramsay, J. G. (1967). "Folding and Fracturing of Rocks," 568 pp. McGraw-Hill, New York.

Methods of graphically recording the longitudinal strains and orientations of strain ellipsoid axes will be found on pp. 134–141.

Talbot, C. J. (1970). The minimum strain ellipsoid using deformed quartz veins. *Tectonophysics* **9**, 47–76.

This paper describes how it is possible to define the strain ellipsoid by determining the surface of no finite longitudinal strain from boudinage and folded competent veins. This paper, together with Ramsay (1976) (p. 71), would provide appropriate background reading to Question 11.3★.

SESSION 12

Progressive Displacement and Progressive Deformation

Principles

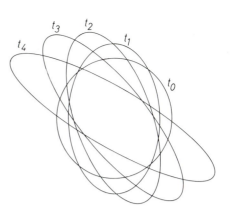

The geometric properties of a deformation sequence are established by envisaging the sequence in terms of displacement increments and strain increments. The build up of a finite displacement field by incremental displacement steps is analysed using matrix multiplication techniques and it is shown how successive states of finite strain can be evaluated from successive matrix products. The concepts of displacement rate and strain rate arise by reducing increment size to that occurring over a period of time, one second in duration, and the relation between these rates and the rates of change of finite displacement and rate of change of the finite strain state are analysed. Various types of constant incremental processes are investigated and it is shown how some sequences lead to progressively increasing states of finite strain, whereas others show pulsating finite strains. Some of these sequences are shown to be common in geological processes and others shown to be rare. The geological implications of increment superposition on previously established finite strain states are investigated: it is shown how fields of active boudinage, active folding, folded boudinage and boudinaged folds can arise. The complexity of geometric forms of small scale structures is discussed in terms of general strain increment sequences.

INTRODUCTION

The concepts of progressive displacement and progressive deformation provide the key for the understanding of many of the observed structural features of naturally deformed rocks. Until now we have mostly developed the idea of finite strain state by comparing the geometry of the end state of a deformation process with geometric features of the initial state without consideration of the geometric features of the intervening stages. In Session 1 we encountered briefly some of the possibilities and complexities of these intervening states during our experiments on simple shear. We saw that lines drawn on the surface of the progressively sheared card deck model sometimes undergo complicated changes in longitudinal strain and before we go further it is a good idea to check back on our conclusions (Answers 1.1, 1.2 and 1.3, pp. 5–6). In this Session we will analyse the changes that take place in different types of deformation sequences in a more general way and discuss the geological implications of these changes.

All deformation processes take some specific length of time to accomplish the total or finite strain that we observe at the end of that process. In geological processes the time can vary from a duration of seconds in the case of earthquakes to millions of years in the case of the formation of folds in orogenic zones. During the process of deformation the rock passes through a series of changing geometrical states and, depending upon the nature, speed and intensity of these changing states, a whole range of mechanical instabilities may arise during the deformation sequence. In the case of the rapid deformations taking place during the propagation of seismic waves through the crust we are aware that a rock element may become deformed and, because the deformation is an elastic and reversible one, these distortions may be completely removed at a later stage. During the slower and larger deformations developed during the ductile processes of rock creep, strains can also grow and then later be partially or completely removed. It follows that the structures that we observe in naturally deformed rocks, for example the fracture orientations, boudin axes and fold directions, reflect the nature of the whole progress of the deformation. A structure may be initiated at a particular stage in the deformation sequence, have its geometry modified by the distortions taking place during the later part of the deformation sequence, or the structure may be superposed by new structures arising during the concluding stages of this sequence. We will now investigate the geometric changes taking place during a general type of deformation so that we will be in a better position to understand the full potential of progressive deformation and be able to interpret the structures we observe in naturally deformed rocks in terms of progressive deformation processes.

Progressive displacement

A convenient and instructive way of starting an investigation into progressive shape changes is to consider what happens to an initially orthogonal grid as a result of progressive changes in the coordinates of different points situated initially at positions (x, y). Figure 12.1 illustrates the progressively changing shape of a small element of material $Op_0q_0r_0$ undergoing a displacement sequence such that the strains at each stage in the sequence are homogeneous. The sequence 0, 1, 2 ... 9 illustrates the shapes of this grid at different stages (or times $t_0, t_1, t_2 ... t_9$), and defines the **progressive displacement** of the geometric changes. These can be envisaged as the stills or individual frames in a continuous movie picture. If we ran them through a movie projector we would get the impression of a continuously moving shape change sequence. This particular sequence was constructed in the following way: the initial rectangular grid at stage zero was subjected to a displacement field and during the time interval between t_0 and t_1 the element took up a parallelogram shape Op_1, q_1, r_1 (stage 1). At stage 1 a new rectangular grid of identical shape to $Op_0q_0r_0$ was constructed (dashed lines). This new grid was then displaced in the time interval between t_1 and t_2 into exactly the same shape as the original grid between times t_0 and t_1. The positions of $Op_1g_1r_1$ become displaced to new positions $Op_2q_2r_2$ at stage 2. This sequential displacement process was repeated nine times to arrive at the **finite displacement field** at stage 9, and the geometry of this total displacement plan can therefore be investigated by determining the geometry arising by the superposition of nine identical **displacement increments**. If the time taken to develop each displacement increment was the same, the process would be one of constant displacement increment rate. In describing in general terms the speed of the displacement process we describe the displacements taking place in some standard time interval, usually one second. The displacements taking place during one second define the **displacement rate** of the process,

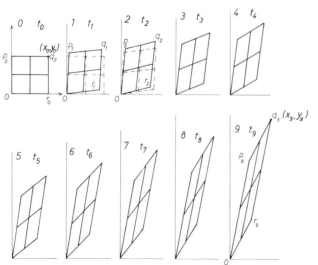

Figure 12.1. *The successive finite displacements arising by the superposition of nine identical displacement increments, Stages 0–9 at times t_0, t_1, $t_2 ... t_9$, illustrate the shape of an initial orthogonal grid. Any general point q_0 with coordinates (x_0, y_0) moves successively to positions q_1, $q_2 ... q_9$ where it has final coordinates (x_9, y_9).*

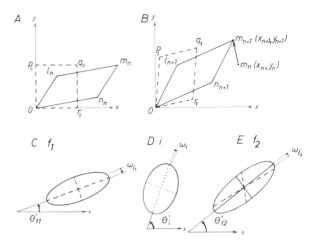

Figure 12.2. *A illustrates the parallelogram $Ol_nm_nn_n$ that has arisen from the finite displacement of an original square element and B illustrates its shape $Ol_{n+1}m_{n+1}n_{n+1}$ after the superposition of the next displacement increment. C is the finite strain ellipse f_1 related to the parallelogram $Ol_nm_nn_n$ and D illustrates the incremental strain ellipse i which, when superposed on the ellipse f_1 transforms it to the finite strain ellipse f_2. For further explanation see text.*

and if the time intervals for the development of each stage in the sequence of Figure 12.1 were each of one second duration the process would be one of *constant displacement rate* known as *steady state flow*.

To analyse the geometric properties of a sequence like that of Figure 12.1 we need to investigate how the total coordinate transformation matrix at any stage $(n + 1)$ is related to that of an earlier stage (n) and that of the displacement increment (i). Figure 12.2 shows the general geometry of the superposition of a displacement increment on a previously attained displacement. The parallelogram in Figure 12.2A defining the previous total displacement $Ol_nm_nn_n$ has been superposed by a square $Op_0q_0r_0$. In Figure 12.2B the new square has been displaced to $Op_1q_1r_1$ and the parallelogram $Ol_nm_nn_n$ is transformed into another parallelogram $Ol_{n+1}m_{n+1}n_{n+1}$, each corner being moved along the appropriate **incremental displacement vector** like that between m_n and m_{n+1}.

The relations between any point (x_n, y_n) and its initial coordinates (x, y) can be expressed in terms of the four components $(a_nb_nc_nd_n)$ of the coordinate transformation matrix for stage n. In terms of the coordinate transformation equations:

$$x_n = a_nx + b_ny$$
$$y_n = c_nx + d_ny \qquad (12.1)$$

The coordinate transformation equations which will transform points (x_n, y_n) to (x_{n+1}, y_{n+1}) for the displacement increment require four components $(a_ib_ic_id_i)$ for their definition.

$$x_{n+1} = a_ix_n + b_iy_n$$
$$y_{n+1} = c_ix_n + d_iy_n \qquad (12.2)$$

Combining 12.1 and 12.2 we obtain the transformation for the total displacement at stage $n + 1$

$$x_{n+1} = (a_ia_n + b_ic_n)x + (a_ib_n + b_id_n)y \qquad (12.3)$$
$$y_{n+1} = (c_ia_n + d_ic_n)x + (c_ib_n + d_id_n)y$$

Table 12.1.
Progressive displacement and strain (Figures 12.1, 12.3).

Increment n	a_n	b_n	c_n	d_n	$1 + e_{1n}$	$1 + e_{2n}$	θ_i'	θ_n	ω_n
0	1·0000	0·0000	0·0000	1·0000	1·00	1·00	—	—	—
1	0·9136	0·0500	0·1000	1·1000	1·13	0·89	71·29	69·87	1·42
2	0·8397	0·1007	0·2014	1·2150	1·27	0·79	71·99	69·18	2·81
3	0·7772	0·1527	0·3055	1·3466	1·43	0·70	72·64	68·53	4·11
4	0·7253	0·2068	0·4138	1·4985	1·61	0·62	73·25	67·93	5·32
5	0·6833	0·2638	0·5277	1·6668	1·81	0·55	73·79	67·38	6·44
6	0·6506	0·3243	0·6488	1·8599	2·04	0·49	74·27	66·91	7·33
7	0·6268	0·3893	0·7787	2·0783	2·30	0·44	74·68	66·49	8·19
8	0·6116	0·4596	0·9193	2·3251	2·59	0·39	75·04	66·14	8·90
9	0·6047	0·5361	1·0724	2·6036	2·91	0·34	75·51	66·02	9·49

Using this mathematical technique of matrix multiplication the successive coordinate transformation components for each of the stages of Figure 12.1 have been calculated (Table 12.1).

The Equations 12.2 defining the displacement increment field in two dimensions require four components for their definition. If the time interval t for the increment is infinitesimally small (in practice one second duration) these terms are:

$$\frac{\partial a_i}{\partial t} \quad \frac{\partial b_i}{\partial t} \quad \frac{\partial c_i}{\partial t} \quad \frac{\partial d_i}{\partial t}$$

and define the **rates of change of the incremental strain matrix** which is related to the **displacement gradient rates**. They are generally written with a dot convention (described as a dot, b dot, etc.).

$$\dot{a}_i \dot{b}_i \dot{c}_i \dot{d}_i$$

We will not be employing this displacement rate notation extensively until Volume 2 of this series but an appreciation of the differences between displacement increments and displacement rates are clearly fundamental in understanding the parameters of a displacement sequence process.

Progressive strain

The geometric changes taking place in the displaced grid of Figure 12.1 clearly imply that the finite strain states at each successive stage are also changing. The changing geometry of these successive strains is known as **progressive deformation** and is mathematically related to the geometry of the progressive displacement process. At each stage (n) in the sequence we can evaluate the characteristic features of a new strain ellipse, and each ellipse requires four strain components for its definition (two principal strains e_{f1n}, e_{f2n}, and the orientations before and after deformation of the strain ellipse axes θ_{fn}, θ_{fn}'). The difference between θ_{fn}' and θ_{fn} defines the finite rotation ω_{fn} at stage n. These strain components may be evaluated from the four components of the coordinate transformation equations using the relations set out in Appendix B (Equations B.14, B.15, B.16, B.19). The appropriate strain components for the displacement sequence of Figure 12.1 are set out in Table 12.1, and these were used to calculate the successive strain ellipses shown in Figure 12.3. Because the displacement increments between any two stages are constant, this implies that the **incremental strain ellipse**

defining the superposed distortion and rotation between any two strain ellipses in the total sequence is constant. This incremental strain ellipse has four components, two **principal incremental longitudinal strains** e_{i1}, e_{i2}, with characteristic orientations before θ_i and after θ_i' the increment and with an **incremental rotation** $\omega_i = \theta_i' - \theta_i$. In the example illustrated in Figure 12.3 the ellipse shown at stage 1 represents the incremental strain ellipse for all stages in the deformation. If the increment of deformation takes place in an infinitesimally small time period t (conventionally one second) the components of the incremental strain ellipse define the **strain rate**. For a complete specification of strain rate we need to define the values of the **two principal strain rates**

$$\frac{\partial e_{i1}}{\partial t} = \dot{e}_{i1} \quad \text{and} \quad \frac{\partial e_{i2}}{\partial t} = \dot{e}_{i2}$$

the direction in which these act (θ), and the **rate of rotation** or **vorticity**

$$\frac{\partial \omega_i}{\partial t} = \dot{\omega}_i$$

During the process of *steady flow*, all these four components remain constant.

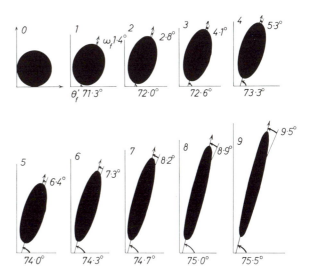

Figure 12.3. The successive finite strain ellipses arising from the displacement increment sequence 0–9 of Figure 12.1. θ' is the orientation of the maximum finite strain axis and ω_f is the finite rotation.

Progressive displacement and progressive strain

Question 12.1

We will now investigate some of the geometric features of Figures 12.1 and 12.3 in more detail and study the relationships between the successive stages of displacement and successive states of strain. Before we do this perhaps we should emphasize that this sequence is one of constant displacement increments and constant strain increments. If these increments each took place in the same time interval it would be a steady state flow process of constant displacement rate and constant strain rate. In general, geological processes are far more complex, but we must first appreciate the complexity of the geometric features arising from a simple, incremental process before we investigate the rules for general progressive deformation.

Determine the actual **movement paths** of the individual particles in Figure 12.1. To do this, place a piece of tracing paper over the stages illustrating the successive displacements keeping the coordinate axes superposed, and note the positions of $p_0, p_1, p_2 \ldots p_9$, and those for q and r. Are these lines straight or curves? How do the *movement paths* of points p, q and r differ from the total *displacement vectors* for points p_9, q_9 and r_9? Draw particle movement paths for all the grid line intersections and determine the **particle movement path field** for this constant displacement increment process. Do movement paths cross?

Question 12.2

Plot graphically the values of the terms a_n b_n c_n d_n of successive displacements set out in Table 12.1. Do these components vary in a linear or non-linear manner? Graph the values of the semi-axes of the successive strain ellipses and their total rotation components ω_n and discuss what these curves imply. What do the changes of values of the initial orientation of the principal axes of the strain ellipses (Table 12.1, values of θ_n) imply? Do you think the results arising from this particular progressive deformation sequence are typical of all progressive sequences?

Question 12.3

Plot the shapes of the successive strain ellipses on a two-dimensional deformation plot (see Figure 4.10 and discussion, p. 65) and determine the *shape deformation path*.

Now proceed to the Answers and Comments section, then proceed to the starred question or to Session 13.

STARRED (★) QUESTIONS

Types of strain arising from steady state flow

Question 12.4★

In this question we wish to investigate the whole spectrum of types of finite strain that can be developed as a result of the superposition of constant displacement and constant strain increments in a way that will cover all the possibilities of two-dimensional steady increment processes. Because the calculations required for the superposition of successive increments are exceedingly tedious and subject to error you will need a programmable desk computer for the solution of this problem.

Any incremental strain can be envisaged as being made up of an incremental irrotational distortion (analysed using an incremental irrotational strain ellipse) and an incremental rotation. To analyse the spectrum of finite strains arising from the accumulation by superposition of successive increments we need to choose particular values for the distortional and rotational part of the increment. Each chosen pair will define a particular type of steady increment process. For simplicity we will restrict ourselves to conditions of plane strain at constant volume, whereby the distortional aspects of the incremental strains are connected by the relationship $1 + e_{i1} = 1/(1 + e_{i2})$. Choosing the principal incremental strains to coincide with the y- and x-coordinates respectively we can derive the matrix defining the incremental irrotational distortion:

$$\begin{bmatrix} \dfrac{1}{1 + e_{i1}} & 0 \\ 0 & 1 + e_{i1} \end{bmatrix} \tag{12.4}$$

where e_{i1} is the maximum principal incremental longitudinal strain. The matrix representing the transformation of a positive (anticlockwise) incremental rotation ω_i is given by (see Appendix C, Equation C.10):

$$\begin{bmatrix} \cos \omega_i & -\sin \omega_i \\ \sin \omega_i & \cos \omega_i \end{bmatrix} \tag{12.5}$$

The strain matrix combining distortional (first) and rotational (second) aspects of the increment is the matrix product of Equation 12.4 and Equation 12.5 which, from Equation 12.3, is:

$$M_i = \begin{bmatrix} \dfrac{\cos \omega_i}{1 + e_{if}} & -\sin \omega_i (1 + e_{i1}) \\ \dfrac{\sin \omega_i}{1 + e_{i1}} & \cos \omega_i (1 + e_{i1}) \end{bmatrix} \tag{12.6}$$

Our problem consists of making successive matrix multiplications of Equation 12.6 with values of e_{i1} and ω_i chosen to cover the whole spectrum of incremental sequences. If $Mf_1, Mf_2 \ldots Mf_n$ represent the successive deformation matrices for successive finite strains $f_1, f_2 \ldots f_n$ then $Mf_1 = M_i$, $Mf_2 = M_i \times Mf_1 \ldots Mf_n = M_i \times Mf_{n-1}$. From each of the matrices Mf_n it is possible to evaluate the orientations θ'_n and values of the principal finite strains e_{fn1}, e_{fn2} and finite rotation ω_{fn} from the equations given in Appendix B (Equations B.14, B.19, B.16). With a short computer program it will be possible from the two input values e_{i1} and ω_i to print out the successive finite strains. A suggested value for e_{i1} is 0·1, and for this distortion a range of incremental rotations is recommended: $\omega_i = 0\cdot0°$, $4\cdot0°$, $7\cdot0°$, $10\cdot0°$, $20\cdot0°$. If your time is limited investigate incremental rotations of $0\cdot0°$, $4\cdot0°$ and $10\cdot0°$.

There must be a particular incremental rotation value which, combined with e_{i1}, gives simple shear increments. This value of ω_i can be obtained by solving Equation 2.7 to determine the value of shear strain γ appropriate to a value of $e_1 = 0\cdot1$, and then replacing this value of γ in Equation 2.6 to obtain the corresponding rotation. Check that this ω_i value is $5\cdot45°$. If you have time determine the finite strain sequence for simple shear terms of successive increments.

Before assessing the geometrical significance of these results we should ask the question "What is the significance of these different increments in terms of overall flow patterns?" The increments differ only in their rotation components. Where the incremental rotation is zero all the successive finite strain ellipses will be coaxial, and at all stages the deformation sequence is without body rotation. This is *progressive pure shear*. Those with incremental rotations lying between zero and $5 \cdot 45°$ are rotational strain sequences between pure and simple shear. As the incremental rotation becomes progressively larger so the sequence becomes closer to that of a body rotation, and sequences with $5 \cdot 45° < \omega_i < \infty$ are intermediate between simple shear and pure body rotation types.

Does the finite longitudinal strain e_{f1} increase progressively for all sequences? If not what is likely to be the value of ω_i which separates fields of progressively increasing strains from those where the strains can decrease? Distinguish those fields where the orientations of the strain ellipses (a) remain static, (b) change, but approach a constant orientation or, (c) continuously change.

To answer these questions it is probably best to plot curves showing (a) principal finite longitudinal strain e_1 as abscissa, orientation of the strain ellipse axes θ' as ordinate and (b) number of increments as abscissa, total rotation of finite strain as ordinate.

ANSWERS AND COMMENTS

Answer 12.1

Figure 12.4A illustrates the **movement paths** of points p, q and r. Point 0 at the origin of the coordinate system does not move. All other points move on curved paths and represent the tracks of particles relative to a fixed point 0 in the body. The **displacement vectors** for the points

p_9, q_9, r_9 would be represented by the straight lines joining p_0 to p_9, q_0 to q_9 and r_0 to r_9 respectively and clearly do not generally coincide with the movement paths. The only deformation sequence where displacement vectors and movement paths coincide is that of simple shear. The incremental displacements along each movement path progressively increase along each path, and these increases relate to increasing distance of the particle from the origin. Figure 12.4B illustrates the movement paths of nine particles and defines part of the movement path field. In this part of the whole field (positive x, positive y) the particle paths converge away from the origin, but if the field for positive x, negative y had been constructed you would see that the particle paths move towards the origin and diverge. This geometry implies progressively increasing values of the principle finite strains with increase in number of increment. The particle movement paths never cross. This feature is characteristic of constant displacement increment (and constant strain increment) geometry. Constant displacement increments imply that the displacement increment vector for a given point is fixed; as a point moves from one place to another its path must therefore always be uniquely defined. If the progressive displacement process is one of varying increment geometry the particle movement paths are not uniquely fixed in this way and movement paths can cross each other.

Answer 12.2

The plots of the values of the components a_n b_n c_n d_n of successive displacement states are illustrated in Figure 12.5. None of these shows growth in any linear manner. This non-linearity is a general feature of all constant increment processes with the exception of simple shear. In simple shear these components take special values ($a_n = d_n = 1$, $b_n = \gamma$ and $c_n = 0$) such that a linear increase in shear displacement (γ) gives a linear increase in component b_n.

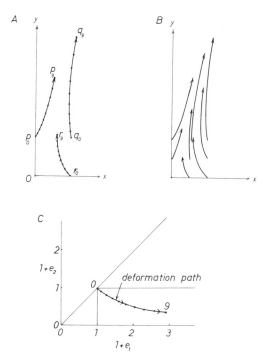

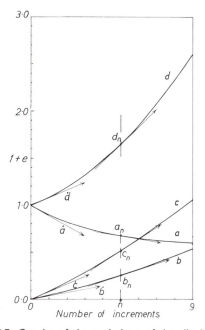

Figure 12.4. A shows the successive positions of p, q, and r through the nine increment stages of Figure 12.1, with reference to fixed x and y coordinate axes. B illustrates the movement paths of all nine points of the original orthogonal intersection grid of stage 0 of Figure 12.1. C shows the two-dimensional deformation plot of successive finite strain ellipses.

Figure 12.5. Graphs of the variations of the displacement components a, b, c and d defining the total displacement after n increments. The slopes of these curves at n = 0 define the displacement rates à, b̀, c̀ and d̀ at the start of the deformation. During a later increment the rate of change of the displacements can be evaluated from the slopes of the four curves (e.g. at n = 5).

If the time interval $t_1 - t_0$ was one second, then the components a_1 b_1 c_1 d_1 would represent the **displacement rates** $\dot{a}_i$ $\dot{b}_i$ $\dot{c}_i$ $\dot{d}_i$ for the first increments relative to the origin in units per second. If this time interval was greater than one second then the actual displacement rates would be given by the tangents to the various curves at time t_0 (i.e. at increment zero) (see Figure 12.5). These correct displacement rate values would be lower than those determined for the first increment by dividing the increment by the time period in seconds $t_1 - t_0$. This has special implications when it is necessary to determine the displacement rate from an incremental displacement process measured over a particular number of seconds. For example, if the time interval $t_1 - t_0$ was 10 seconds then the average displacement rate a_i would be less than $0.9136/10$ per second (i.e. less than $9.136 \times 10^{-2}\,\mathrm{s}^{-1}$). The value can be calculated by using a determination of the tangent slope and is $(8.82 \times 10^{-2}\,\mathrm{s}^{-1})$.

At any period during the progress of finite displacement we could describe the **rates of change of the finite displacement** in terms of the rates at which the values of a_n b_n c_n and d_n were changing per second (i.e. $\dot{a}_n$ $\dot{b}_n$, $\dot{c}_n$ and $\dot{d}_n$), and these displacement rates of the finite displacement could be obtained directly from the tangents to the curves at a particular increment value if we know the time scale involved (Figure 12.5). It should be clear from Figure 12.5 that these rates of change will differ from those of the incremental displacement rate at that particular time.

The values of the semi-axes of the successive strain ellipses and the rotation ω_n are plotted in Figure 12.6. All increase in a non-linear fashion with respect to linear increment values. This non-linearity is a general feature of all incremental strain processes including that of simple shear. In the example plotted here the curve for the values of major semi-axes of the finite strain ellipse is concave upwards, implying that the finite strains arising from two increments are always more than double those from one increment. This accelerating growth of finite strain state is characteristic of many geological strain sequences, but it is not an inevitable result of all constant incremental flow processes, as we will see later in the starred question of this Session. If, in our example, the time interval t_0 to t_1 was one second we could define the **principal strain rates** as $\dot{e}_{i1} = 1.13\,\mathrm{s}^{-1}$, $\dot{e}_{i2} = 8.9 \times 10^{-1}\,\mathrm{s}^{-1}$, and the **vorticity** $\dot{\omega}_i = 1.42°\,\mathrm{s}^{-1}$. If, however, this first increment was achieved over a period of 10 seconds these rates would be slightly different. The rates could be constructed from the tangents to the three curves of Figure 12.6 determined at time t_0. From these tangents the average principal strain rates (assuming a constancy of strain rate over the time interval $t_0 - t_1$) would be $\dot{e}_1 = 1.11/10$ per second $(1.11 \times 10^{-1}\,\mathrm{s}^{-1})$, $\dot{e}_2 = 0.87/10$ per second $(8.7 \times 10^{-2}\,\mathrm{s}^{-1})$, $\dot{\omega} = 1.63° \times 10^{-1}\,\mathrm{s}^{-1}$. The changes of the finite strains that

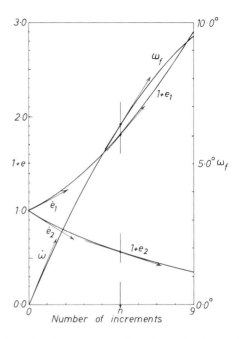

Figure 12.6. Graphs of the variations in semi-axis lengths of finite strain ellipse $(1 + e_1, 1 + e_2)$ and of the finite rotation ω_f. The slopes of the curves at $n = 0$ give the strain rates $\dot{e}_1$ $\dot{e}_2$ and vorticity $\dot{\omega}$. The tangents to these curves at increment n can be used to determine the rates of change of the principal finite strains e_{f1}, e_{f2} and the rates of change of finite rotation $\dot{\omega}_f$.

are taking place at any period during the deformation sequence, describing in mathematical terms the degree of non-linearity of these changes, can also be derived from the curves of Figure 12.6. At any stage during the strain history we can establish the tangents to the finite strain curves, and knowing the tangent slope and time interval it is possible to define the **rates of change of finite strains** $\dot{e}_{f1}, \dot{e}_{f2}$ and **rates of change of finite rotation** $\dot{\omega}_f$. It will be clear from the graphs that these values are in no way *simply* connected to the strain rates $\dot{e}_{i1}$ $\dot{e}_{i2}$ $\dot{\omega}_i$ taking place at the same time.

To illustrate the complexity of the geometric relationships of successive stages of finite strain we will return to the example we chose to illustrate the effect of the superposition of two displacements (Figure 12.2). The numerical values of displacements and strains for an initial state of strain f_1, an incremental strain i and the finite strain product f_2 is set out in Table 12.2 and illustrated as the three strain ellipses f_2, i and f_2 in Figure 2.2 C, D and E. Ellipse f_1 has principal axes indicated by perpendicular dashed lines, its orientation is given by θ'_{f1} and it has a negative finite rotation ω_{f1}. The incremental strain ellipse has principal axes indicated by dotted lines, and has a small positive incremental rotation ω_i. The effects of superposing these two strains produce ellipse f_2. The principal axes of this

Table 12.2.
Incremental displacement and strain (Figure 12.2).

	a	b	c	d	$1 + e_1$	$1 + e_2$	$1 + \Delta_A$	θ	θ'	ω
Finite displacement and strain f_1	1.5000	0.5000	0.2500	0.7500	1.66	0.60	1.00	25.67	19.33	−6.34
Incremental displacement and strain f_i	0.9000	0.1000	0.2000	1.2000	1.26	0.84	1.06	66.14	68.86	2.72
Finite displacement and strain f_2	1.3750	0.5259	0.6000	1.000	1.78	0.60	1.06	34.88	36.69	1.81

new ellipse lie between those of the two component strains. The total rotation ω_{f2} is positive, yet shows no simple relationship to the two rotations ω_{f1} and ω_i. The dashed lines which once occupied the principal directions of ellipse f_1 are rotated in opposite senses and do not coincide with the axes of ellipse f_2. This effect whereby the principal finite strain directions are continually adjusting to new positions in the body is a general one in progressive deformation and we will see later that it has very important geological consequences. You will recall that we discovered this effect when we conducted experiments on card deck models in simple shear (Session 2, Question 2.9).

Returning to the data of Table 12.1 values of θ_n are continually changing during our deformation sequence. This accords with the conclusion outlined above implying that the lines marking the positions of successive finite strain ellipse axes are continually moving to new positions in the body.

Answer 12.3 See Figure 12.4C

The general progressive deformation process

So far we have investigated only processes where the deformation and strain increments are of constant type. Such deformation sequences do occur, but in natural deformation systems the incremental strains and strain rates vary during the progress of deformation.

Although the general geometric effects of superposing an incremental strain on an already established strain are quite complex, they do not significantly differ in principle from those effects we have previously considered under constant processes. A general analysis is illustrated in Figure 12.7. Figure 12.7A shows an initial circle and the finite strain ellipse f derived from it. This ellipse has four strain components, two principal finite longitudinal strains e_{f1} and e_{f2} with characteristic orientations θ' derived from directions θ in the undeformed circle. The strain is rotational $\omega_f = \theta' - \theta$. Figure 12.7B shows the incremental strain ellipse and a circular marker from which it was derived. This ellipse also has four components, two principal incremental longitudinal strains e_{i1} and e_{i2} with characteristic orientations and an incremental rotation ω_i. The superposition of this increment on the ellipse f can be considered in two parts: a distortional part and a rotational part. The distortional part moves points lying on the periphery of the ellipse f by amounts proportional to the strains taking place in the incremental strain ellipse, and moves the peripheral points towards the long axis of the incremental strain ellipse. The maximum ellipse axis of f is increased in length, but not as much as that along a line oriented in a clockwise sense, likewise the minimum axis of f is decreased but not as much as that along a line oriented in a clockwise sense. The ellipse f therefore takes up a new form with new axes oriented more clockwise than those of the original ellipse, these new axes located along new material lines in the body. Because of the change in position of the new ellipse axes, the finite rotation of the new ellipse is modified. This change of rotation due to incremental distortion may be termed the *distortion increment component of the rotation*. The values taken by this rotation component depend not only on the orientations and values of the strain increments, but also on the ellipticity of the strain ellipse on which the increment is superposed. The greater the initial ellipticity of the existing finite strain the smaller is the effect of the distortion increment component of the rotation. The incremental strain has its own rotation component, and this *body rotation increment component* is numerically added to the distortion increment component of the rotation. In some incremental systems these two components changing the finite rotation work in the same sense, in others they work in opposition. It should now be realized why the relationships between finite rotation and incremental rotation is generally complex.

Figure 12.7A indicates the two directions where the original finite strain ellipse intersects the unit circle from which it was derived, known as lines of no finite longitudinal strain ($e_f = 0$). These lines are symmetrically disposed about the principal axes of the finite strain ellipse with orientations depending upon values of the principal finite strains (see Appendix D, Equations D.9, D.10) and they divide the finite strain ellipse into sectors with positive and negative longitudinal strains (Figure 12.7A, $e_f +$ve, $e_f -$ve). In a similar way lines of no incremental longitudinal strain ($e_i = 0$) are symmetrically disposed about the principal axes of the incremental axes of the incremental strain ellipse and divide this ellipse into sectors with positive and negative incremental longitudinal strains (Figure 12.7B, $e_i +$ve, $e_i -$ve). When we consider the possibility of a general case of increment superposition on a pre-existing strain state there are four combinations of these strain sectors (see Figure 12.7C)

$$(1)\ e_f + \text{ve} \qquad e_i + \text{ve}$$
$$(2)\ e_f + \text{ve} \qquad e_i - \text{ve}$$
$$(3)\ e_f - \text{ve} \qquad e_i - \text{ve}$$
$$(4)\ e_f - \text{ve} \qquad e_i + \text{ve}$$

These various combinations have important geological consequences and can be related to differences in form and style of structures that are found in different orientations within a deformed rock. We will now examine these four possible combinations from a geological viewpoint, beginning the discussion by considering in a simple way the effects on a cross-section of a competent layer, where this cross sectional line lies in one of the four strain combinations set out above.

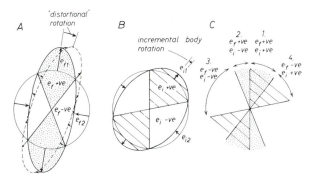

Figure 12.7. *Geometry of the superposition of incremental strain ellipse i (B) on an initial finite strain ellipse f (A). C shows the four sectors of possible combinations of positive and negative finite longitudinal strain e_f with positive and negative incremental strain e_i.*

1. e_f + ve, e_i + ve field of active boudinage: Lines lying in these sectors of the finite and incremental strain ellipses have suffered a finite elongation and as a result of the superposition of the next increment they will continue to elongate. Competent layers so positioned will have suffered boudinage, and these boudins will be actively developing. The field geologist will see simple boudins.

2. e_f + ve, e_i − ve field of folded boudins: Competent layers in this orientation will have developed boudin structures, but during the next deformation increment the boudinaged layers undergo shortening. It is mechanically most unlikely, if not impossible, that the already extended and separated competent layer boudins will simply reverse the boudinage process and, in so doing, come to lie adjacent to each other with the form of the undeformed layer. Boudin necks are either filled with incompetent material or crystalline vein material and therefore generally act as sites of mechanical weakness during the shortening increment. What happens in practice seems to depend upon the aspect ratio or length to width ratio of the boudin cross section. If this aspect ratio is large (greater than 5:1) shortening often leads to the development of folds in the boudinaged competent blocks. Figure 12.8 illustrates an example of such folded boudinage: the more competent layers consist of amphibolite, and it is clear from the relationships of pale quartz-feldspar areas in the initial neck zones to the fold forms that the amphibolite sheet was first stretched, then shortened. Figure 12.9 is another example of compressed boudins in competent metasandstone and incompetent mica schist. The initially stretched sandstone layers were compressed and developed folds with a number of different wavelengths. In situations where the aspect ratio of the

boudins is small (less than 5:1) and where the competent layers are inclined to the principal incremental strain axes the competent layers are obliquely pressed together to develop an imbricate structure, like the overlapping tiles on a roof (Figure 12.10). In Figure 12.10 the total overall longitudinal strain along the competent dark amphibolite layers is negative (an overall shortening) even though the early phases of strain history led to the formation of extensional boudin structure under a positive longitudinal strain. This implies that much of the later part of the deformation history gave rise to layer shortening.

3. e_f − ve, e_i − ve field of active folding: Layers lying in this location have been shortened and during the next increment of deformation undergo further shortening. In competent layers this gives rise to actively forming folds. These are folds which, during the strain increment, show a decrease in wavelength and an increase in amplitude. In the field one sees folds with stable geometry.

4. e_f − ve, e_i + ve field of boudinaged folds: Competent layers in this sector have been initially shortened and formed folds, and subsequently stretched. In the discussion of field 2 it was pointed out that boudins cannot "reseal" on compression. However, in field 4 it is possible for the folds in competent layers to become unfolded, and in laboratory experiments carried out with ductile materials this is often what does happen. There are, however, a number of other geometric modifications that are possible, especially where the orientations of the principal incremental strains are oblique to the competent layers. Under these circumstances it is common for the limbs of early

Figure 12.8. *Folded boudins of competent amphibolite in a less competent banded biotite gneiss, Lower Pennine nappes, Ticino, Switzerland.*

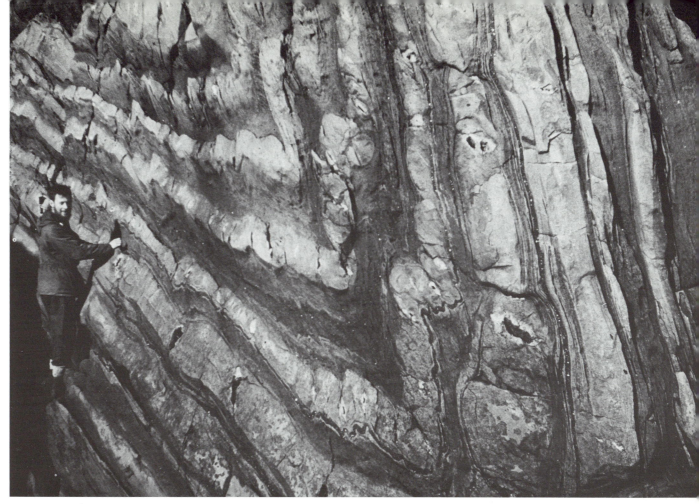

Figure 12.9. Folded boudins in Moinian metasediments, Isle of Mull, Scotland. The paler, more competent layers are metamorphosed sandstones, the less competent dark layers are muscovite–biotite metapelites.

Figure 12.10. Imbricated boudins in amphibolitic gneisses, Lower Penninic nappes, Ticino, Switzerland. The necks between the original boudins have acted as weak zones: when the boudins were later compressed they guided the development of the imbricate structure.

Figure 12.11. *Intrafolial folds in competent, pale, quartz–felspar gneiss bands surrounded by less competent biotite schist, Lower Penninic nappes, Cristallina, Ticino, Switzerland. The limbs between the fold hinges have been strongly thinned as a result of layer parallel stretching after layer parallel folding.*

Figure 12.12. *Intrafolial folds in competent calc–silicate rock surrounded by incompetent marble, Khan Gorge, Namibia.*

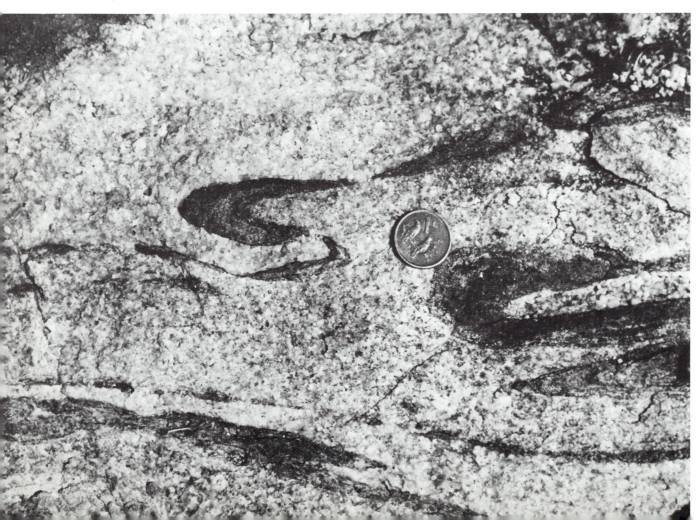

formed folds to be preferentially stretched, thinned and perhaps to develop boudin structure. In strongly folded terrains of metamorphosed rocks it is quite common to find remnants of fold hinges in competent layers contained in a matrix of incompetent material. These are known as **intrafolial folds** and they indicate a period of fold formation followed by a period of strong extension leading to a rupture of the fold wave train. The folds are sometimes more or less connected by stringers of strongly thinned competent rock (Figure 12.11) or they may be isolated so that the fold hinges are disconnected and apparently free swimming in the incompetent matrix (Figure 12.12).

We will now extend our discussion of some of the geological consequences of progressive deformation by considering the reaction of a competent layer to changing two-dimensional incremental strains. Figure 12.13 shows a plan view of part of a competent layer undergoing progressive strain, and A illustrates the effect of the first increment of deformation. The competent layer shows a maximum shortening in one direction and a maximum elongation perpendicular to the shortening. The layer is therefore likely to develop boudinage structure with the boudin axes aligned in direction b_1 and to form folds with hinges in direction f_1 according to the principles we established in Session 4 (Field 2 strain ellipse, Figure 4.11). Figure 12.13B, C and D illustrates the changing geometry of the finite strain axes (firm lines) and the changing orientation of the boudin necks and fold hinges resulting from a non-coaxial superposition of further strain increments. The initially perpendicular boudins and folds progressively lose their perpendicular relationship as both directions rotate towards the principal axes of successive incremental strains, and both take up positions which do not coincide with the axes of the finite strain ellipse. The movement of the boudin axes from the direction of the minor axis of successive finite strain ellipses (0° to 7° to 16° to 27°) is more marked than the movement of the fold axes away

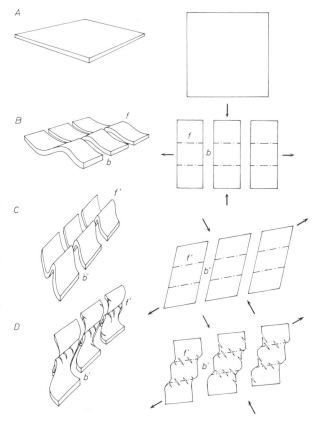

Figure 12.14. *Geological implications of a changing strain increment sequence: A, original state of competent layer; B, first part of strain history with the development of orthogonal boudins b and fold hinges f; C, later part of strain history with distortion of early formed structures to new non-orthogonal orientations b' and f'; D, later part of strain history where the folds and boudins are still active but take up incremental strains in an oblique fashion.*

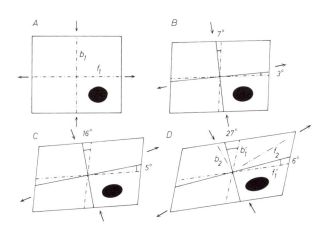

Figure 12.13. *Distortion of the axial directions of an early incremental strain ellipse (A, dashed lines), by successively changing incremental strains (B, C and D). The incremental strain directions are indicated with arrows, the finite strain for each successive increment is shown, with finite axial directions indicated by the continuous lines. Directions b_1 and f_1 refer to initial boudinage and fold axis directions, b_1' and f_1' show these directions at the end of the strain sequence. Direction b_2 and f_2 show potential late stage boudins and folds forming as a result of the last increment of deformation.*

from the direction of the major axis of the finite strain ellipses (0° to 3° to 5° to 6°). Although Figure 12.13 shows the geometric implications of one particular incremental system, the effects seen here hold true for many of the types of incremental strain systems arising during the contraction of competent layers. The geological consequences of these geometric changes are:

1. Folds and boudins initiated early in a deformation sequence are unlikely to show perpendicular relationships at the end of the sequence. Figure 12.15 illustrates an example of a naturally deformed rock with such geometry. The outcrop surface consists of a layer of more competent limestone enclosed in a less competent shale. The layer is folded and also shows the development of quartz–carbonate-filled extension fissures. The traces of the fold axes and extension fissures on the bedding surface are not perpendicular because of a clockwise change of incremental axes during the deformation sequence.

2. Fold hinge lines are likely to be fairly closely oriented to the major axis of the finite strain ellipse on the folded surface, although there is likely to be some departure from true parallelism. The major and minor axes of the two-dimensional strain ellipse may lie close

Figure 12.15. *Folded bedding surface of competent limestone, with fold axes inclined from upper right to lower left, cross-cut by quartz–carbonate extension fissures. The two structural features are not perpendicular because of the super-position of non-coaxial strain increments (cf. Figure 12.14C). Curtain Rock, Ilfracombe, N. Devon, UK.*

Figure 12.16. *Boudins in competent sandstone beds with zones of en-echelon extension fissures aligned along early initiated boudin necks. Culm Series, Bude, Cornwall, UK.*

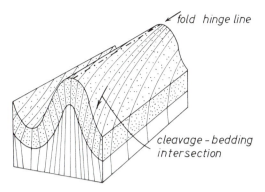

Figure 12.17. *Geological implications of Figure 12.13D with the principal XY plane of the strain ellipsoid, marked by cleavage, obliquely crossing the fold. The cleavage–bedding intersection lineation is no longer parallel to the fold hinge line (cf. Figure 10.21).*

to the Y and Z directions of the three-dimensional finite strain ellipsoid. In Session 10 we saw that if the finite strain shows sufficient shortening (>30%) along the Z-direction slaty cleavage or schistosity forms parallel to the XY plane of the strain ellipsoid. As a consequence of the non-alignment of the Y-direction with the fold hinge direction, the relationships between cleavage orientation and fold geometry will be more complex than those discussed in Session 10 and illustrated in Figure 10.21. The cleavage planes will be slightly skewed across the fold and the cleavage–bedding intersection line will no longer be parallel to the fold hinge. This effect (Figure 12.17) is not uncommon in strongly compressed terrains. If we examine closely Figure 4.12B, an example used previously to illustrate structure forming in a field 2 strain ellipse, it will be seen that the main extension veins are not absolutely perpendicular to the fold hinge lines (see 1, above) and that the trace of the cleavage on the folded bedding surface (seen best on the sunlit upper surface of the fold) is not parallel to the fold hinge line.

3. The folds initiated during the early stages of deformation (Figure 12.13A, f_1) may become so oblique to the main shortening directions of late increments that they are unable to continue to actively fold the layer (Figure 12.13D, f_1'). In such circumstances new folds may be initiated with axial directions oriented perpendicular to the directions of maximum incremental shortening (Figure 12.13D, f_2). For such superposed folding to develop the competent layer should not possess a close array of early folds because such a well-formed fold system gives rise to significant increase in the strength of the layer along the fold axis direction. This strengthening effect is like that produced in metal sheets by forming regular corrugations. It is very difficult to buckle corrugated steel in a direction transverse to the corrugation axes.

4. Extension fissures or boudin necks formed early in a deformation sequence tend to be rotated away from the directions of the minor axis of the finite and late incremental strain ellipses at a rather rapid rate. They may subsequently be superposed by or intersected by late developing fissures or boudins with directions governed by the late strain increments (Figure 12.13D, b_2). Extension vein systems are particularly sensitive indicators of changing increment systems,

and in Section 13 it will be shown how their geometry can be used to evaluate the incremental strain sequence. The degree of rotation of the early formed extension fissures illustrated in Figures 12.13 and 12.14C has been determined using the assumption that the rock layer behaves in a completely homogeneous manner during the superposition of late strain increments. This assumption does not generally hold true, because the heterogeneities induced by the early formed extension structures and folds localize the later strains in such a way that they are heterogeneously dissipated along the earlier formed instabilities. The early formed extension structures may continue to be active during later strain increments, but the direction of late wall rock separation becomes obliquely inclined to that of the initial structure (Figure 12.14D). This oblique opening can also lead to the development of new extension systems propagating obliquely from the walls of the initial fissure and showing an en-echelon arrangement (Figures 12.14D, 12.16). The oblique contractions taking place during the superposition of late strain increments can be accommodated by continued closure of the early fold hinges, but with a maximum shortening aligned in a non-perpendicular direction to the fold axes (Figure 12.14D). This oblique shortening can lead to the initiation of new en-echelon fissure systems localized along fold hinge lines.

Answer 12.4★

To produce satisfactory answers to this question relatively rapidly it is necessary to program the matrix multiplication of successive increments with a reiterative loop so as to produce a data print out for each successive increment. The results are presented graphically (Figures 12.18, 12.19, 12.20) and in a form which shows the changing finite strain ellipse shapes and orientations (Figure 12.21). There are a number of different types of strain sequence, depending upon the value of the incremental rotation ω_i which accompanies the fixed distortion ($e_{i1} = 0.1$).

1. Progressive pure shear, $\omega_i = 0$: The increments are superposed coaxially and all states of finite strain are irrotational (Figure 12.19). The finite strain distortion grows very rapidly, and, of all the sequences to be discussed, this shows the most rapid growth of ellipticity of the strain states (Figure 12.18). The major semi-axis of the finite strain ellipse increases as a power series, such that after n increments

$$1 + e_1 = (1 + e_{i1})^n$$

and the ellipticity R of the ellipse is given by

$$R = \frac{1 + e_1}{1 + e_2} = (1 + e_{i1})^{2n}$$

2. Sequences between progressive pure shear and progressive simple shear, $0 < \omega_i < 5.45°$: The finite strains grow rapidly, although not as rapidly as those of pure shear (Figure 12.18). The orientations of successive finite strain ellipses change, but the rate of change decreases as the finite strain increases (Figure 12.20). This is because the effect of the negative distortional rotation increment is greater than that of the positive body rotation increment. As deformation proceeds the positive and negative values

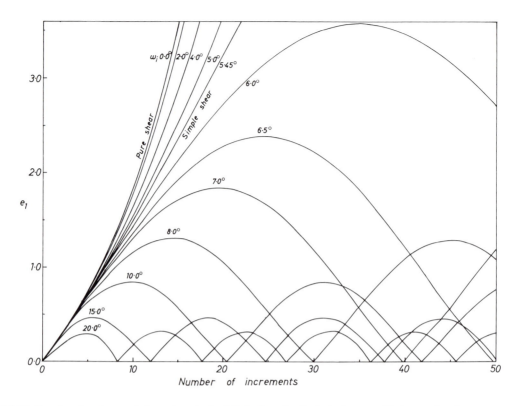

Figure 12.18. Graphs showing variations in principal finite longitudinal strains e_1 with increasing increment number for different values of incremental rotation ω_i.

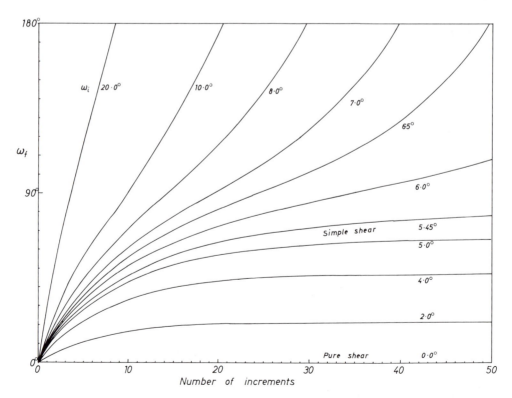

Figure 12.19. Graphs showing variation in finite rotation ω_f with increasing increment for different values of incremental rotation ω_i.

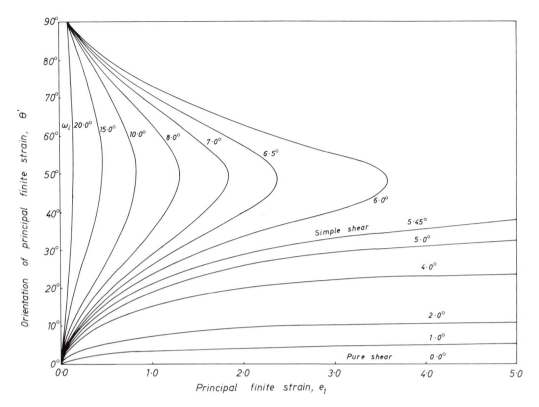

Figure 12.20. Graphs showing relationships between values of principal finite strain e_1 and orientations of the principal strains for different values of incremental rotation ω_i.

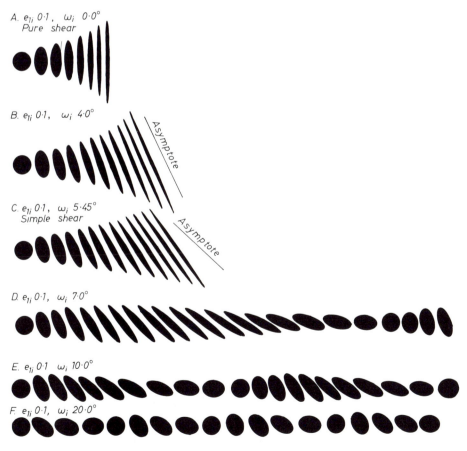

Figure 12.21. Representation of finite strain ellipse sequences for varying amounts of incremental rotation ω_i.

of these components approach numerical equality and the long axis of the finite strain ellipse becomes progressively more stabilized towards a direction asymptote. With small increments of body rotation (ω_i) this asymptote lies close to the direction of the maximum elongation of the strain increment, whereas as the rotation increment becomes closer to a value of 5·45°, the asymptote becomes closer to a 45° orientation.

3. Progressive simple shear, $\omega_i = 5\cdot45°$: The finite strains grow quite rapidly and the ellipticity of the finite strain ellipse after $2n$ increments is always more than double that for n increments. We have already met and graphically recorded this effect in the simple shear experiments performed in Session 2 (values of R in Figure 2.10). With increase in increment number the maximum finite rotation approaches a limit of 90° (Figure 12.19), and the directions of the principal finite strains asymptotically approach a value of 45° (Figure 12.20). In terms of the simple shear experiments carried out in Session 2, it can be seen that this asymptote corresponds with the direction of the card surfaces. These card deck experiments showed that the long axis of the finite strain ellipse initiated in a direction at 45° to the card surface, the strain axes became closer to the card surfaces with increasing shear, but could never pass through the card surface direction.

4. Sequences between progressive simple shear and body rotation, $5\cdot45° < \omega_i < \infty$: These strain sequences show a surprising geometric property not seen in those we had discussed so far. This property is that of periodic straining and unstraining and produces a pulsation in the strain state with increase in increment number. In each sequence the ellipticity of the finite strain progressively increases, the rate of increase gradually declines to zero where a maximum distortion is reached, and then the ellipticity decreases to its initial unit value. On reaching this unit value the sequence recommences and the strain cycle is repeated (Figure 12.18). The frequency of period and value of the peak strain are both functions of the values of ω_i. At values slightly greater than $\omega_i = 5\cdot45°$ the cycle length is long and the peak strain (e_1) of the strain ellipse is high, whereas at high values of ω_i the cycle length is short and the peak strain is low (Figures 12.18, 12.20). The directions of principal strains show a continuous positive motion, and the peak strain is attained where the principal strains become oriented at 45° to the incremental strain axis (Figure 12.20). The finite rotation increases continuously but with varying rate, and is unbounded (Figure 12.19).

5. Progressive body rotation, $\omega_i \approx \infty$: At very large values of incremental rotation the deformation effects of the progressive strain sequence are completely outweighed by the rotational effects, and the initial circle effectively rolls in constant body rotational motion.

Particle path line fields

Figure 12.21 gives a summary of the various finite strain ellipses arising from constant incremental distortion, and for each of the types A–F the particle path line fields have been computed (Figure 12.22). Because the incremental displacements and strains remain constant, these fields

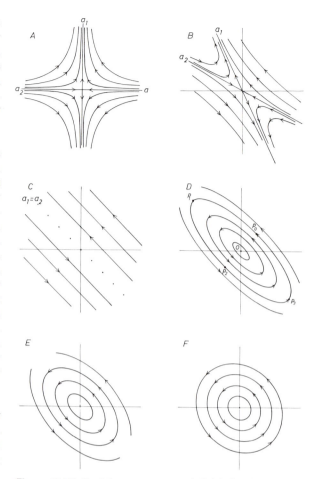

Figure 12.22. Particle movement path fields for the six deformation sequences illustrated in Figure 12.21.

remain constantly fixed at all stages of the deformation sequence.

The particle path line fields for pure shear consist of rectangular hyperbolae (Figure 12.22A). The two perpendicular asymptotes are parallel to the principal strain directions of the system.

Deformation sequences between those of pure shear and simple shear also have hyperbolic path fields, but the asymptotes are not perpendicular and do not coincide with the x- and y-coordinate directions (Figure 12.22B). There is always a convergence of particles towards one asymptote (a_1) and a divergence away from the other (a_2). You will see that the particle movement path fields we constructed from Figure 12.1 and illustrated in Figure 12.4B is of this type. The asymptote a_1 represents the line towards which the major axes of the finite strain ellipse converge (cf. Figures 12.21B, 12.22B). The bisector of the acute angle between the two asymptotes is always inclined at $+45°$ to the y-axis, and the value of the acute angle is a function of the incremental rotation component. The angle is 90° where $\omega_i = 0$ and decreases to zero as ω_i approaches 5·45°. In simple shear (Figures 12.21C, 12.22C) the two asymptotes converge to form a common line along which no displacement occurs (the movements towards and away from the origin along a_1 and a_2 respectively are cancelled out) and this line becomes parallel to the simple shear plane orientation of the system. All the previously hyperbolically curved displacement paths now degenerate into a series of straight lines. The shear plane now takes on the properties of the asymptote a_1 in that it forms the finite strain ellipse maximum strain asymptote oriented at 45° to the axis of the incremental strain ellipses.

Where the incremental rotation exceeds that for simple shear it has been shown that the strain sequences show a pulsatory growth and decline. The particle movement path fields for the sequences illustrated in Figure 12.21D, E and F are shown in Figure 12.22D, E and F. The fields all show elliptical particle motion. These imply that any point initially positioned at p_0 (Figure 12.22D) undergoes a movement cycle through positions p_1, p_2, p_3 and then returns to its initial position. All points undertake a similar cycle and the geometric result is that, after a given number of increments, all points have returned to their initial position. As the point moves from p_0 to p_1 the length of the line Op_0 increases to a maximum length Op_1. The direction Op_1 therefore coincides with the direction of maximum finite longitudinal strain in the sequence. In a similar way any point initially located at p_1 moves to p_2, length Op_1 decreases to Op_2 and becomes the direction of minimum finite longitudinal strain. It should be evident from this discussion that the shapes of the particle path ellipses have ellipticities related to that of the maximum finite strain ellipticity in the form of $(1 + e_{f1})/(1 + e_{f2}) = R = (Op_1/Op_2)^2$.

As the rotation increment increases the ellipticity of the particle paths decreases and approaches a circular form as the movement approaches that of a body rotation (Figure 12.22F).

It is remarkable how all the different types of constant incremental strain patterns fit so beautifully into mathematically continuous and elegant scheme, with particle path fields forming a continuous spectrum of classic conic section shapes. To a geologist, the obvious question that arises is "Are all these strain sequences possible and, if so, are some more common than others?". In principle all are possible but those that are bounded by pure shear and simple shear seem to be the most common. Although sequences lying between simple shear and body rotation do occur, and in Session 14 we will provide good evidence for the existence of these types, the strain pulsations rarely, if ever, pass through more than one cycle. Most large scale crustal geological processes are dominated by types of displacement close to pure shear (crustal thickening and thinning by predominantly surface parallel displacements) or by simple shear (underthrusting of one plate beneath another during subduction processes of sideway movement along transverse shears). Although local displacements can perturb these systems to give rise to vorticities greater than simple shear, such effects do not appear to have general or regional significance.

KEYWORDS AND DEFINITIONS

Intrafolial folds Folds in competent layers that have been subjected first to layer shortening, and then to a later layer elongation. They show highly thinned or boudinaged limbs and are found as discontinuous isolated hinges in the incompetent matrix (Figures 12.11, 12.12).

Particle movement path The line traced out by the movement of a single point in a body undergoing progressive displacement. The lines traced out by all the points in the body define the **particle movement path field**.

Progressive deformation The changes taking place in the internal distortions and rotations of a body as a result of progressive displacement of its constituent particles. The incremental distortions and rotations are recorded by the shape and rotations of the **incremental strain ellipse**: this is the ellipse derived from any circle of unit radius drawn on the body at the start of the increment in question. The major and minor axes of this ellipse have semi-axis lengths $1 + e_{i1}$ and $1 + e_{i2}$, where e_{i1} and e_{i2} are the **principal incremental strains** and the difference in orientation between the initial and final position of these axes defines the **incremental rotation** ω_i. The rates of change of these components define the **principal strain rates** $\dot{e}_{i1}$ and $\dot{e}_{i2}$, and the **vorticity** $\dot{\omega}$.

Progressive displacement The changes taking place in the displacement of points in a body undergoing a change of spatial position. The **incremental displacements** are recorded by the **incremental displacement vectors** linking any point at one stage in the displacement with its position at some later stage. The **rates of change of the components of the incremental strain matrix** are $(\dot{a}_i, \dot{b}_i, \dot{c}_i, \dot{d}_i)$, and define the **displacement rate** of the system.

KEY REFERENCES

Elliott, D. (1972). Deformation paths in structural geology. *Geol. Soc. Am. Bull.* **83**, 2621–2638.

The concepts of progressive deformations are developed and some practical ways of measuring incremental strains from field observations are established. This paper would form good bridge reading from Session 12 to the last two Sessions of this book.

Flinn, D. (1978). Construction and computation of three-dimensional progressive deformations. *J. Geol. Soc. Lond.* **135**, 291–305.

This paper discusses the various graphical techniques for representing finite and progressive deformation concluding that the orthogonal logarithmic ratio plot provides the most convenient method for recording data. It considers, from a theoretical viewpoint, different types of deformation paths.

McKenzie, D. (1979). Finite deformation and fluid flow. *Geophys. J. R. Astr. Soc.* **58**, 689–715.

This paper discusses the problems of different types of flow involving differing amounts of vorticity, and relates the resulting total strains to the development of anisotropy in mantle material undergoing convective flow.

Means, W. D., Hobbs, B. E., Lister, G. and Williams, P. F. (1980). Vorticity and non-coaxiality in progressive deformations. *J. Struct. Geol.* **2**, 371–378.

The geometry of deformation sequences with changing directions of rate of strain tensor is described with emphasis on mathematical analysis.

Pfiffner, O. A. and Ramsay, J. G. (1982). Constraints on geological strain rates, arguments from finite strain states of naturally deformed rocks. *J. Geophys. Res.* **87**, 311–321.

This work describes strain paths involving different amounts of distortion and vorticity and establishes the complete spectrum of steady state flow possibilities. It then proceeds to analyse the most likely types of geologically important flow systems to evaluate strain rates from observed finite strains. This paper would provide a development to the conclusions arising from Question 12.4★.

Ramberg, H. (1975). Particle paths, displacement and progressive strain applicable to rocks. *Tectonophysics* **28**, 1–37.

In this work the possibilities of defining a general two-dimensional progressive plane strain in terms of combinations of pure shear and simple shear are described. Particle path fields are investigated and lead to the possibility that finite strain states may undergo pulsatory change. This paper would form a good background to Question 12.4★.

Ramsay, J. G. (1967). "Folding and Fracturing of Rocks", 568 pp. McGraw-Hill, New York.

Several sections of this book will provide good general background to this Session, especially pp. 114–120, 174–177, 322–332.

Stringer, P. and Treagus, J. E. (1980). Non axial planar S_1 cleavage in the Hawick Rocks of the Galloway area, Southern Uplands, Scotland. *J. Struct. Geol.* **2**, 317–331.

The cleavage planes described here are consistently oblique to the axial planes of the major folds, and the phenomenon is related to the non-orthogonal relationships of fold axes to the bulk strain axes. The authors suggest that the non-orthogonal relationships could arise as a result of a sequential strain history of non-coaxial increments, or be the result of folds generated on surfaces obliquely inclined to the principle finite strain axes. They favour the latter explanation.

SESSION 13

Measurement of Progressive Deformation

1. Extension Veins

The geometric forms of extension fissure walls and internally crystallizing vein fibres developing during an incremental deformation are established. The special features of the four commonest types of vein structure (syntaxial, antitaxial, composite and stretched crystal) are described. The fibre patterns of veins can be used in a general way to evaluate changing incremental extension directions and in some examples it is possible to compute values and orientations of the successive incremental strain ellipses. Methods are proposed for computing finite strains from measurements of incremental strain. The geometric features of chocolate tablet boudinage is related to incremental strain sequences. The vein and fibre geometry of shear veins is compared and contrasted with that of extension veins.

INTRODUCTION

In the last Session we established the concepts of progressive deformation from a theoretical viewpoint and gave some indications as to how changing incremental strains might be observed in the field as complex successions of superimposed structures. The next two Sessions aim at establishing practical methods whereby we can determine the sequential changes of incremental strains during a continuous phase of deformation; how we can determine the directions of successively imprinted strain increments and also, in certain terrains, how we are able to determine the actual incremental principal strain values for the differently oriented strain increments. In making these measurements, the whole deformation history of a region is revealed, and it becomes possible to appreciate tectonic evolution in a way that one is not able to do from a study of finite strain state alone.

The measurement of progressive strain sequences is a comparatively recent development in structural geology. However, the principles now seem well established, are accepted as standard, and we believe that they are sufficiently practical and important as to justify devoting two Sessions of this book to their exposition. Although some of the more elaborate analytical methods which will be described in the next two sessions are complex and time consuming, many of the basic principles are of extreme simplicity and can be used in the field to provide valuable data during the course of a regional structural exploration. Like many of the techniques described in these books, before starting any research investigation it is always necessary to consider very carefully what are the particular problems which need to be solved, and to select the technique that is appropriate to obtain solutions.

In Session 12 we found that the geometric forms of extension structures, such as boudin necks and extensional veins, can provide very sensitive indicators for establishing progressive strain increment sequences. Session 13 will be aimed at following up this approach. In Session 12 we discussed how the geometry of an extension fracture system could be rotated as a result of the superposition of increments; now we wish to discuss this feature in more detail, and look into the way a fissure progressively opens, propagates sideways and becomes filled. We should first contrast the types of fracture and fissure systems that develop during a progressive deformation sequence and those fractures in rocks known as joints. **Extension fissures** developing as a result of mechanical rupture during deformation usually initiate in a direction perpendicular to the maximum incremental longitudinal strain acting at the time of initiation. As a result of further increments of strain they generally continue to open, either in a direction normal to their walls, or obliquely to their walls, depending upon the orientation of the incremental strain ellipse. In most tectonically active systems they are fluid filled. Because the fissure is a site of low pressure and because the fluid phase contains material in solution which can be deposited in a low pressure zone, the fissure is filled with crystalline material at the same time as it is progressively opening, and so forms a **vein**. In contrast, **joints** are usually very fine cracks which develop generally as a result of crustal uplift and the associated release of elastically stored stresses.

Because the strains involved in this process are very small and, because the stored stresses are related to a unique set of principal stress axes it follows that joints are usually very fine cracks without the crystalline infilling typical of vein systems, and that joint sets are geometrically rather simply and constantly oriented.

Figure 13.1 illustrates schematically the progressive evolution of a tectonically active extension system and shows the typical geometric forms of the extension fissures. The earliest formed fissures are oriented perpendicular to the direction of maximum incremental longitudinal strain (Figure 13.1A) and the wall separations are sub-perpendicular to the walls (a–a'). If the directions of subsequent increments do not coincide with those of early increments three geometric features arise (Figure 13.1B): first, the initial fissure tips propagate sideways but in a direction controlled by the later strain increment axes; second, the dilation across the early formed fissure continues to increase, but the new dilation direction is parallel to the maximum strain of the new increments; third, the blocks between the opening fissures show a tendency to rotate. With further change of the incremental strain the fissure tips continue to propagate sideways, but in a direction more oblique to that of the initial fissure (Figure 13.1C). The earliest initiated vein now has an orientation that is highly inclined to the latest developed principal increment directions. It may continue to open in an oblique way, but often it appears, from observations of natural systems, that new veins are initiated which cross-cut and displace the earlier formed system (Figure 13.1C). The overall geometry of fissures developed during a changing increment sequence is characterized by sigmoidally curving systems often with curving splays at the fissure tips (Figures 13.2, 13.3), with later developed cross-cutting fissures oriented in directions of the tips of the sigmoidal veins. We have already met one such sequence in Session 2 (Figures 2.11, 2.13) when we discussed the evolution of vein systems developed during progressive simple shear and it might be instructive to return to this example and offer an explanation of this geometry in terms of the incremental distortions and incremental rotations arising during progressive simple shear. It should be stressed that the intersection of fissure systems only indicates unambiguously the time relationships between the successive systems and is not always connected with a sequential progressive deformation. The different systems of extension could be formed during completely unrelated deformation events. For the extension fissures to be related in a progressive sequence there should be a geometric spectrum of orderly intersections, the orientations linked by the sub-parallelism of late fissures with the tip geometry of earlier fissures.

The progressively opening fissures become infilled with crystalline material, the commonest mineral species being quartz, calcite, chlorite, epidote and albite. A particularly striking aspect of this material which develops the mineral *vein* is its *fibrous crystal morphology*. This fibrous habit is quite distinct from the normal habit shown by a crystal growing freely into an open space. Although freely growing crystals are elongated they rarely show the very elongated form characteristic of vein fibres and which commonly show extreme length to width ratios (see Figure 13.4). Another important feature of vein fibres which contrasts with that of free growing crystals is that the length of the fibre generally has no special orientation with respect to the crystallographic directions of the crystal making up the fibre. With some types of fibrous veins, the crystallographic orientations of the fibre can be shown to be directly inherited from those of the wall rock crystals (Figures 13.10, 13.11). The fibrous crystals in a vein may be straight or curved. If the fibres have a curved form the crystallographic directions remain fairly constant through the curved fibre, and do not change with the fibre orientations (Figure 13.5). This shows that the curved fibre form is not generally the result of deformation of a previously straight fibre: it must be an original growth feature. The geometry of the fibres of veins can often be linked very convincingly with the displacements and dilations that have taken place across the vein. Figure 13.6 shows various sets of intersecting veins in a limestone from the Helvetic Alps. The latest set

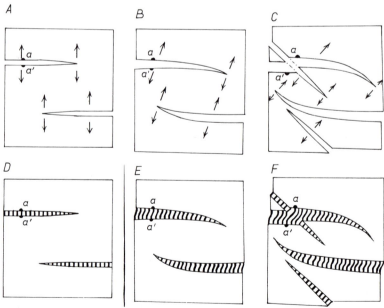

Figure 13.1. A, B *and* C *show the geometry of progressively opening extension fissures;* D, E *and* F *illustrate the forms of crystal fibres forming syntectonically with the progressive opening. Points* a *and* a' *were once in contact and become separated during the progressive strain—the vector joining* a *to* a' *gives the local displacement vector of the vein walls—compare this with the fibre geometry.*

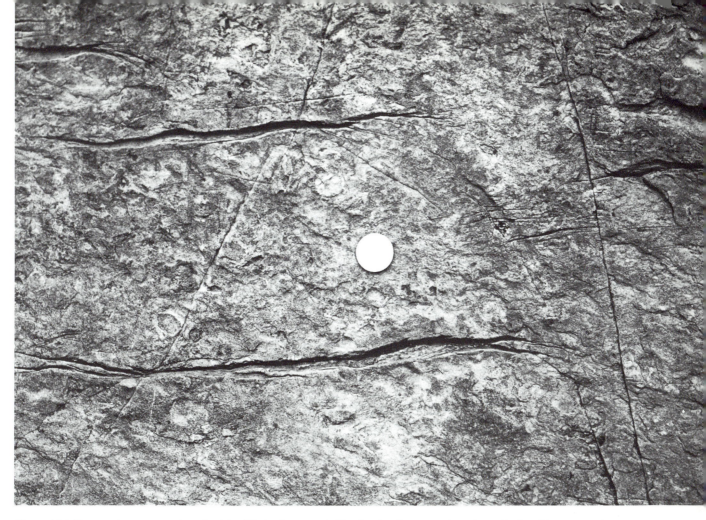

Figure 13.2. Progressive propagation of calcite filled veins revealed by the development of curving vein tips. The progressive maximum extensions rotated in a clockwise sense. Wildhorn nappe, Helvetic Alps, Switzerland.

Figure 13.3 Sequential planar fracture development indicating a clockwise rotational sense. Calcite veins in marly limestone, Wildhorn nappe, Switzerland.

Figure 13.4. Thin section of fibrous calcite (crossed polars, × 150) in a composite vein in slate, Llanberis, North Wales. The vein shows a thin marginal development of chlorite and a well-developed line of inclusions along the median surface. The decreasing length to width ratio of the calcite fibres on the left centre of the photograph arises because the fibres are curving out of the plane of the section.

Figure 13.5. Detail of curved calcite fibres (crossed polars, × 200) in a thin section of calcite–chlorite composite fibre vein, Dinorwic, North Wales. The optic orientations are unaffected by the change of direction of the fibre axes. The fibre curvature is therefore the result of growth and not subsequent deformation.

of veins runs horizontally across the block and shows well-developed fibrous crystals of quartz and calcite. The dilations that have taken place during the formation of these late veins can be inferred from the missmatch relationships of the early formed veins. The fibres link points that were once in contact. In this case the fibre gives the displacement vector across the vein. The same conclusion holds true if the fibres have a curved form. Figure 13.7 shows a fibrous quartz–calcite vein cutting a sandy limestone. A sandstone layer in the lower middle part of the photograph is displaced upwards and to the left across the vein, and the fibres link points which were once in contact.

We can now review the sequence of fissure propagation in the light of the geometry of the fibres filling the fissure space and building the vein. Figure 13.1D, E and F illustrates the sequential formation of fibres based on the geometrical rules discussed above, that (1) fibres link points on either wall that were once in contact and that (2) they grow in the direction of displacement: if the displacement pattern is simple the fibres are straight, whereas if the displacement is the result of a complex sequence of changing oblique extensions across the vein the fibres will be curved. The further constraint placed on the resulting fibre geometry of Figure 13.1 has been to allow new fibres to grow in the central part of the vein. We will see later that this type (known as syntaxial fibre growth) is just one of several possibilities. From the result of combining all these constraints it will be seen from Figure 13.1F that the final fibre geometry is complex, that there are clear geometric links between fibre geometry and vein form and that these features provide a key to determining the incremental strain history of the block.

Interpretation of vein increments

Question 13.1

From Figure 13.7 determine, in general terms, the changes of orientations of the principal axes of the incremental strains leading to the vein geometry. Suggest relative values for the incremental longitudinal strains in different directions.

Question 13.2

Figure 13.8 shows a thin section of a complex calcite vein array developed in a fine grained argillaceous sandstone. How does the fibre geometry differ from that used to develop the model of Figure 13.1F? Suggest a likely incremental strain sequence to account for this vein pattern.

Types of fibrous vein systems

Fibrous vein systems are normally developed in rocks of differing competence undergoing deformation in regions of low to middle grade metamorphism. They are especially common in environments where deformation takes place under anchimetamorphism to middle greenschist facies metamorphism. Fibrous veins can sometimes be found in terrains of metamorphism above greenschist grade but they are uncommon and, where found, they generally show fabrics that suggest that the fibre structure was developed in a lower grade metamorphic environment. For example, the fibres sometimes apparent in hand specimen size pieces of rock of overall amphibolite grade metamorphism are

Figure 13.6. *Fibrous quartz–calcite veins in limestone. The intersection relationships and vein offsets can be used to establish a development sequence. The fibres that are developed in the last, sub-horizontal vein set connect points on the vein walls which were originally in contact.*

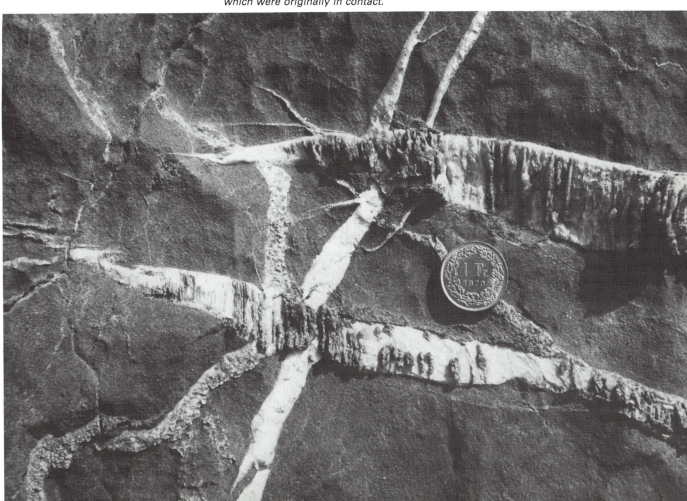

Figure 13.7. Curving quartz–calcite fibres in a siliceous limestone, autochthonous cover of the Aar massif, Gemmipass, Switzerland. Note how the displacements across the vein of the vertical, more massive, sandy bed in the middle of the photograph relate to the fibre shapes.

Figure 13.8. Thin section (crossed polars, ×100) of calcite veins cutting an argillaceous limestone, North Spain. The deformation increment sequence is complex. See Answers and Comments, p. 253.

seen in thin sections to be totally recrystallized into equi-dimensional polygonal crystals. In grades of metamorphism where temperatures in excess of 350°C occur fibrous crystal forms, showing high ratios of crystal surface area to volume, are not thermodynamically stable.

Fibrous vein systems are generally most abundant in the more competent layers of a lithologically stratified succession of rocks. In sedimentary assemblages they usually form preferentially in competent sandstone or limestone layers. Vein systems are only developed strongly in incompetent materials where these materials have suffered especially high local strains (and perhaps high local strain rates). For example, differential slip between competent layers in certain types of folds, termed flexural slip folds, may be localized in the incompetent strata, and these incompetent strata might show a higher frequency of veins

than is seen in the more competent strata. However, if two rock types have been subjected to more or less the same total strain, it is the more competent layer which will show the greatest extent of vein formation.

There are a variety of different types of vein systems with differing fibre geometry, and in order to correctly interpret the strain sequences which led to different geometric organization of the fibres it is necessary to be able to differentiate these types illustrated schematically in Figure 13.9.

A. Syntaxial fibre veins

The general geometric features of these veins are illustrated in Figure 13.9A. The material filling syntaxial veins is compositionally identical or very close to that of the vein

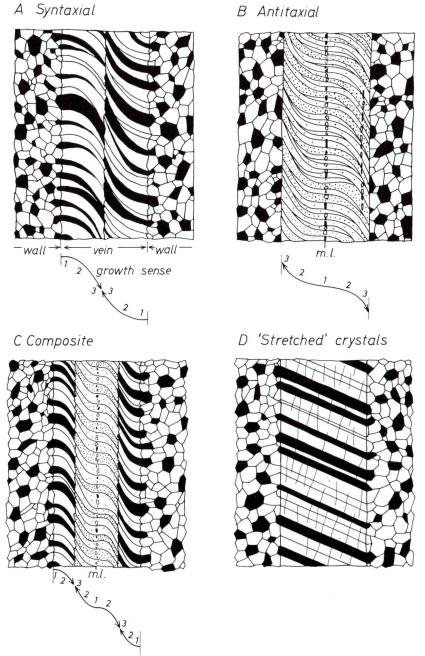

Figure 13.9. *Principal features of the four main types of fibrous vein systems showing the relationship of the composition and crystallographic orientations of vein and wall rock crystals. The black and white unstippled areas are crystals of the same species but with differing crystallographic orientations. The stippled areas are crystals of another species, and their crystallographic orientations are indicated by differing intensities of stippling.*

walls; quartz veins in sandstones and calcite veins in limestones are good examples of this compositional link. Where the wall rock has several components it is usual for the syntaxial vein that develops in this environment to show fibres of several compositions growing together in sub-parallel fashion. The fibres occur in two distinct groups attached to either wall of the vein respectively, and each group is crystallographically related to the corresponding wall rock crystals (Figures 13.9A, 13.10). In thin section, the optical orientations of fibres are seen to be parallel to those in the walls (Figure 13.11). The contact between the two groups of vein fibres is a more or less centrally located suture, and is a feature which can often be seen in the scale of a field outcrop (Figure 13.12). Where the fibre groups have a curved form the two sets show more or less mirror image curvatures. The individual curved fibre axes are always perpendicular to the vein walls at the vein–wall contacts and lose this perpendicular relationship towards the central suture. In some syntaxial veins certain of the crystal fibres become wider towards the central suture at the expense of their neighbours. This change in fibre width is best interpreted in terms of progressive growth sense from wall to centre of the vein (hence the name **syntaxial** as a result of certain crystals being crystallographically more favourably oriented for growth than their neighbours. Generally the crystals forming syntaxial veins are relatively inclusion-free (except at their contact with wall rock crystals—see Figure 13.11) and form clean vein crystals.

It sometimes occurs that more than one crystal species exists in the vein wall, and yet the vein developed is a syntaxial type and of monomineralic composition. In such examples the adjacent syntaxial vein crystals may show differing geometric features. Figure 13.13 shows an example of a vein of fibrous gypsum (so-called satin-spar gypsum) developed from a wall containing large gypsum crystals together with small (dark) calcites. Where the syntaxial overgrowth of fibrous gypsum is on a wall gypsum, the crystallographic orientation of the new fibres is very close to that of the wall rock parent, whereas the gypsum overgrowth developed on wall rock calcite shows a more extreme fibrous form with very variable crystallographic orientation.

The interpretation of progressive incremental strain directions of syntaxial veins accords with that used to develop the scheme of Figure 13.1. Although we often tend to think of most vein fillings as being governed by syntaxial geometry this conclusion is not supported from the internal geometry of the vein fibres, and the types described below are of equal abundance.

B. Antitaxial fibre veins

The geometry of these veins is illustrated schematically in Figure 13.9B. The crystal material making up the veins has a different composition from that of the vein walls, although usually there is a source for the vein filling near at hand. Thus antitaxial veins of calcite in shale often have a probable source of their calcite from nearby (0.1–50 m distance) limestones or marls. The geometry of antitaxial fibres shows a number of significant differences from that of syntaxial fibres, and these differences suggest a totally different type of vein growth. The vein crystals in antitaxial veins in contrast to those of syntaxial types can have no crystallographic relationships to the crystals of the vein walls, because the crystal species are not the same. The fibres do not therefore form two wall-related groups, but

Figure 13.10. *Thin section (crossed polars, ×60) of a syntaxial calcite vein in limestone from the Vanoise area of the W. Alps, France.*

Figure 13.11. *A detail from Figure 13.10 showing the crystallographic correspondence between vein fibres (right-hand side) and wall rock crystals (left). The vein crystals at the vein wall are heavily impregnated with foreign material. This material was probably brought into the initial crack during the influx of hydraulically active fluids creating the fissure, and was trapped by subsequent crystallization of first formed vein material. The later developed vein fibres are comparatively clean, containing much less included material.*

Figure 13.12. *A syntaxial quartz–albite–chlorite vein with well-developed suture line between the two groups of wall rock controlled crystals. Grandes Rousses, W. Alps, France.*

Figure 13.13. *Syntaxial vein growth of fibrous gypsum (satinspar) on wall rock gypsum and wallrock calcite (dark coloured). Thin section, crossed polars, × 100.*

Figure 13.14. *Antitaxial calcite veins in a matrix of micaceous phyllite, Col de Croix de Fer, W. Alps, France. The median line is well developed in the centre of the thickest part of the vein with symmetrically disposed calcite fibres on either side. The progressively developing incremental strains determined from fibre geometry show changes of maximum longitudinal strains with clockwise sense, and the veins intersection and propagation is in accord with this interpretation of fibre geometry.*

Figure 13.15. *Thin section (crossed polars) of antitaxial calcite veins in argillaceous flysch deposits of the Morcles nappe, one of the Helvetic nappes of the Valais, Switzerland. The sandstone at the top of the photograph shows no vein development. The median line in the vein on the right hand side is centrally and symmetrically disposed in the vein, whereas that in the vein in the middle of the photograph is not located midway between the vein walls. Enlargement ×20.*

run with crystallographic continuity from wall to wall (Figures 13.14, 13.15, 13.16). A line of wall rock fragments usually runs along a central or near central line of the vein, these pieces being incorporated as solid inclusions within the vein crystals (Figures 13.14, 13.15, 13.16). This structure is known as the median line (Figure 13.9B, *m.l.*) and appears to mark the debris derived from the walls of the first fissure or from material brought in during the influx of fluid (hydraulic fracturing) during crack initiation. In many examples of antitaxial veins the individual fibres keep a remarkably constant width, but where competitive fibre growth occurs it always produces an increase in fibre width away from the median line and towards the vein walls (Figure 13.17). In some antitaxial veins inclusion bands of wall-derived material in addition to that seen at the median line may be observed, these bands being often arranged in remarkably regular wall-parallel lines (Figures 13.9, 13.18). These trails appear to mark successive openings along the vein–wall contact with progressive evolution and often impart a "cloudy" or "dirty" appearance to antitaxial vein crystals. They suggest a periodic opening (with ripping off of wall rock particles) and crack filling, a process called the **crack–seal mechanism**. We will discuss later the mechanical implications of this process. All the geometric features of antitaxial veins support a vein accretion process by addition of new material along the vein–wall contact (Figure 13.9B, growth sense 1 to 2 to 3).

C. Composite fibre veins

Composite veins consist of two component crystal species zonally arranged so that one species forms a central zone of the vein, the other forming the two marginal zones of the vein (Figures 13.9C; 13.19, 13.20). The crystal species forming the marginal zones are the same species as those predominant in the vein walls (Figures 13.5, 13.9C, 13.19, 13.20) and are crystallographically attached to root crystals in the vein walls. In contrast, the crystal species forming the central part of the vein are those which are either of minor volumetric importance in the walls or which are only found in rock types outside the immediate environment of the composite vein. The composite veins of Figure 13.19 show marginal zones of white calcite fibres, attached to the calcites of the wall rock limestone, and central zones of quartz probably, derived from the isolated quartz sand grains or chemical breakdown of clay minerals in interbedded marls. Figures 13.4 and 13.5 illustrate composite veins with central calcite, and with marginal chlorite attached optically to the wall rock phyllosilicates. The central zone of composite veins consists of crystallographically continuous fibres containing a more or less centrally located median line of included rock fragments (Figures 13.4, 13.9C, m.l., 13.19). The geometric characteristics of this central zone are identical to that of an antitaxial vein system, whereas those of the two marginal zones is comparable with the two fibre groups of a syntaxial vein system.

Figure 13.16. Thin section (crossed polars) of an antitaxial calcite vein developed in a diagenetically cemented quartzite, Donegal, Eire. Note the centrally located median line, the optical continuity of the calcite fibres across the vein and the connection of the fibre ends with originally like points in the walls. Enlargement ×35.

Figure 13.17. Thin section (crossed polars) of the central part and median line of a quartz filled antitaxial vein, Parys Mountain, Anglesey, North Wales. Note that certain of the quartz crystals grow larger away from the median line at the expense of certain neighbouring grains. Enlargement ×100.

Figure 13.18. *Thin section (crossed polars) of a fibrous antitaxial vein containing elongate crystals of quartz and calcite, Windgällen, N. Aarmassif, Switzerland. The fibres contain regular inclusion bands which are parallel to the vein wall and contain crystals of wall rock affinity. In the centre the inclusions are of calcite and are derived from calcite in the wall, whereas the darker inclusion crystals are of chlorite and connect with chlorite rich parts of the wall rock (chamositic oolite material).*

Figure 13.19. *Composite veins of central quartz and marginal calcite developed in a deformed limestone, Leytron, Morcles nappe, Helvetic Alps, Switzerland. A centrally disposed median line is well developed in the zone of quartz.*

Figure 13.20. *Composite calcite–quartz veins developed in the neck zones of boudinaged belemnites, from the same locality as Figures 6.8 and 13.18. The tectonically produced vein calcite is crystallographically overgrown on the organically formed calcite of the belemnite.*

If the crystals curve as a result of progressive growth under conditions of changing incremental strains, this curvature shows a double sigmoidal form with mirror symmetry reflected along the median line. The vein crystal fibres show a perpendicular relationship to the vein walls along the fibre–wall contacts and along the median line, and they are oblique to the vein trend along the two sutures separating the central and marginal components. The geometry of the internal structure of these veins is in accord with the addition of new material along the two contact sutures between the two crystal species groups (see growth sense 1 to 2 to 3 in Figure 13.9C).

D. "Stretched" crystal fibre veins

The crystals in this type of vein system have a similar overall chemistry to those found in the vein walls, and in this respect have the same vein–wall compositional control as that found in syntaxial veins. In contrast to the double fibre grouping of syntaxial veins, however, "stretched" crystal fibres form a crystallographically continuous link from wall to wall of the vein (Fibres 13.9D, 13.21, 13.22). The original rock grains appear to be broken by the vein and the two separated parts are connected by an optically continuous seal. This relationship is shown especially well in Figure 13.22, where two parts of the same sand grain showing dark grey polarization tints are tied together by a quartz fibre. The origin of the fibre link between the two walls can be deduced from the presence of internal structures in the fibres which are absent in the parent grain.

The new material contains many completely or partially sealed cracks marked by zones of clear quartz or zones with bands of fluid inclusions respectively. The contacts between adjacent quartz fibres are, in detail, not planar, but show a saw-tooth form (Figures 13.23), the teeth-like irregularities showing a spacing of some 30–60 µm. The quartz within the fibres shows a general crystallographic continuity with that of the fragmented parents, however, in detail the fibre quartz shows small briquette-like elements generally elongated perpendicular to the fibre axis and showing widths of 30–60 µm (Figures 13.22, 13.23). These special geometric features shown by the vein fibres suggest a growth by the **crack–seal mechanism**. The fibres are built up by the development of successive small openings, followed by a period of cementation of the walls of these cracks by material derived from intergranular fluids. In stretched crystal veins, and in contrast to anti-taxial veins, this episodic activity of cracking and sealing appears to be rather haphazardly localized and rarely do successive cracks occur in exactly the same place. This effect may be related to the fact the the newly crystallizing quartz is particularly strong because it contains fewer dislocations than are found in earlier crystallized and slightly deformed quartz. Because of the geometrically irregular localization of the cracking process and the resulting lack of any systematic addition of new fibre material on particular surfaces, the curves in individual fibres cannot be simply related to the changes in the orientation of the principal strain increments producing the vein opening. The joins of the ends of any fibre gives the total displace-

Figure 13.21. "Stretched" crystal fibres of quartz in a quartz sandstone, Coombe Martin, Devon, UK. These fibres are of almost constant width, and their straight form indicates a constant incremental extension process.

Figure 13.22. "Stretched" crystal fibres in thin section (crossed polars) from a flysch type sandstone, Chaines Subalpines, W. Alps, France. The fibrous quartz shows a fine banded structure, the bands parallel to vein walls, and the optic axes of the fibres are clearly continuous with those in the broken and separated grains now situated in the vein walls. Enlargement × 100.

Figure 13.23. Detail of the internal structure of "stretched" quartz fibres. Note the characteristic saw-tooth contacts between adjacent fibres and the briquette-structure inside the fibres caused by the slight crystallographic misfits resulting from imperfect addition of new quartz and during the cementation part of the crack-seal process.

ment vector across the vein, but this vector cannot be separated into its individual progressive displacement components.

Measurement of incremental strains from fibrous vein systems

From a study of the curved patterns of the fibres in syntaxial, antaxial and composite veins it is possible, using a fibre growth sequence interpretation appropriate to the vein, to show how the incremental longitudinal strains change during a progressive deformation. An interpretation of the changes in direction of the maximum incremental longitudinal strain follows directly from the changing fibre orientation. The fibre lengths in a particular direction are related to the amount of incremental extension in that direction. In any vein it is possible to measure the fibre length δl developed between any two chosen fibre orientations (Figure 13.24A, directions a, b; $\delta l = \delta l_1 + \delta l_2$) and record how δl varies with orientation. If the exposed surface of a competent layer containing extension veins is sufficiently large, it may be possible to extend these calculations to evaluate the actual values of maximum principal incremental strain e_{i1} for each increment. To make these calculations it is necessary to have a measure of what length l of material existed in a certain direction at the time of the incremental increase of length δl. We can then compute e_{i1} from:

$$e_{i1} = \frac{\delta l}{l}$$

Figure 13.24B shows schematically several sub-parallel syntaxial veins and illustrates the technique for measuring the total length l existing at the time of a particular increment. A line is drawn across the surface with the average orientation $(a + b)/2$ where a and b are the values of angles used to delimit the incremental fibre length δl. In a syntaxial vein system the value of l is determined by summing the lengths in this direction of the walls together with the length across vein material of previously crystallized vein material and which existed at the time of the formation of the increment we are evaluating. In syntaxial veins this length does not include the central part of the vein, because this crystallized after the increment we are evaluating. For each average orientation it is possible to calculate the incremental longitudinal strain e_{i1} and record this graphically to give a clear visual representation of the changing strain history of the rock (Figure 13.24C). The measurements which are necessary to compute values for the existing length l at the time of development of any increment vary with the type of vein and Figure 13.24 (D, E and F) illustrates the different types of length summations required for syntaxial, antaxial and composite veins. In practice the most convenient way to undertake such a succession of measurements is to first construct isogonal lines on the curving fibres (lines of equal fibre orientation). It is then possible to obtain more accurately the average values for both l and δl and, by making a sum of measurements over several parallel cross-section lines taken at different positions in the system, any local variations in fibre development and vein orientation can be smoothed out. The main problem arising from the application of this

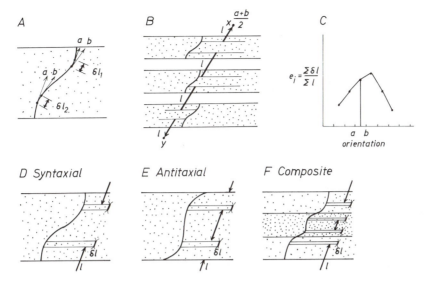

Figure 13.24. *Techniques for the measurement of existing length l and incremental length δl for a sector of fibre vein between orientations a and b.*

technique is that the ends of the measurement lines (Figures 18.24B, x and y) have to be fixed at some specific positions, and this choice of end points may be open to a certain degree of personal interpretation. As the area of the surface used for measurement gets larger, and as the number of veins used in the computation increase, so the values of incremental longitudinal strain become more accurate.

Measurement of strain increments

Question 13.3

Figure 13.25 shows a deformed Liassic sediment from Leytron, Valais (locality shown in Figure 11.9), geologically situated in the core of a large recumbent fold known as the Morcles nappe. The surface shown in this figure consists of a coarsely crystalline competent layer of limestone (A). This limestone is embedded in a matrix of less competent marl: it has been subjected to a complex sequence of extensions and has formed a type of chocolate tablet microboudinage. The spaces between the boudinaged limestone are now occupied by extension veins with marginal calcite (lightest material in the photo, B in diagram) and central quartz (C, stippled in the diagram). In the centre of each quartz vein a well-developed median line (m.l.) is developed, marking the location of the earliest deposited quartz in the earliest formed crack. The calcite and quartz are fibrous crystals and show a complex form.

Discuss the general significance of the vein development in terms of a progressive deformation sequence and determine if the increments of strain accumulated in a clockwise or anticlockwise sense.

Using the techniques discussed earlier (p. 250, Figure 13.24) determine the values of the principal maximum incremental extensions for orientation changes of 10°. To make the measurements of l and $δl$ with maximum accuracy it is probably best to construct orientation isogons at 10° intervals on the fibre directions. Plot variations of e_{i1} (ordinate) against orientation of the incremental maximum longitudinal strain direction (abscissa) in graphical form.

There are several ways of determining the finite strain ellipse produced by the extension vein system. One simple way is to reconstruct the original connections of the walls, draw a circle on the reconstructed fragments, then determine the fragmented form of this circle in the deformed state. Make a tracing of the wall–vein contacts of Figure 13.25. Cut along these contacts and reassemble the fragments into the best fit you can make of the jigsaw puzzle. To assist reconstruction, remember that fibres link points which were once in contact. Stick the reassembled walls on to a base and construct one or more circles across the reassembly. Trace the positions of fragmented circular arcs on Figure 13.25 and then find an average elliptical shape which fits the separated pieces of the original circle. This will be the finite strain ellipse: determine the axial lengths and orientations. What is the type of this ellipse (Field 1, 2 or 3)?

If time is available determine the successive strain ellipses by reconstructing, for any incremental stage, the intermediate shape of the vein system at the time of the increment and the locations of initial fragmented circular markers at the period of a specific increment.

Now proceed to the Answers and Comments section, then go on to the starred questions below, or to Session 14.

STARRED (★) QUESTIONS

Calculation of successive finite strain states

Question 13.4★

The geometry of the fibre patterns of Figure 13.25 suggests that the incremental strain ellipses had values of e_{i2} close to zero. For example, the fibre patterns show a form which is consistent with a progressive stretch across the veins, there has been no folding of the veins or fibres in a direction perpendicular to the extension directions, and there are no indications of pressure solution structures cutting the

Figure 13.25. *Fibrous quartz–calcite vein array cutting limestone from Leytron, core of the Morcles nappe fold, W. Switzerland. The diagram illustrates the distribution of quartz C (stippled), calcite B, and wall rock carbonate A, and shows the detail of the fibre lengths of the crystals. At the centre of the quartz veins is a well-developed median line, m.l.*

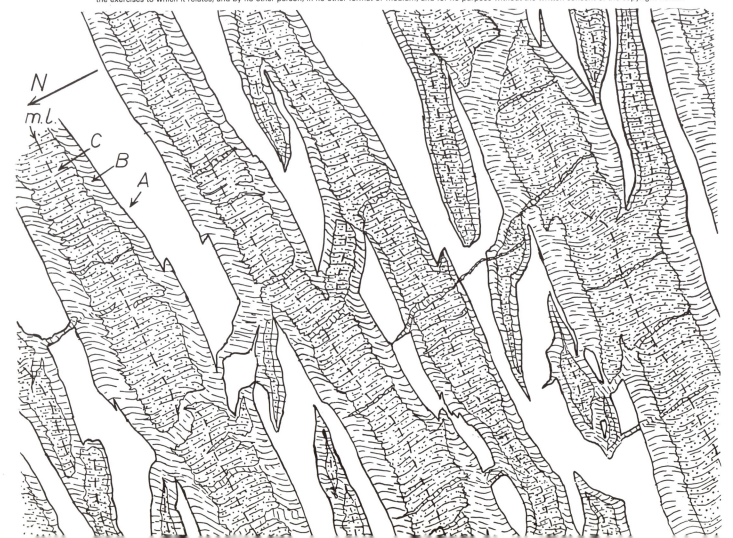

vein walls. In a system such as this it is possible to calculate the successive states of finite strains by a matrix multiplication technique. From the equations established in Appendix B (Equation B.16) an incremental strain ellipse with principal strains e_{i1} and $e_{i2} = 0$, with orientation from a reference axis (the y-direction is the north azimuth in the specimen Figure 13.25) given by θ' has a deformation matrix given by:

$$\begin{bmatrix} a_i & b_i \\ c_i & d_i \end{bmatrix} = \begin{bmatrix} (1 + e_{i1})\cos^2\theta' + \sin^2\theta' & e_{i1}\cos\theta'\sin\theta' \\ e_{i1}\cos\theta'\sin\theta & \cos^2\theta' + (1 + e_{i1})\sin^2\theta' \end{bmatrix}$$

Record the components of the incremental deformation matrices for successive increments in tabular form, and then use the equations for matrix multiplication (Appendix C, Equation C.22) to evaluate the matrix components $a_n b_n c_n$ and d_n defining successive finite displacements. From these matrix components compute the strain components of the successive finite strains (Appendix B, Equations B.14, B.19).

Changing proportion of mineral components in a composite vein

Question 13.5★

From an analysis of the isogonal lines drawn on the quartz and calcite fibres respectively, determine how much quartz fibre δl_q and how much calcite fibre δl_c was accreted during each deformation increment. For each of the deformation increment directions determine δl_q and δl_c and compute the proportion $\delta l_q/\delta l_c$.

Graph these values as ordinate against incremental orientation as abscissa. Is the change in proportion $\delta l_q/\delta l_c$ systematic? Discuss what might be the reasons for the observed changes.

ANSWERS AND COMMENTS

Interpretation of extension vein increments

Answer 13.1

The crystals at the edge of the large vein are narrower than those in the centre. The angles between crystal fibres and the vein walls are 90° at the vein–wall contact. These features imply that the fibres initiated with a vertically oriented maximum incremental longitudinal strain, and that an anticlockwise rotation of the incremental strain axes took place. This change of orientation of the principal incremental strains led to the development of smaller, obliquely oriented veins which cross-cut the main vein (cf. the geometric features of Figure 13.1). The lengths of the vertically oriented fibres are much smaller than those of the obliquely oriented late fibres, and this feature shows that the positive incremental longitudinal strains of the early increments were smaller than those taking place during the later increments.

Answer 13.2

The calcite fibres in the centre of the widest veins show a sub-perpendicular relationship to the trend of the main vein walls and they are oblique to the walls along the

vein–wall contact. This geometry does not accord with that set out in Figure 13.1. Two explanations are possible: first, the thin section might not be a truly perpendicular profile of the vein, so that the angular relationships of vein and wall are influenced by the section orientation; second, the fibre growth sense might be different from that of the syntaxial scheme. In this example the first explanation is incorrect. This can be deduced from the fact that the fibres close to the vein wall are optically continuous and show high length to width aspect ratios, features indicating the the fibre axes lie in the plane of the section (compare with the changing fibre orientation seen in Figure 13.4). The relationship of the fibres and displacements in the small, cross-cutting veins to the fibres in the main veins is clear (see bottom right): fibre lengths in the late veins are sub-parallel to fibres at the *edges* of the main veins. These geometric features imply that the growth sense in the widest veins was not from edge to centre, but from centre to edge. Such veins are termed **antiaxial**. The progressive strain sequence indicated by this example is first, a relatively small extension perpendicular to the walls of the main veins, followed by a strong positive elongation in an anticlockwise sense leading to the formation of the obliquely inclined veins (from top left to bottom right of the photograph). The few small veins running obliquely from upper right to lower left could be either later or earlier than those of the main sequence. The lack of obviously simple displacement relationships to the main veins and fibres (cf. Figure 13.1, a–a') is suggestive that they are probably later than the main sequences and are perhaps the end product of a general anticlockwise stretch sequence.

Measurement of strain increments

Answer 13.3

The vein system is of composite type. This implies that the earliest extensions are recorded by fibre orientations along the vein–wall contacts and along the median line and that the latest extensions are given by the direction of fibres along the quartz–carbonate sutures. An inspection of Figure 13.25 shows that the deformation sequence initiated in a NNW direction with relatively large incremental extension. The principal extensions of the later incremental strains rotate in a clockwise sense and the values of the extensions decrease in intensity. The sequence is concluded by relatively small incremental extensions in an ENE direction. This strain increment sequence produces a vein pattern with oblique splays propagating off the quartz–calcite suture zone of the earlier veins and in some places cross-cutting and displacing them, and the sequence concludes with the development of narrow calcite veins, preferentially developed in the quartz sectors of the previously formed veins. It is worth emphasizing how quickly a general interpretation of directions and relative proportions of incremental strains can be established from a simple analysis of the overall vein and fibre geometry. Although the determinination of the fine details of this sequence are fairly time consuming, the speed at which general conclusions can be drawn means that a direct analysis of the main incremental features is feasible as a practical field technique.

For a more detailed analysis fibre orientation isogons

Figure 13.26. *Contoured diagram of fibre isogons derived from an analysis of Figure 13.25. The sectors are oriented as follows: black 10°W to 10°E, close stipple 10°E to 30°E, fine stipple 30°E to 50°E, very fine stipple 50°E to 70°E. Vein walls left unornamented.*

were constructed at 10° intervals. The pattern has been simplified into 20° orientation difference sectors in Figure 13.26. This diagram brings out very clearly the progressive evolution of the veins and the sequential development of the splay and cross-cutting veins. For example veins with black margins were initiated in the development increment stages with maximum incremental longitudinal strains oriented between 10°W and 10°E, those with heavy stippled margins in the phases oriented between 10°E and 30°E, and so on. The incremental extension values δl for a particular orientation range were measured by placing a ruler across the contoured figure in the mean direction of the orientation range and measuring the lengths between the two limiting orientation contours. The existing length l was measured from the wall lengths plus lengths across previously crystallized vein material. These measurements were repeated across several parallel traverses, the results averaged, and a value of $e_i = \delta l/l$ computed and graphed (Figure 13.27). The systematic and continuous changes in values and orientations of the successive strain increments is very clear from this graph.

To determine the finite strain resulting from the development of the vein array, the wall rock areas were cut away from the vein, and reassembled like a jigsaw puzzle into a best fit reconstruction (Figure 13.28A) using fibre traces to guide the wall particles back to their original

starting points. On this reassembly two circles were constructed. The wall fragments were then placed in their final position and, in a crude way, the form of the finite strain ellipse was established from the detached parts of the circle (Figure 13.28D). These two ellipses had long axes oriented 13°E, and gave major and minor semi-axial lengths of $1 + e_1 = 3\cdot3$, $1 + e_2 = 1\cdot1$. The finite strain ellipse therefore lies in Field 1 (Figure 4.10). Two intermediate stages of

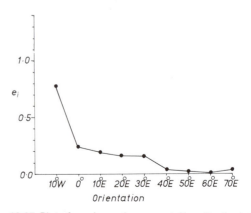

Figure 13.27. *Plot of maximum incremental longitudinal strain e_i against orientation for the vein array of Figure 13.25.*

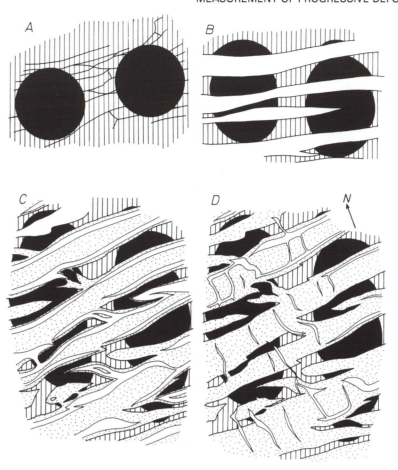

Figure 13.28. *Sequential reconstruction of various stages of the evolution of the vein systems of Figure 13.25. Wall rocks have vertical line ornament. The existing vein material at the time of the increment is stippled, and vein material developed during the increment is unornamented.*

the deformation sequence were established in a similar way, (Figure 13.27B and C), by restoring the original circles to intermediate displacement positions determined from the contour lines of Figure 13.26.

Chocolate tablet structure

the overall finite strain e_{f2} becomes positive.

Chocolate tablet structure is not uncommon in deformed sedimentary rocks. It can form in individual competent sediment layers and is often exceedingly well formed in pyrite sheets of diagenetic origin contained in the sediments (Figures 13.00, 10.00).

appears to be very close to the X and Y directions of the finite strain ellipsoid, and the fact that the wall rock pieces all lie in the same planar surface suggests that this surface was close to the X and Y directions of the intermediate stage finite strains. The fibres all lie exactly in the surface of the competent layer and indicate that the X_i and Y_i directions of the incremental ellipsoids were all coplanar. The regular patterns of changes of fibre directions recorded in Figure 13.26 strongly suggest that the three-dimensional incremental deformations were of plane strain type with $e_{i2} = 0$, but that, as a result of rotation of the increments,

veins may be very fast, whereas that in others may, at the same time, be slow or even static. The fibre growth rates depend upon the orientation of the fissure in question, and its relationship to the directions of the principal incremental strain axes. Figure 13.31 shows the formation of antitaxial fibres between four rigid plates of pyrite undergoing a changing strain increment history. The geometry has been reconstructed by allowing the centre of gravity of each of the four plates to move according to the requirements of a homogeneous increment of strain, and then to discover the separations along the plate contacts in order to compute the fibre directions. In the first increment, note how the

Figure 13.29. *Chocolate tablet structure developed in a rigid pyrite layer (dark) contained within a less competent marl. Calcite fibres with complex curvatures are located between the pyrite fragments, the complex geometry being the result of a sequential rotational separation history of the fragments. Root zone of the Morcles nappe, Switzerland.*

Figure 13.30. *Thin section of chololate tablet structure developed in a pyrite sheet. The antitaxial fibres consist of quartz and calcite and show a well-developed median line, growth zoning and curved forms as a result of a changing incremental strain sequence. Wildhorn nappe, Morge valley, Switzerland. Enlargement ×150.*

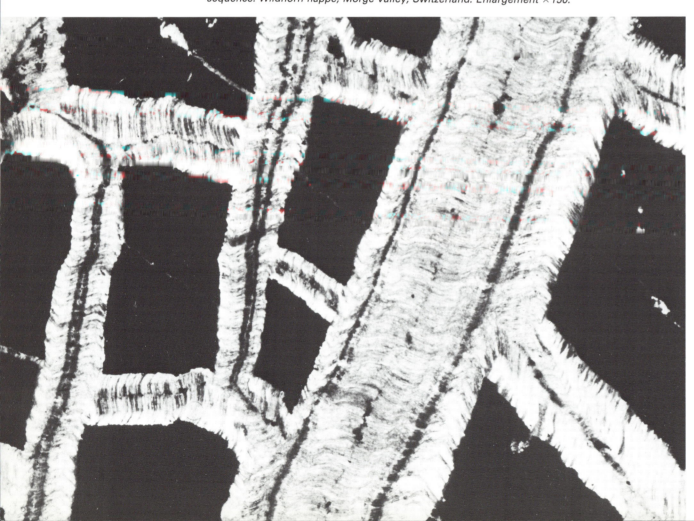

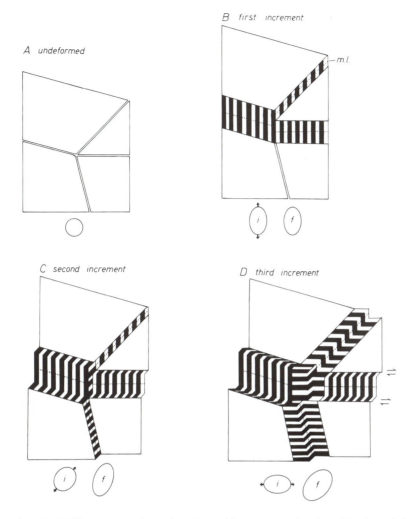

Figure 13.31. *Predicted antitaxial fibre patterns from chocolate tablet structure developed by the relative displacements of four rigid plates (e.g. pyrite). The successive incremental and finite strain ellipses are indicated i and f respectively. The median line marking the initial opening of each fissure is marked with a row of points (m.l.).*

fibre lengths along each plate boundary are not the same. The reasons for this variation are, first, that the total displacements are taken upon along one contact on the left hand side and along two contacts on the right hand side. Second, the displacements are functions of the centre of gravity points and they, therefore, are a function of relative positions of the plate centres. The second increment (Figure 13.31C) was arranged to produce no extension across one of the veins, and fibre addition therefore takes place only along three boundaries to produce sigmoidal curved fibres along two of these. The final increment (D) gives rise to opening along three boundaries and sets up a simple shear displacement along the pyrite–vein fibre contacts along the fourth. It will be seen that the strain sequences are all preserved in the fibre geometry, but that the interpretation of the strain history from this geometry requires careful thought.

Shear vein systems

The geometry of many extension fissure systems and the vein structure developed in them is closely controlled, at least in the initial stages of growth, by the initial perpendicular relationship of the first fissure to the direction of maximum incremental longitudinal strain. Although we have emphasised this direction in detail in the previous types of fissure where the opening increases direction in quite a marked way, we should in many, if not most, natural systems th not always make such variations, and ma tion fissures show fibres that are more or to the vein walls. There are, however, veins where the predominant displacement from the start of the deformation, sub-parallel to the fissure walls. Such systems, termed **shear veins**, arise in two principal ways. They can form by the superposition of a strain system on rocks containing pre-existing fractures or weaknesses geometrically and mechanically unrelated to the new strains, or they can form from shear systems developing during the new deformation and inclined obliquely to the principal strain directions. During the predominant shear displacements taking place along the surface there are generally dilational components. These dilations might result from the presence of surface irregularities along the fracture walls or because there is a natural dilation component in the system acting normal to the

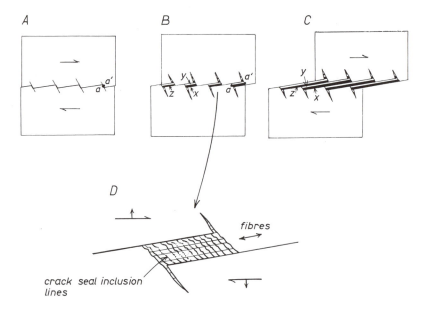

Figure 13.32. *Features of shear fibre veins. A shows an initial fracture surface with irregular form due to alternating shear and extension sectors. B shows the type of fibre geometry induced by sliding movements on the shear sectors with fibres connecting points originally in contact (a–a'). There is often a geometrical link between the fibres developed in en-echelon extension fissures and the shear fibres. C illustrates the build up relatively thick shear fibre packets with increased shear displacement. D shows the type of inclusion structure which may be developed as a result of progressive crack–seal activity.*

walls. The evolution of sliding movement along such surfaces where displacements are predominantly those of shear can produce a type of fibrous crystal growth which has many affinities with that we have discussed above. Figure 13.32 illustrates the type of vein geometry arising in such shear systems. The shear movement along an initially irregularly inclined shear surface produces spaces behind morphological irregularities. Fibrous crystals can grow in these spaces as the walls progressively displace past each other, and the fibre axes are constrained by exactly the same principles as we have discovered for fibres in extension fissures: the axes join points which were once in contact. The fibre growth can be syntaxial, antitaxial, composite or stretched crystal type. The fibres form a series of overlapping sheets slightly oblique to the average orientation of the walls of the shear. In the field the appearance of such shear veins is highly characteristic. If the displacement across the walls is not great (0–10 cm) the shear surface shows a series of small step-like fibre treads separated by areas with wall to wall rock contact (Figures 13.32B, 13.33). In thin section, the obliquity of the fibre axes to the main shear and the control of fibre dimensions by wall rock irregularities or wall rock crystals is related to the differential shear sense (Figure 13.34). With increased shear, the thickness of the fibre sheets accumulating along the shear surface increases and wall to wall contacts become overgrown by developing fibres. The progressive development of growing fibres often takes place by the crack–seal mechanism, and inclusion trails may be preserved inside the individual antitaxial fibres marking successive detachments of the fibre and its wall (Figures 13.32D, 13.35). The grains of wall material are successively plucked from the wall-fibre contact during the cracking stage, and are subsequently incorporated into the crystal fibre during the sealing process.

One important aspect of fibre growth process concerns the relative ages of adjacent fibres. The progressive increase of fibre length brings different groups of fibre sheets into contact in a rather interesting way. In Figure 13.32B fibres labelled x and y are growing side by side with the same growth rate, and fibre z is initiated at a different irregularity and is quite separate from fibres x and y. On continued sliding (Figure 13.32C) fibre z has increased in length and has now come into direct contact with the fibre packet xy. The central part of crystal z is in contact with the end of crystal y, and between the crystals is an active transform slide. In fact the overall geometry of this process has its analogy in plate tectonic theory: shear veins arise by the type of displacements sometimes termed "leaky transform faults". The transform like discontinuity between crystals x and z means that the overall shear fibre mass is subdivided into a number of packets of individual fibres, each crystal group separated by a shear discontinuity. The discontinuities bring into contact fibres formed at different times in the overall shear displacement history and impart a layered appearance to the fibre mass (Figure 13.37A). The discontinuities separating the fibre layers can often be seen in field outcrops as surfaces of crushed fibre, or concentrations of clay minerals. If the shear displacement vector is changing with time, the fibre bundles on either side of the discontinuity have markedly differing orientations because the adjacent fibres on either side the discontinuity were formed at different times (Figures 13.36, 13.37B).

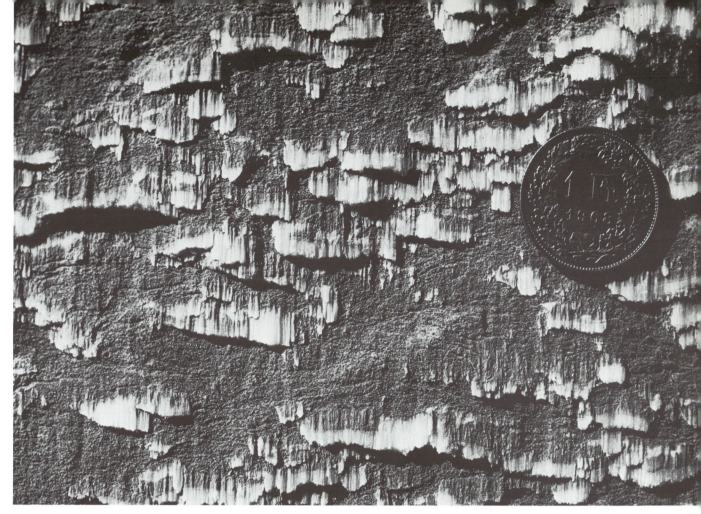

Figure 13.33. *Typical calcite fibre steps produced along irregularities in the walls of a shear fibre vein in limestone, Wildhorn nappe, Switzerland. The surface of rock exposed here moved upwards relative to the observer. Fine light and dark bands crossing the crystal fibres in a perpendicular manner are traces of the crack-seal accretion structure.*

Figure 13.34. *Thin section (crossed polars) of an antiaxial calcite shear fibre vein in a cleaved siltstone, Chaines subalpines, W. Alps. The top part of the vein has been displaced to the right relative to the lower part. Enlargement ×100.*

Figure 13.35. Wall rock inclusions preserved in quartz fibres in a shear vein as a result of growth by the crack–seal process (cf. Figure 13.32D). Barberine, Aiguilles Rouges Massif, W. Switzerland. Thin section, enlargement ×200.

Figure 13.36. Overlapping sheets of curved shear fibres developed on a bedding surface in sandstones, Culm Series. N. Cornwall, UK. Between each fibre sheet is a discontinuity of transform shear aspect.

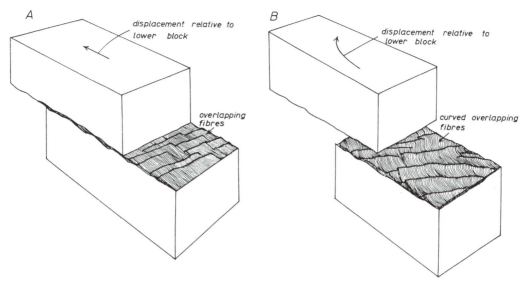

Figure 13.37. *The geometry of fibre sheets developed on shear surfaces (A) where the relative shear displacement vector is constant and (B) where this displacement vector changes direction with progressive evolution of the shear.*

Calculation of successive finite strain states

Answer 13.4★

In Question 13.3 it was shown how it is possible to reconstruct, in a graphical way, the shape of the finite strain ellipse and the shapes of successive finite strain ellipses arising during a progressive deformation (Figure 13.28). Using the data from Figure 13.27 (and set out in Table 13.1) it has been possible to evaluate successive finite strains by making successive matrix multiplications of the incremental strain component. The total finite strain has values $1 + e_{f1} = 3.37$, $1 + e_{f2} = 1.15$ and the maximum strain direction has an orientation θ' of $9.1°E$. These values are in good general accord with those established graphically, and are probably more accurate because they rely on an integration of features measured through the whole vein array, and not just local displacement determined from a single circular marker. The nine successive strain ellipses are plotted as a deformation path in Figure 13.38.

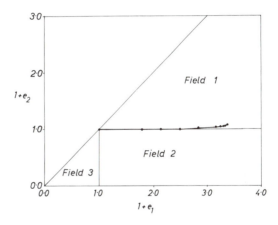

Figure 13.38. *Two-dimensional deformation path of the successive strain ellipses of the progressive deformation sequence which developed the vein array of Figure 13.25.*

Table 13.1.

Increment	θ'	R	a_i	b_i	c_i	d_i	a	b	c	d	$1 + e_1$	$1 + e_2$	θ'
1	10	1·8	1·7759	0·1368	0·1368	1·0241	1·7759	0·1368	0·1368	1·0241	1·80	1·00	+10.00
2	0	1·2	1·2000	0·0000	0·0000	1·0000	2·1311	0·1642	0·1368	1·0241	2·15	1·00	+7.35
3	−10	1·18	1·1746	−0·0307	−0·0307	1·0055	2·4990	0·1614	0·0721	1·0247	2·51	1·02	+3.78
4	−20	1·15	1·1324	−0·0483	−0·0483	1·0176	2·8264	0·1333	−0·0473	1·0349	2·83	1·04	+0.03
5	−30	1·15	1·1125	−0·0650	−0·0650	1·0375	3·1474	0·0810	−0·2328	1·0650	3·16	1·07	−4.21
6	−40	1·05	1·0293	−0·0246	−0·0246	1·0207	3·2453	0·0572	−0·3150	1·0851	3·26	1·09	−5.86
7	−50	1·03	1·0124	−0·0148	−0·0148	1·0176	3·2902	0·0418	−0·3686	1·1034	3·31	1·10	−6.92
8	−60	1·02	1·0050	−0·0087	−0·0087	1·0150	3·3099	0·0324	−0·4028	1·1196	3·34	1·11	−7.60
9	−70	1·05	1·0058	−0·0161	−0·0161	1·0442	3·3356	0·0146	−0·4739	1·1686	3·37	1·15	+9.08

Changing proportions of mineral components in a composite vein

Answer 13.5★

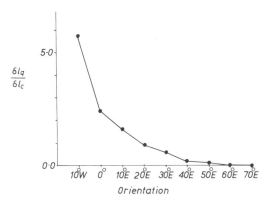

Figure 13.39. Changing proportions of quartz to calcite fibre length for different incremental orientations of the vein array of Figure 13.25.

From the fibre isogon plot of Figure 13.26 the incremental extensions have been subdivided into quartz δl_q and calcite δl_c components, and a plot of changing ratio of quartz to calcite with changing increment orientation is shown in Figure 13.39. The changing ratio is very systematic. At the start of the deformation nearly six times as much quartz as calcite was deposited, whereas at the end of the sequence the veins contain only calcite. The changing quartz–calcite proportion probably arises from changes in the fluid composition filling successive openings or to temperature controlled differences in deposition rate. By determining the homogenization temperature of primary fluid inclusions David Durney has been able to show that the changes in the quartz–calcite proportions are associated with a decrease in environmental temperature of about 200°C. In this system it appears that the fluid composition chemistry is very closely dependent upon temperature. This systematic and large change in temperature is consistent with the geometrical studies we have made suggesting vein evolution over a long time period, probably of the order of millions of years duration.

KEYWORDS AND DEFINITIONS

Antitaxial veins	Veins with filling material grown from vein centre towards the vein walls (Figure 13.9B).
Composite veins	Veins with two dominant crystal species arranged in zones parallel to the vein walls (Figure 13.9C).
Crack-seal mechanism	A mechanism whereby a vein filling is built up by successive development of microcracks, followed by successive periods of cementation. It is especially common in anti-taxial veins (Figure 3.18) where it often produces a series of wall-parallel inclusion bands and is also characteristic of **"stretched" crystal fibre veins** (Figure 13.9D).
Extension fissures	Crack-like discontinuities formed perpendicular to the direction of maximum incremental stretching as a result of brittle or semi-brittle failure. **Extension joints** are small cracks on which little movement has taken place generally developed as a result of uplift and release of stored stress. **Extension veins** are wider than joints, and the dilational displacements produce significant (>1%) longitudinal strain in the rock. The space between the walls of a vein is filled with crystalline material usually deposited more or less synchronously with the progressive vein opening (Figure 13.1).
Shear veins	Crystal filled fissures in which the differential movement across the fissure is sub-parallel to the fissure walls (Figure 13.32).
Syntaxial veins	Veins with filling material grown from walls towards the vein centre (Figure 13.9A).
Vein fibres	Crystals with fibrous morphology developed in extension- and shear veins. The fibrous form is related to the differential displacement taking place across the vein (Figures 13.1, 13.32).

KEY REFERENCES

Adams, S. F. (1920). A microscopic study of vein quartz. *Econ. Geol.* **15**, 623–664.

Geologists in the field are sometimes overheard to say "Oh, it's only a quartz vein!" This paper should be read to correct the impression of those who think that vein systems are unimportant. It describes different types of veins relating them to differences in geological environment and comments on some of the fibrous vein types discussed in this Session.

Casey, M., Dietrich, D. and Ramsay, J. G. (1983). Methods for determining deformation history for chocolate tablet boudinage with fibrous crystals. *Tectonophysics* **92**, 211–239.

This describes the geometry of several examples of chocolate tablet structure found in naturally deformed rocks and would form good background reading for Questions 13.3 and 13.4★. It also discusses various mathematical models for the progressive formation of the structure in terms of velocity gradients and strain rates, showing how the rigid plates may be both separated and differentially rotated.

Durney, D. W. and Ramsay, J. G. (1973). Incremental strains measured by syntectonic crystal growths. *In* "Gravity and Tectonics" (K. A. De Jong and R. Scholten, eds) 67–96. Wiley, New York.

This was the first paper to distinguish between several types of tectonic extension veins and to show how an analysis of fibre geometry could be used to evaluate successive strain ellipses in a progressive deformation sequence.

Ramsay, J. G. (1980). The crack–seal mechanism of rock deformation. *Nature* **284**, 135–139.

This paper describes the internal geometric features of tectonic veins formed by successive development of microfracture followed by cementation.

Wickham, J. S. (1973). An estimate of strain increments in a naturally deformed carbonate rock. *Am. J. Sci.* **273**, 23–47.

The theoretical background and practical details of methods of determining incremental dispacement and incremental strain history are discussed, and the orientations of tremolite fibres are used to evaluate a strain path. The deductions as to the strain history are checked by comparisons with finite strains determined from deformed ooids.

SESSION 14

Measurement of Progressive Deformation

2. Pressure Shadows

The geometric forms of pressure shadows developed around deformation resistant objects can be used to determine the orientations and values of the principal incremental strains. Three main types of pressure shadows are recognized: pyrite, crinoid and composite types, each showing different types of growth sequences of the crystal fibres developed in the pressure zone. Two principal measurement techniques are used to evaluate the shapes and orientations of successive incremental strain ellipses, depending upon whether the fibre masses growing during deformation behave in a rigid or ductile way. Graphical and cartographic methods of representing strain history are presented with reference to strain information taken from an area some 700 square kilometres in extent from the Helvetic nappes of Switzerland.

INTRODUCTION

Many rocks contain objects which are resistant to deformation, and these rigid particles set up special types of deformational heterogeneity which may be used to evaluate both the finite strain and the progressive strain history. This deformational heterogeneity leads to the formation of **pressure shadows**: symmetrically disposed regions of low strain developed on opposite sides of a rigid object where the rock matrix has been protected from the full effects of deformation. In the pressure shadow region the rock matrix may be undeformed and still attached to the object, but often the matrix becomes detached from the object along the mechanically weak contact. This detachment sets up extension fissures on either side of the object and these fissures, like those discussed in Session 13 become infilled with fibrous crystalline material (Figures 14.1, 14.2). The geometrical forms of the fissures and of the fibrous crystal infilling depend upon the nature of the progressive extension history of the rock, and in this Session we will investigate the methods whereby a progressive strain sequence may be evaluated from this geometry.

The most frequently found deformation resistant objects in rocks are isolated crystals or crystal aggregates made up of ore minerals, pyrite and magnetite being especially common. In naturally deformed rocks in a metamorphic condition up to greenschist facies, both pyrite and magnetite appear to retain their original pre-deformed shape, whereas in higher metamorphic environments the external crystal form often becomes modified by chemical activity during metamorphism. Pyrite is common in many sedimentary rocks and often develops during early diagenesis associated with the activity of anaerobic sulphate reducing bacteria. This diagenetically formed pyrite may have a

euhedral form (Figure 14.1), but often it is found as sub-spherical masses of small particles of about 1–20 μm diameter aggregated together to form a globule with a raspberry-like external form known as **framboidal pyrite** (Figure 14.3). Euhedral pyrite often originates after diagenesis, perhaps during the course of low grade metamorphism accompanying a tectonic deformation. Analysis of the pressure shadows formed around such euhedral crystals can be used to evaluate that part of the deformation subsequent to the mineral growth, but not the total strain history. In limestones and marls certain clastic fossil fragments or sand grains may act as rigid masses on account of differences of composition (Figure 14.5) or of differences in grain size between the particle and its matrix (e.g. a single large calcite plate of a crinoid ossicle embedded in a fine grained limestome). In metamorphic terrains certain porphyroblastic minerals may act as rigid bodies (Figure 14.6) and develop pressure shadow structures.

Types of pressure shadows

There are several different ways by which the progressively developing extension fissure is infilled with fibrous crystalline material. There are three main types which show geometrical correspondence with certain of the features of extension veins described in Session 13. The three types are termed pyrite-type, crinoid-type and composite-type.

1. Pyrite type

This is probably the most frequently met type of pressure shadow and its main features are shown in Figure 14.7 A

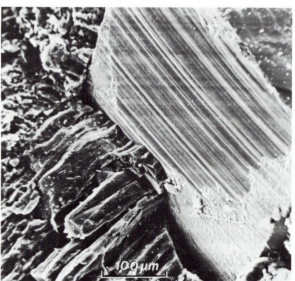

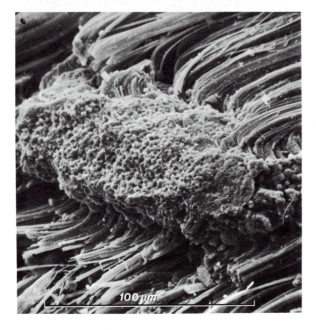

Figure 14.1 (top of page). Quartz–chlorite fibres developed in a complex pressure shadow zone around euhedral pyrite, N. Pyrenees.

Figure 14.2 (middle left). Calcite fibres developed in a sigmoidally shaped pressure shadow zone around a pyrite mass embedded in marl. Grande Mulveran, Morcles nappe, Switzerland.

Figure 14.3 (bottom left). Sigmoidally shaped quartz fibres developed around framboidal pyrite, SEM image. Leytron, Morcles nappe, Switzerland.

Figure 14.4 (middle right). Quartz fibres developed in a pressure shadow around euhedral pyrite, SEM image. Panixer Pass, Infrahelvetic nappes, E. Switzerland.

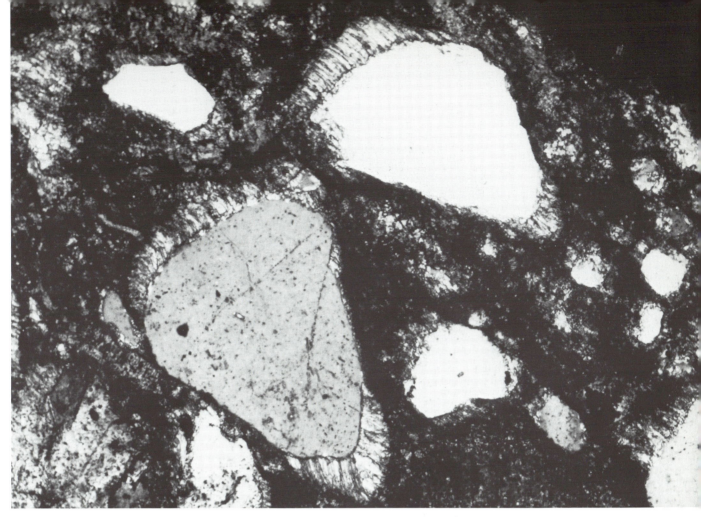

Figure 14.5. Quartz sand grains with calcite filled pressure shadows embedded in a deformed marl. Flysch, Wildhorn nappe, Engelberg, Switzerland. Thin section enlargement ×100.

Figure 14.6. Large crystal of porphyroblastic chloritoid with composite chloritoid–calcite pressure shadow in a fine grained metamorphosed marl. Termignon, W. Alps. Thin section enlargement ×100.

and D. It is termed the pyrite type because of its prevalence around resistant pyrite bodies but this type of growth can be observed around many deformed resistant objects wherever there is a marked mineral species contrast between the object and its matrix (Figures 14.1, 14.3, 14.5). The fibres grow with crystallographic continuity on wall rock crystals of the same species (Figures 14.7, 14.9) or from wall rock crystals which have some chemical affinity with the composition of the fibres (e.g. quartz fibres overgrowing phyllosilicates in the wall). If the fibres are more or less straight and sub-parallel, the fibre width sometimes increases towards the object–fibre contact as a result of preferential growth on crystals which have a more favourable crystallographic orientation for growth than their neighbours (Figure 14.9). Although the first formed fibres have a width which is often controlled by the grain size of the wall rock crystals, the fibre width may be controlled by the grain size of the individual particles making up a pyrite framboid (Figure 14.3). Where the fibres curve, individual fibres generally retain a constant crystallographic orientation showing that the curvature is a growth feature and not the result of a late deformation superposed on previously straight fibres (Figures 14.1, 14.7D). All the geometric features set out above are consistent with a progressive growth of the fibre from the wall towards the resistant object, and such fibres are **displacement controlled fibres** which can be used to evaluate certain features of the progressive strain history.

The fibres around euhedral crystals may show somewhat different morphological relationships with respect to the walls of the resistant crystal. Where the fibres are very narrow or where the fibres consist of two or more species growing in a sub-parallel fashion then the fibre geometry generally obeys the features set out above for displacement controlled fibres. However, where the fibres are mono-mineralic and fairly wide ($>30\ \mu m$), the fibre axes form perpendicular to the face of the resistant crystal irrespective of the displacement directions and they are known as **face controlled fibres** (Figures 14.4, 14.10). Although the factors controlling the two types of fibres are not fully understood, the geometric contrasts of the two types are so clear that, in practice, no confusion arises in deciding which relationship is the correct one.

With displacement controlled fibres the shapes and orientations of the pyrite faces influences the fibre pattern but not the fibre orientation (Figure 14.8). Where the displacement vector makes an angle with a pyrite face, the fibres grow with this orientation. Changing incremental extensions may result in the displacement direction becoming parallel to a crystal face, and where this occurs the earlier formed fibres are slid by transform motions along the face (Figure 14.8, a to a'). With continued movement, the earliest fibres may be completely slid away from the face on which they originated to form an isolated group showing discordant relationships to later formed fibres (Figure 14.8, a'', and Figure 14.1).

Although strain history cannot be determined from the directions of face controlled fibres, it can be determined from the geometry of the contact suture which separates

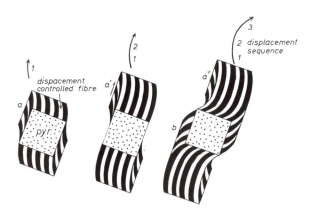

Figure 14.8. Development of a pyrite type pressure shadow around an euhedral rigid crystal. The fibres are displacement controlled. Those originally formed along the wall of the crystal at a become slid along the wall to a' and at the end stages are isolated from the wall on which they formed at a''.

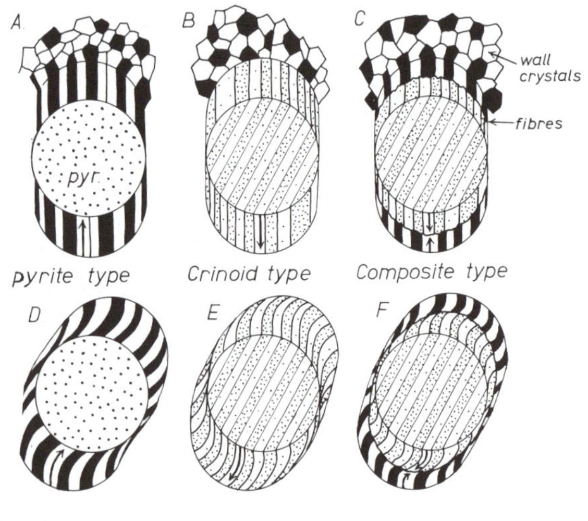

Figure 14.7. Three main types of pressure shadow. A, B and C show the appearance of the pressure shadows and the relationship of crystal fibres to object and wall under a coaxial deformation sequence; D, E and F show the same types under conditions of non-coaxial increments. Arrows show fibre growth sense. The black and unornamented areas are crystals of the same species but differing crystallographic orientations, the stippled areas are crystals of a second species.

Figure 14.9. Quartz fibres growing towards the face of a euhedral pyrite. The optical orientation of the fibres is linked with crystals in the sandstone making up the rock matrix. Thin section enlargement ×50.

different fibre groups. This contact suture (Figure 14.10, *s.l.* and Figure 14.11) forms as the separate face oriented fibre groups are pulled away from adjacent walls of a euhedral crystal. The suture therefore traces out the locations of points which once lay at the edge between the two crystal faces, and therefore records the progressive displacement history of this point. Face controlled fibres sometimes contain straight or curved inclusion trails made up of circular particles of chlorite, haematite or limonite (Figures 14.10, 14.11, 14.12). These inclusion trails, known as **"ghost" fibres** cross cut the face controlled fibres and show forms which are parallel to the contact suture. They appear to mark out displacement paths between the pressure shadow wall and the rigid object, and their geometric forms are equivalent to those of displacement fibres.

Figure 14.10. *Typical geometric forms of face controlled fibres. The suture line (s.l.) between differently oriented face controlled fibre groups can be used to determine the displacement history. The fibres may contain inclusion-trails of "ghost" fibres which also are related to the displacement.*

2. Crinoid type

The characteristic features of this type of pressure shadow are diagrammatically shown in Figure 14.7B and E. The rigid particle is generally of a different composition or grain size from that of the walls. A typical situation for the development of such a pressure shadow would be a deformation resistant crinoid ossicle in a marl matrix. The fibres filling the pressure shadow are the same species as found in the resistant object and grow with crystallographic continuity on the object wall. The fibrous texture is either inherited from crystallographic variations in the parent object (e.g. lamellar twinning in calcite), or arises from the incorporation of lines of foreign inclusions. All the geometric relationships observed are consistent with progressive fibre growth from the object surface *towards* the displaced matrix.

3. Composite type

This type of pressure shadow incorporates features which are consistent with both pyrite and crinoid types, and is closely related to composite extension veins. The fibres form two well defined species groups which crystallographically or chemically relate to the object and to the rock matrix respectively, and these fibre groups meet together along a common suture (Figures 14.6, 14.7C and F). The interpretation of fibre growth direction is in accord with progressive growth of both fibre sets simultaneously, one group from the object to the suture, the other group from the matrix wall towards the suture.

Figure 14.11. *Face controlled quartz fibres containing abundant "ghost" fibre inclusion trails. The "ghost" fibres are parallel to the well developed suture line between the two groups of face controlled crystals. Mica schist, S. Sudan. Thin section enlargement ×100.*

Thin section preparation

To investigate the geometric features of pressure shadow fibres in sufficient detail to establish the progressive strain history of the rock it is generally necessary to prepare thin sections of the rock. The general features of the pressure shadows and fibres are first determined with hand lens and this initial inspection directs the choice of section plane. The plane is selected so that it is parallel to all or most of the fibre axes. Fibres sometimes show very complex curved forms in three dimensions and it may not be possible to find a single plane containing all the fibre directions. If no fibres can be observed in the hand specimen it is probably best to choose section planes which correspond to the principal planes of the finite strain ellipsoid (XY parallel to cleavage, and XZ perpendicular to cleavage containing any "stretch" fabric). Clearest interpretations of fibre geometry come from sections passing through the centres of resistant objects, and in such sections pressure shadow zones should have a more or less mirror symmetry about the object centre. To find such sections, the first polishing of the rock slice must be carried out very carefully, the rock being slowly ground away and frequently inspected with a hand lens. Once a satisfactory surface has been reached, the rock slice is mounted on a glass slide and prepared in the normal way. Because pyrite is very brittle and often chemically weathered it is advisable to impregnate any pyrite originating pressure shadows with Araldite before mounting on the glass slide and, when the thin section is in a transparent stage, to complete the last stages of thinning with exceptional care. If normal preparation methods are used it is common for the pyrite mass and attached fibres to break away and be plucked from the section surface.

Measurement techniques

The directions and lengths of the fibres relate to the directions and values of the maximum incremental longitudinal strains. Because the shape of the rigid object generating the pressure shadow remains constant during the progress of deformation it is used as a base length to standardize the increments determined from the fibre lengths.

To determine the incremental strains it is necessary to record fibre orientations, to measure the lengths of fibres between chosen angular limits and to record the maximum diameter of the resistant object in the mean direction of the fibre sector. The interpretation of the progress of changing increment directions depends upon pressure shadow type (growth sense arrows in Figure 14.7). There are two principal models used for the calculation of incremental strains, these models forming end members of a behavioural spectrum. The first is known as the **rigid fibre model**, and it assumes that, for any particular incremental fibre length, any previously formed fibres are undeformable and act in a rigid way like that of the central resistant object. This model implies that any early formed fibres must be combined with the object radius in order to determine a base length for strain determination. The second model, the **deformable fibre model**, assumes that, once a fibre is formed, it deforms in a manner identical to that of the rock matrix surrounding the rigid object. The early

Figure 14.12. Detail of part of Figure 14.11 showing face controlled fibres and acicular "ghost" fibres of chlorite.

formed fibres therefore play no part in determining the values of later incremental strains, but they do suffer length changes as a result of later increments of deformation. The geometry of quartz fibres mostly appears to accord with the rigid fibre model, whereas that of chlorite and calcite fibres can relate to either model. If the incremental strains are non-coaxial and if a separate method is available to compute the finite strain it is possible to check which of the two fibre development models is correct.

Rigid fibre model

Figure 14.13A shows the principal features of this type of fibre geometry appropriate to the pyrite type of pressure shadow undergoing *coaxial incremental deformation*. The pyrite, with radius l develops an initial set of fibres of length δl_1. The incremental strain e_{i1} is given by

$$e_{i1} = \frac{\delta l_1}{l}$$

During the next stage of strain, a fibre of length δl_2 is formed. The initially formed fibres, being rigid, now combine with the rigid body to form the base line of length $l + \delta l_1$. The second principal incremental extension e_{i1} is given by

$$e_{i1} = \frac{\delta l_2}{l + \delta l_1}$$

and in general for the nth increment

$$e_{i1} = \frac{\delta l_n}{l + \delta l_1 + \delta l_2 \ldots + \delta l_{n-1}} \quad (14.1)$$

The total finite elongation e_{f1} can be derived by multiplication of the semi-axes of successive incremental strain ellipses $(1 + e_{i1})$

$$1 + e_{f1} = \left(1 + \frac{\delta l_1}{l}\right)\left(1 + \frac{\delta l_2}{l + \delta l_1}\right)$$
$$\times \left(1 + \frac{\delta l_3}{1 + \delta l_1 + \delta l_2}\right) \ldots \left(1 + \frac{\delta l_n}{1 + \delta l_1 + \ldots \delta l_{n-1}}\right)$$

This simplifies to

$$e_{f1} = \frac{\delta l_1 + \delta l_2 + \delta l_3 \ldots + \delta l_n}{l} \quad (14.2)$$

Where the *increments are non-coaxial* certain modifications must be made to this analysis because the effective rigid body length in the direction of later increments is a function of the difference in angle ϕ between an early

formed fibre and the direction of the later increment. Figure 14.13B shows a situation: the length of the rigid body base line for the increment of length δl_2 is the sum of the rigid body radius plus $\delta l_1 \cos \phi$. This gives the incremental strain for the second increment

$$e_{i1} = \frac{\delta l_2}{l + \delta l_1 \cos \phi}$$

and, in general, for the nth increment

$$e_{i1} = \frac{\delta l_n}{l + \Sigma_1^{n-1} \delta l_n \cos \phi_n} \quad (14.3)$$

If ϕ_n exceeds 90° and the fibre becomes reflexed (Figure 14.14), $\cos \phi_n$ becomes negative. It follows from this geometry that only increments with positive $\cos \phi_n$ values should be incorporated into the denominator of Equation 14.2.

Deformable fibre model

In a *coaxial deformation* of this type the length of any fibre increment is increased in proportion to the total strain (or product of later incremental strain ellipse semi-axes) after its formation. The first increment with a fibre length δl_1 becomes modified during the formation of the second increment of length δl_2 to a new length $\delta l_1'$ given by

$$\delta l_1' = \delta l_1 \left(1 + \frac{\delta l_2}{l}\right)$$

and after $n - 1$ subsequent increments giving fibres of initial length $\delta l_2, \delta l_3 \ldots \delta l_n$, the first fibre now has a length

$$\delta l_1' = \delta l_1 \left(1 + \frac{\delta l_2}{l}\right)\left(1 + \frac{\delta l_3}{l}\right) \ldots \left(1 + \frac{\delta l_n}{l}\right) \quad (14.4)$$

The correct incremental strain e_{i1} for a measured fibre length $\delta l_1'$ is then given by

$$e_{i1} = \frac{\delta l_1}{l} = \frac{\delta l_1'}{l\left(1 + \frac{\delta l_2}{l}\right)\left(1 + \frac{\delta l_3}{l}\right) \ldots \left(1 + \frac{\delta l_n}{l}\right)} \quad (14.5)$$

If the *increments are non-coaxial*, early formed fibres are *modified both in length and in direction* by subsequent strain increments. An original fibre length δl_1 modified by the next incremental strain with fibre length δl_2 and oriented at an angle ϕ to the direction of δl_1 according to Equation D.3 to a length $\delta l_1'$

$$\delta l_1' = \delta l_1 \left[\left(1 + \frac{\delta l_2}{l}\right)^2 \cos^2 \phi + \sin^2 \phi\right]^{1/2} \quad (14.6)$$

and its orientation is modified to ϕ' given in Equation D.13

$$\tan \phi' = \tan \phi \bigg/ \left(1 + \frac{\delta l_2}{l}\right) \quad (14.7)$$

From Equations 14.5 and 14.6 it follows that incremental strains can be determined from deformed fibres if the calculations are made stepwise. The last strain increment is determined from a length of fibres in contact with the rigid object. The deformational effects of this last increment are then removed from earlier fibres. The next to last strain increment is then determined and its effects removed from earlier fibres, and similar calculations are made stepwise along the length of the fibre until its total length has been covered by the computations.

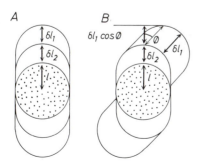

Figure 14.13. *Relationship between incremental extensions δl_1 and δl_2 in the rigid fibre model for pyrite type pressure shadows.*

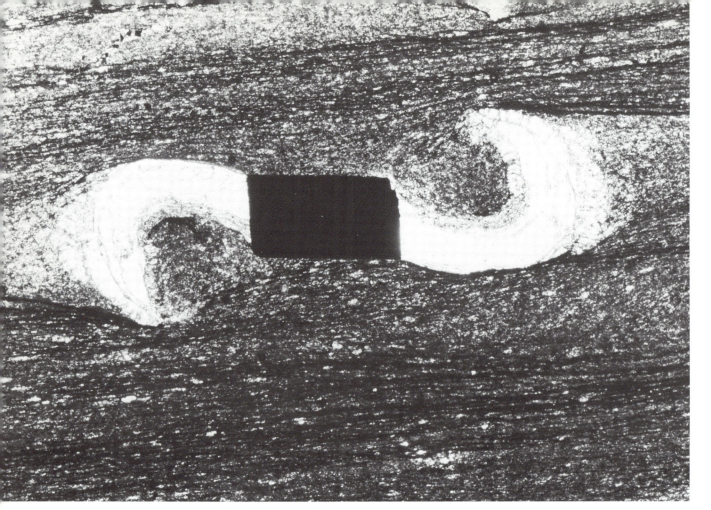

Figure 14.14. *Reflexed fibre pressure shadows around pyrite from the N. Pyrenees. See Choukroune (1971).*

If accurate values of incremental strains are to be established it is clearly important to be able to decide which of the two models is appropriate to solve the problem in hand. There are a number of differences in pressure shadow form which depend on the deformation model, and these are useful in making a choice of a particular computational model. Figure 14.15 shows the differences between the two models in the case of coaxial incremental deformation developed by superimposing 10 increments, each with principal strain $e_{i1} = 0.1$. The positions of time lines for increments with numbers 2, 4, 6, 8 and 10 are shown transecting the fibres. The rigid fibre model shows greatest fibre lengths per increment during the later stage of growth, whereas the deformable fibre model shows greatest fibre lengths per increment at the pressure shadow end. The shapes of the pressure shadow boundaries differ considerably in the two models. In the rigid fibre model the shapes of the ends of the pressure shadows must be identical to the shape of the wall of the rigid object generating the shadow, whereas in the deformable fibre model the boundary of the shadow zone becomes modified by strains superposed on the fibres, and becomes sub-elliptical. The same general geometrical features are observed where the increments are non-coaxially superposed (Figure 14.16, $e_{i1} = 0.1$ and $\phi = 10°$ each successive increment) although the internal geometry is more complex than that produced during a coaxial deformation. For example, the angular relationships between fibre directions at the outermost parts of the shadow differ in the two models (Figure 14.16, cf. B and C). In both models there are convergences and divergences of the boundaries of adjacent fibres as they are followed towards the rigid body contact, and these result from the changing angular relationships of fibre axis to growth surface produced by changes in the principal incremental strain direction. In the deformable fibre model it is to be expected that the first formed fibres will show more signs of internal deformation than those formed at the end of the sequence. When a fibre is traced from the pyrite contact to the edge of the shadow zone there should be an increase in dislocation density, an increase in the development of sub-grain microfabric and perhaps an increase in the features suggestive of grain boundary sliding and microstylotite formation along the fibre contacts.

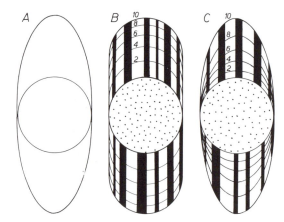

Figure 14.15. *Geometric contrasts of the rigid fibre model (B) and deformable fibre model (C) in coaxial deformations for pyrite type pressure shadows. (A) shows the form of the finite strain ellipse. The numbers refer to successive increments and link fibres formed at identical times.*

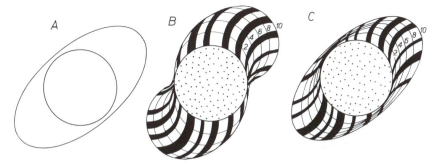

Figure 14.16. *Geometric contrasts of the rigid fibre model* (B) *and deformable fibre model* (C) *in non-coaxial deformations for pyrite type pressure shadows.* A *shows the form of the finite strain ellipse.*

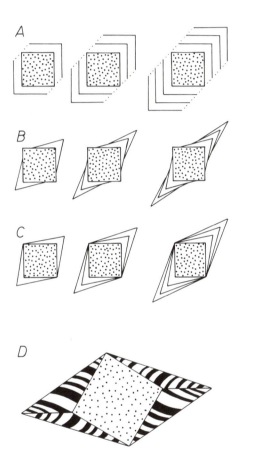

Figure 14.17. *Pyrite type pressure shadows developed according to various models: A, rigid fibre; C, deformable fibre; B, intermediate between A and C with shortening perpendicular to main elongation direction. D shows the type of fibre geometry associated with deformed face controlled fibres.*

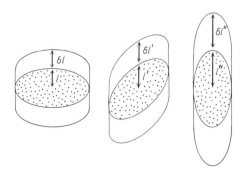

Figure 14.18. *Effect on fibre length of orientation of an inequidimensional rigid object; rigid fibre model.*

The geometric differences between the rigid- and deformable-fibre models where the rigid body is euhedral are very striking (Figure 14.17). Because the angular relationships of the rigid body crystal faces are exactly controlled by the crystallographic structures of the crystal, the shape of the pressure shadow boundary gives a clear indication of whether the fibres were rigid (A) or ductile (C). Certain types of intermediate geometric types have been described (B) where wall generated fibres are successively slid along the wall of the rigid mass; this type generally occurring where the extension in one direction goes together with shortening in a perpendicular direction. If the fibres have a face controlled orientation (Figure 14.10), any tendency for fibre deformation to take place will be accompanied by changes in the perpendicular relationships of adjacent fibre groups and the formation of curved fibre forms (Figure 14.17D).

Inequidimensional rigid objects

If the rigid object is not sub-spherical the length of fibres forming at any given incremental stage will be a function of the orientation of the object with respect to the direction of incremental extension (Figure 14.18, cf. different fibre lengths δl, $\delta l'$, $\delta l''$ with the lengths l, l' and l''). With progressively changing increment directions it is necessary to change the standard base line length with orientation of the object when computing values of incremental strains.

Strain increment calculations from a pressure shadow

Question 14.1

Figure 14.19 shows a thin section (crossed polars) of a pyrite crystal embedded in a marly limestone from the deformed Jurassic sediments of the Wildhorn nappe of the Helvetic Alps. The pyrite is surrounded by a pressure shadow containing fibrous quartz. The fibres are quite strongly curved but retain constant continuity through the curved portions showing that the crystallographic directions of each fibre remain constant through the curved section. There are no indications of changes in internal structure of the quartz along the fibres that might suggest the crystals have been deformed. All the features of this pressure shadow suggest that it was formed according to the rigid fibre model.

From a preliminary inspection of Figure 14.19, deter-

Figure 14.19. Pyrite type pressure shadow with fibrous quartz crystals developed in the shadow zone. Wildhorn nappe, Switzerland. The locality is illustrated on the map of Figure 14.21.

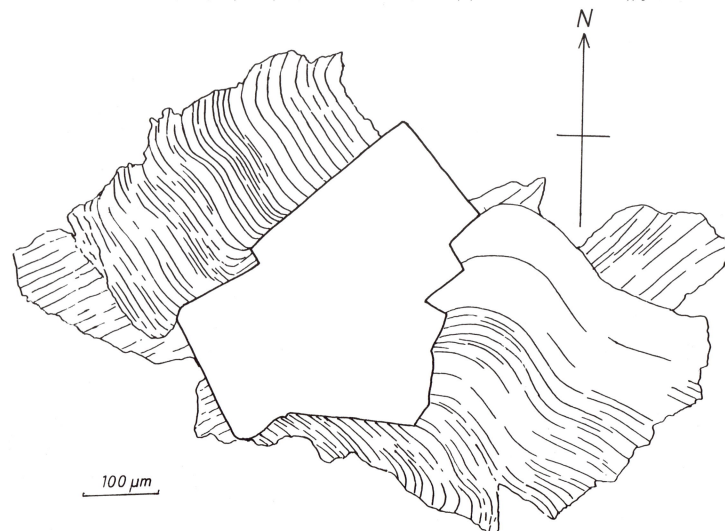

mine the directions of the main incremental maximum longitudinal strains and determine whether the changes in the directions of the principal incremental strains have a clockwise or anticlockwise rotational sense.

Make an isogon plot of the fibre directions with contour interval 10°. From this diagram determine the lengths of fibres δl between 10° spaced contours. Evaluate the incremental longitudinal strain e_{i1} for each interval using Equation 14.2. The pyrite is not equidimensional, and so you will need to correct the base length for successive increments as well as for lengths of earlier formed fibres. Graph e_{i1} against mean orientation of the increment.

Now proceed to the Answer and Comments section and, if you have time, return to the starred questions below.

STARRED (★) QUESTIONS

Progressive evolution of a pressure shadow zone

Question 14.2★

Using the isogons as time lines, make reconstructions of the changing shapes of the pressure shadow zone as it progressively developed.

Calculation of successive total strains from incremental strain

Question 14.3★

From the computations made in Question 14.1 of incremental value e_{i1} and orientation of principal incremental strains compute successive values of the principal finite strains (e_{f1} and e_{f2}) and orientations of the successive strain ellipse long axes θ'. Graph the strains (ordinate) against orientation as abscissa. Discuss the significance of the results.

ANSWERS AND COMMENTS

Strain increment calculations from a pressure shadow

Answer 14.1

The geometric features of Figure 14.19 suggest that the rigid fibre model is most appropriate for the detailed analysis of the strain history. The general correctness of this model is also supported by the fact that the trends of the fibre isogons run sub-parallel to the faces of the pyrite.

The preliminary interpretation of the strain history from the main mass of fibres abutting against the pyrite is reasonably clear: there is a major incremental longitudinal extension directed southeastwards, and this is followed by a rather abrupt anticlockwise rotation to develop an incremental extension in an ENE–WSE direction. The fold structures in the Wildhorn nappe at this locality are generally oriented ENE–WSE, so this sequence shows a strong elongation sub-perpendicular to the fold axes, followed by an extension sub-parallel to the fold axes. The main problem in this initial interpretation concerns the fibre masses

Table 14.1

Increment number	Orientation θ_i'	Pyrite l	δl	$l + \Sigma_1^{n-1} \delta l_n \cos\phi_n$	e_{i1}
1	210	8·0	0·60	8·00	0·08
2	200	8·0	0·65	8·59	0·08
3	190	8·0	0·45	9·20	0·05
4	180	8·0	0·35	9·57	0·04
5	170	8·0	0·40	9·79	0·04
6	160	7·5	0·80	9·50	0·08
7	150	7·5	0·40	10·03	0·04
8	140	7·5	4·60	10·08	0·46
9	150	7·5	2·00	14·96	0·13
10	160	7·5	1·00	16·98	0·06
11	150	7·5	1·80	17·85	0·10
12	140	7·5	1·00	19·36	0·05
13	130	7·5	0·70	19·60	0·04
14	120	8·0	0·60	19·70	0·03
15	110	8·0	1·00	18·87	0·05
16	100	8·0	0·30	18·24	0·02
17	90	8·5	0·50	17·18	0·03
18	80	9·5	0·30	16·60	0·02
19	70	9·5	1·20	14·67	0·08
20	60	9·5	5·00	13·78	0·36

which have fibre axes ENE–WSW and which are discordant to the fibres connected to the pyrite faces. These fibres could be interpreted in two ways: they could have formed early in the sequence and later detached from the pyrite face on which they originated by transform-like sliding (cf. Figure 14.8, fibres a, a', a''), or they could have formed late and complete the obvious anticlockwise rotation sequence of the main fibre mass. Because their basal width is larger than that of the pyrite faces oriented perpendicular to the fibres, and because their orientation and fibre length accords in a more continuous fashion with the last formed fibres of the main fibre mass the second explanation is most satisfactory.

The details of the measurements of fibre lengths and base line lengths for each increment are set out in Table 14.1. The pyrite mass is not equidimensional and so the pyrite component of base line measurement changes as the incremental strain direction changes (cf. Figure 14.18). Because the changing incremental strain axes process through an angle greater than 90° we must be careful not to incorporate early incremental fibre masses (where ϕ_n exceeds 90°) into the rigid base line of calculations for late increments. The shape of the fibres indicate that, although the general sequence of incremental extensions shows an anticlockwise sense, there is a short section which shows clockwise rotations.

The incremental strain data for orientation changes of 10° is graphed in Figure 14.20. The two main incremental peaks (with respect to direction) appreciated previously from a general interpretation are clearly marked, as is the slight rotational reversal in the general anticlockwise directional change of the maximum incremental extension directions.

Inset in Figure 14.20 is a scheme whereby we can represent the incremental changes with respect to a standard circular object. The stippled circle represents a unit circle (a circular mass of rigid pyrite of unit radius), and the two mirror image lines represent a graphical expression of the changing maximum incremental strains as would be recorded by an ideal fibre in this standardized pyrite pressure shadow. These data can be simplified even further for

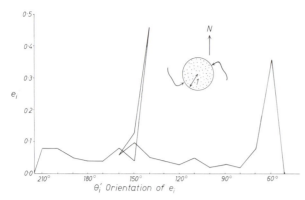

Figure 14.20. *Graphical representation of the changing incremental strains e_i with changed increment orientation θ'_i. Inset shows a method of illustrating the incremental strain history in terms of a standardized rigid object of unit radius. The two mirror symmetric lines with arrow heads illustrate the fibre lengths and directions which would be formed in an ideal pressure shadow.*

plotting strain increment sequences on to a map. Figure 14.21 shows a map of part of the Helvetic nappes of Switzerland. The region is the same as that previously discussed in Session 11 under the general topic of finite strain interpretation. Figure 14.21 includes data from the incremental strain analyses of Questions 13.3 and 14.1 so that the relationships between the specific determinations we have made may be fitted into a perspective of regional geology. Although strain measurements can often be of

interest in themselves from the point of view of technique one should not loose sight of the idea that one of the main reasons for making such measurements is the investigation of regional tectonics. We recommend students to compare the features of the finite strain variations in the XY and XZ planes shown in Figures 11.9 and 11.10 with the incremental strain data of Figure 14.21. There are many problems of interpretation, and we hope that a deeper appreciation of these will be helpful in trying to formulate what additional data should be sought to resolve the geometric problems.

The Helvetic nappes are slices of Mesozoic and Tertiary cover sediments (mostly limestones and marls) which are separated by zones of strong ductile deformation, accompanied by brittle overthrust faults. In the region shown in Figures 11.9, 11.10 and 14.21 the general tectonic plunge is to N 60°E and successively higher nappes are exposed at the surface in a northeasterly direction. The deepest elements are exposed in the southwest and consist of a Pre-Carboniferous basement of crystalline gneisses with infolded Permo-Carboniferous clastic sediments: the Aiguilles Rouges massif. This is overlain by an autochthonous sedimentary cover, and this autochthon is itself tectonically overlain by the Morcles, Diablerets, Wildhorn and Ultrahelvetic nappes. The nappes are strongly affected by folds with wavelengths of kilometre size, and the whole nappe pile is arched into a broad antiformal structure (see profile, Figure 11.10). The general direction of transport deduced from the overall fold geometry, displacements of palinspastic lines was from southeast to northwest. Studies of the relationships of folds and nappe contacts show that

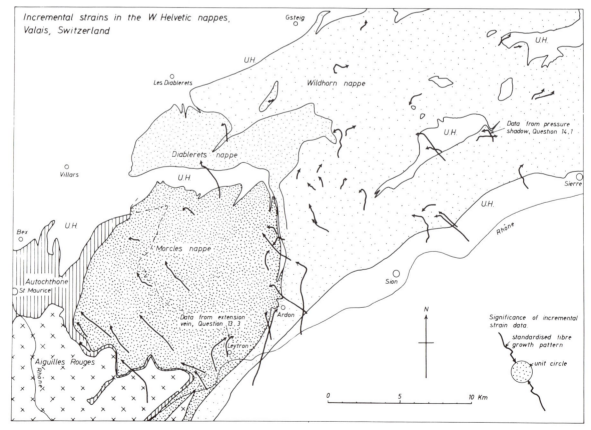

Figure 14.21. *Incremental strain directions in the Western Helvetic nappes of the Valais, Switzerland. Data from Durney and Ramsay (1973) with additional information from D. Dietrich.*

the nappe emplacement sequence was first the transport of the Ultrahelvetic nappes over undeformed Helvetic sediments, followed by the thrusting of the Wildhorn and Diablerets nappes (with piggybacked Ultrahelvetic nappes) over the, as yet undeveloped, Morcles nappe. The development of the Morcles folds was comparatively late, and its northwestward displacement carried all the higher tectonic elements further forward. The overall sequence of nappe evolution was therefore in the reverse order of that seen in the present day pile.

The incremental strain history shows striking variations from nappe to nappe, and within each nappe. We will first set out the main features of the data and then discuss their possible interpretation. Although there are transitions between different areas there are five main types of strain pattern.

1. **Lower limb of the Morcles fold nappe.** The trace of the main axial plane of this fold is marked by a dashed line in the central area of the nappe, and the lower limb lies to the west of this line. The finite strains (Figures 11.9, 11.10) are approximately of plane strain type ($e_2 = 0$) and are generally strong to very strong particularly in the southeastern sector where the nappe approaches the root zone in the Rhône valley. The X-directions of the finite strain ellipsoids are fairly constantly oriented NW–SE and the incremental X-axis strains are sub-parallel to the finite strain X axes. The overall pattern of strain in this zone shows some geometric affinities with the type of shear zones discussed in Session 3 (cf. Figure 11.10 and Figure 3.2), although the shear in the profile is clearly not constant, but weakens from southeast to northwest. This shear zone structure is most strongly developed in the root zone, and the zone of high deformation passes into an intensely deformed part of the Mont Blanc crystalline basement about 4 km southwest of Leytron.

2. **Upper limb of the Morcles fold nappe.** This is located East of the dashed line trace of Figures 11.9 and 14.21. The finite strains are generally of low value although they increase towards the southeast. The incremental strains show maximum extensions oriented initially NW–SE, trending perpendicular to the fold hinges, to a final NE–SW orientation parallel to the fold axes.

3. **Root zone of the upper limb of the Morcles nappe** (between Leytron and Ardon). The finite strains are of a strong flattening type and chocolate tablet structure is very well developed in competent layers and pyrite sheets (Figures 13.19, 13.25, 13.29, 13.30). The analysis of the incremental strains suggests that this overall flattening deformation was built up by successive increments, of more or less plane strain type, initially oriented with incremental X-directions NW–SE to NNE–SSW, followed by maximum elongations oriented NE–SW.

4. **Middle and frontal parts of the Diablerets and Wildhorn nappes.** The finite strains are generally low even though there are well developed large scale buckle folds in the competent limestone layers, and the X-axes are sub-parallel to the fold axes. This weak fold axis parallel stretch is associated with fabrics of **pencil structure** type (Figure 10.26). The incremental strain data are unfortunately sparse, but those sequences which have been established show great irregularity of pattern, with some dominance of extensions in NE–SW directions.

5. **Root zone of the Diablerets and Wildhorn nappes.** The finite strains are quite strong and generally of the flattening type with X-axes oriented mostly NW–SE. The technique of determining finite strains by integrating the incremental strain ellipses derived from an analysis of pressure shadow fibres has been a particularly valuable one in this area. The incremental strains show the maximum incremental X-axes initially oriented NW–SE (sub-perpendicular to the folds) followed by an abrupt transition into a NE–SW direction parallel to the folds (Figure 14.20). The abrupt change in direction of the incremental strain X-axes sometimes is in a clockwise sense, sometimes anticlockwise.

Interpretation of strain data

1. The earliest strain increments are found only in the highest nappe elements, whereas late strain increments are recorded throughout the nappe pile. The Morcles nappe was developed at a comparatively late stage under a tectonic overburden of 5–10 km of sedimentary rocks. Its free movement was therefore constrained and confined to give rather constant local strain gradients and strains.

2. The strongest finite strains of the Morcles nappe were mostly concentrated into a narrow, shear zone-like structure emerging from a shear zone in the underlying basement (Mont Blanc massif), and the shearing movements became less intense and dissipated over a wider zone towards the northwest. The predominantly simple shear strain in this zone produced deformations with characteristics close to those of plane strain ($e_2 = 0$), and the strongest shears led to the formation of the overturned fold limb of the Morcles together with intense thinning of the individual sedimentary formations (thickness reduced by a factor of 0·1). The strong strain developed an intense XY schistosity and X stretching fabrics. The high finite strain probably signifies a relatively high strain rate, and plastic flow by mechanical twinning is the typical deformation mechanism in this shear zone (Figure 7.24A). In contrast, the finite strains in the upper part of the Morcles fold are much lower: the strain rates were lower and the deformation mechanisms were predominantly those of pressure solution and the formation of extension veins (Figure 7.24B). The rock fabrics which dominate are a weak slaty cleavage, accompanied by sub-parallel pressure solution surfaces, associated with weakly developed stretching lineations.

3. The frontal parts of the Morcles, Diablerets and Wildhorn nappes were very weakly strained. The incremental strain sequences are highly irregular because deformation took place with a relatively thin overburden (1–5 km). The nappes were probably forming quite close to the Earth's surface and individual parts of the nappe sheets were free to move differentially with respect to each other (perhaps in response to

gravitational gliding on local slope irregularities), setting up irregular displacement gradients and therefore irregular incremental strains.

4. Positive extensions sub-parallel to the fold axes developed in all the higher tectonic units at a comparatively late stage. These longitudinal strains are perhaps related to transport of nappe pile over underlying irregularities (ramp structures) running in a general NW–SE direction which could be, in part, due to irregular thickening of the late developing Morcles nappe. This late NE–SW stretch was superimposed on rocks which had previously been stretched in a dominant NW–SE direction and led to the formation of an overall flattening type of three-dimensional strain.

5. The rotation sense of the changeover from dominant NW–SE incremental extensions to SW–NE extensions is not systematic. The reason for this is not completely understood. It could have come about by differential advance of different parts of each nappe. A slight preferential northwestward advance of the northeast side of any element would lead to an overall clockwise increment sequence, whereas a preferential advance of the southwest side would lead to an anticlockwise rotational sense (Figure 14.22).

From this discussion it will be seen that this nappe pile, which at first glance looked to have a rather simple overall structure, is, in fact, of great geometrical complexity. However, with the techniques of strain analysis currently available we are in a position to establish what these complexities are and postulate displacement patterns that could account for them.

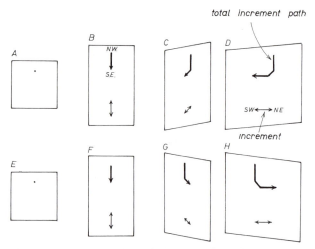

Figure 14.22. *Explanation of changes of rotation sense of incremental strains of the Helvetic nappes. A, B, C and D show an initially sub-horizontal rock element undergoing an early NW–SE stretch and differential advance leading to a clockwise incremental sequence into the final fold axis parallel incremental deformation. E, F, G and H show a sequence leading to anticlockwise rotation of the increments.*

Progressive evolution of a pressure shadow zone

Answer 14.2★

In a rigid fibre model, such as seems appropriate for the analysis of Figure 14.19, the fibre isogons should more or less reflect time correlation lines for the formation of the fibres. It is therefore possible to make a reconstruction of the growth sequence of the pressure shadow (Figure 14.23).

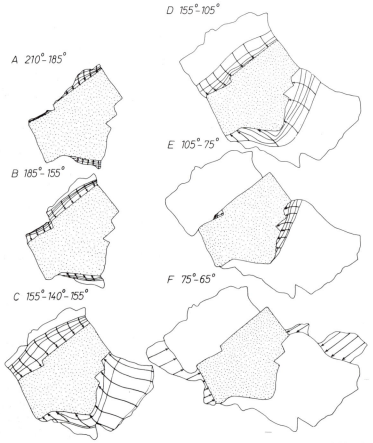

Figure 14.23. *Successive growth stages of the pressure shadow zone of Figure 14.19.*

This figure illustrates very well how the fibres are successively accreted at the pyrite face, and during the late stages of growth it reveals very clearly the sinistral sliding of the early formed quartz fibres along the faces of the pyrite.

Calculation of successive total strains from incremental strain

Answer 14.3★

Using Equations C.16 the strain matrix for each increment was computed, and the successive matrix products giving the strain matrix for each stage in the sequence were evaluated (Equation C.22). The values of the successive principal strains and their orientations were computed from Equations B.19 and B.14. A programmable desk computer was used to make the evaluations, which are set out in Table 14.2.

The data from successive strain states are graphed in Figure 14.24. The successive strains are all of flattening type (both e_1 and e_2 positive), but the main development of the flattening occurs during the last stages of the strain history. It is clearly possible that this late incremental

elongation in a N 60°E orientation, if it proceeded further, might lead to a complete reversal of the positions of major and minor axis of the finite strain ellipse, and it is possible that some of the finite strain states with major axes oriented NE–SW on Figure 11.10 might have been produced in this way.

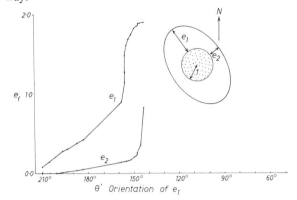

Figure 14.24. Graph showing progressive change of values and orientations of the principal strains e_1 and e_2 resulting from the progressive deformation sequence illustrated in Figure 14.20.

Table 14.2

Increment number	e_{fi}	θ_i'	a	b	c	d	e_{f1}	e_{f2}	θ'
1	0·08	210	1·0600	0·0346	0·0346	1·0200	0·08	0·00	210
2	0·08	200	1·1358	0·0633	0·0622	1·0304	0·16	0·00	205
3	0·05	190	1·1914	0·0746	0·0725	1·0325	0·22	0·00	201
4	0·04	180	1·2391	0·0764	0·0735	1·0324	0·26	0·00	197
5	0·04	170	1·2868	0·0707	0·0664	1·0331	0·30	0·02	194
6	0·08	160	1·3761	0·0466	0·0359	1·0409	0·38	0·04	187
7	0·04	150	1·4169	0·0237	0·0171	1·0504	0·42	0·05	183
8	0·46	140	1·7949	−0·2127	−0·2973	1·2455	0·90	0·15	158
9	0·13	150	1·9774	−0·3587	−0·3717	1·3088	1·14	0·15	156
10	0·06	160	2·0877	−0·4112	−0·4072	1·3265	1·27	0·15	156
11	0·10	150	2·2625	−0·4968	−0·5094	1·3769	1·49	0·15	156
12	0·05	140	2·3396	−0·5534	−0·5707	1·4195	1·61	0·15	155
13	0·04	130	2·3867	−0·6015	−0·6234	1·4666	1·69	0·16	154
14	0·03	120	2·4091	−0·6388	−0·6599	1·5112	1·75	0·17	152
15	0·05	110	2·4302	−0·6799	−0·7192	1·5921	1·83	0·20	150
16	0·02	100	2·4273	−0·7095	−0·7255	1·6325	1·85	0·21	149
17	0·03	90	2·4245	−0·7191	−0·7406	1·6844	1·87	0·24	148
18	0·02	80	2·4196	−0·7265	−0·7377	1·7185	1·88	0·26	148
19	0·08	70	2·4214	−0·6957	−0·7226	1·8232	1·89	0·35	146
20	0·36	60	2·5237	−0·4901	−0·5262	2·2104	1·90	0·84	144

KEYWORDS AND DEFINITIONS

Displacement controlled fibres Fibrous crystals growing in an extension fissure with long axes parallel to the displacement taking place in the fissure.

Face controlled fibres Fibrous crystals developed in pressure shadow zones with long axes controlled by the orientations of the faces of the rigid object (generally euhedral pyrite or magnetite), and forming perpendicular to these faces.

"Ghost" fibres A linear trail of inclusions cross-cutting face controlled fibres and marking out a progressive displacement path of the pressure shadow walls.

Pressure Shadow A region of low strain protected from deformation by a rigid or competent object in a rock of lower competence. The rock matrix and object are frequently detached along their contact, the space being filled with crystals showing fibrous growth forms.

KEY REFERENCES

Choukroune, P. (1971). Contribution à l'étude des mécanismes de la déformation avec schistosité grâce aux cristallisations syncinématiques dans les "zones abritées". *Bull. Soc. Géol. France Ser.* **13**, 257–271.

This is an outstanding paper from the viewpoints of descriptive excellence, the quality of the diagrams and photographs and the theoretical analysis of the data. It describes the very complex geometry of pyrite pressure shadows and of the quartz–chlorite fibres from a region in the northern Pyrenees. Remarkable examples of highly curved and reflexed pressure shadow zones are figured and the forms related to strong rotations occurring during the development of fold phases of regional significance. A good discussion is presented of the relation of pressure shadow geometry in two dimensions and its relation to different types of three-dimensional strain.

Durney, D. W. and Ramsay, J. G. (1973). Incremental strains measured by syntectonic crystal growths. *In* "Gravity and Tectonics" (K. A. De Jong and R. Scholten, eds) 67–96. Wiley, New York.

This paper sets out a detailed synthesis of the different types of pressure shadow geometry, comparing the various types of fibrous crystal infilling with the structure of extension vein fibres. The descriptive section, containing good photographs and diagrams, is followed by an account of the methods for numerical analysis of incremental strain. Incremental strains calculated from vein geometry and pressure shadows at the same locality are compared, and the work concludes with a discussion of the regional significance of incremental strains in the Helvetic nappes of Central Switzerland.

Mugge, O. (1930). Bewegungen von Porphyroblasten in Phylliten und ihre Messung. *Neues Jahrb. Mineral Geol. Palaeont.* **61**, 469–520.

This classic paper describes the features of quartz and chlorite crystals found in pressure shadows around euhedral magnetite crystals of the Ardennes. It relates overall pressure shadow geometry to the tectonic features of schistosity and "longrain" (*X*-direction of the finite strain ellipsoid) and illustrates the changing patterns of growth forms. The paper was the first to show how pressure shadows grow during a progressive deformation, and how the resulting geometry can be used to determine the relative rotation between porphyroblast and rock matrix. Mugge examined various growth models for producing the crystalline infill, these models based on differing rates of separation of wall and object and rates of crystallization.

Pabst, A. (1931). "Pressure-shadows" and the measurements of orientation of minerals in rocks. *Am. Mineral.* **16**, 55–61.

This discusses the geometric features of pressure shadows developed around pyrite, emphasizing the control exerted by the pyrite crystal faces on the growth of quartz fibre. Using a universal stage microscope Pabst investigated the optic axis orientation of the quartz fibres, showing that the orientations were random and concluding that no crystallographic control on the fibres had been exerted by either the pyrite faces or the principal tectonic strain directions.

White, S. H. and Wilson, C. J. L. (1978). Microstructure of some quartz pressure fringes. *Neues Jahrb. Mineral. Geol. Palaeont.* **134**, 33–51.

In Session 13 and 14 we noted that curving fibres may be interpreted in a number of ways. This paper discusses the problem of how it may be possible to determine if the curvature is a primary growth feature or due to later deformation. The authors describe, using transmission electron microscope techniques, sub-microscopic features of curving quartz crystals that suggest that some curving fibres may result from deformation of previous straight fibres.

Appendix A
Strain Parameters

Internal strain in a body leads, in general, to changes of lengths of lines and changes in the angles between intersecting lines. The extent of these changes enables us to measure strain, and it should be noted that all the parameters described below refer to a particular direction in the strained body, and that they generally vary with orientation.

Extension

If the initial length of any line in direction A is given by l_A, and its final length after deformation (in direction A') is l'_A, then the **extension** (sometimes termed **engineers' extension**) is given by:

$$e = \frac{l'_A - l_A}{l_A} \tag{A.1}$$

Extension may have either a positive or a negative sign depending upon whether the line has increased or decreased in length. Its numerical range is $-1 < e < +\infty$. The physical significance of the extension is that it is a measure of the change of length of a line of initial unit length. This parameter is used extensively in analyses of the small deformations arising during elastic behaviour (hence the term "engineers'" extension).

Quadratic extension

This is a length parameter of particular use in analyses of large finite strains where its use simplifies the appearance and practical use of the strain equations. It is designated by the Greek letter "lambda" (λ) and defined as:

$$\lambda = \left(\frac{l'_A}{l_A}\right)^2 \tag{A.2}$$

The parameter always has a positive value lying in the range $0 < \lambda < +\infty$. The quadratic extension and extension are linked by the relationship

$$\lambda = \left(\frac{l'_A}{l_A}\right)^2 = \left(\frac{l_A}{l_A} + \frac{l'_A - l_A}{l_A}\right)^2 = (1 + e)^2 \tag{A.3}$$

In some geological problems it is sometimes convenient to use the reciprocal of this parameter, the so-called **reciprocal quadratic extension** (lambda prime, λ'):

$$\lambda' = \frac{1}{\lambda} \tag{A.4}$$

Logarithmic strain

Logarithmic strain is usually denoted by the Greek "eta" (ε) and defined as:

$$\varepsilon = \log\left(\frac{l'_A}{l_A}\right) \tag{A.5}$$

The logarithmic base can be either 10, or the base for natural logarithms 2·1415. The parameter is sometimes called **natural strain** when the logarithmic base is 2·14. This is perhaps an unfortunate term in that it has in the past given the impression that this parameter is a more "natural" parameter than other measures, and this is not the case. Logarithmic strain is a parameter most suited to the examination of different types of progressive deformation sequences (particularly coaxial paths) and for the graphical plotting of strain data. Logarithmic strain has a numerical range $-\infty < \varepsilon < +\infty$, and it is related to extension and quadratic extension by

$$\varepsilon = \log(1 + e) = \frac{1}{2}\log \lambda \qquad\qquad (A.6)$$

Shear Strain

The parameter used for defining angular changes describes the angular modification of any two initially perpendicular directions. In general (but not always) two initially perpendicular lines loose their orthogonal property, and the angular deflection from the perpendicular defines the **angular shear strain** (Figure 1.4). This is designated by the Greek letter "psi" (ψ), and a sign convention (positive or negative) is required to distinguish left- and right-handed shear sense respectively. For example, relative to a line A', line B' moves in an anticlockwise sense (left-handed shear), so the angular shear strain is positive, whereas relative to line B', line A' moves in a clockwise sense, so the shear strain for B' has the same numerical value, as that for A' but the sign is negative (Figure 8.3).

In many of the equations describing finite strain the parameter which simplifies their appearance is the **shear strain**, designated by the Greek letter "gamma" (γ) and defined as

$$\gamma = \tan \psi \qquad\qquad (A.7)$$

The sign convention for γ follows that for ψ. The physical significance of the shear strain is that it is a measure of the distance "sheared" by the perpendicular to the reference direction at unit distance from the reference line (Figure 1.4).

Another measure of angular deflection used in strain analysis combines both angular and length parameters. This has no special name, it is termed **gamma prime** (γ') and defined as

$$\gamma' = \frac{\gamma}{\lambda} \qquad\qquad (A.8)$$

For direction A', for example, it would be

$$\gamma'_{A'} = \gamma_{A'} \left(\frac{l_A}{l'_A}\right)^2$$

Dimensions of strain parameters

The parameters of strain are all numbers or scalars. This should be contrasted with the parameters of normal- and shear-stress which have the dimensions of force per unit area ($ML^{-1}T^{-2}$), where M is mass, L length and T time.

Appendix B
Strains from displacement

The aim of this appendix is to set out all the basic proofs for the principal features of the two-dimensional finite homogeneous strain state, and to show how they are dependent on the four numerical parameters of the displacement.

1. Displacement

The finite displacement of any point in a two-dimensional surface is defined as the straight line joining the initial position (x, y) and the final position (x', y') (see Figure B.1). It is a vector quantity which has length and orientation, and it can be expressed in terms of two components u and v measured parallel to the x- and y-axes respectively.

The **finite displacement vector** represented by this line does not record the actual **movement path** of the point (x, y) to its final position, it relates only to a connection between the start and the finish of the displacement process.

The field of vectors representing the displacements of all points is known as the **displacement vector field**, and generally has a quite complicated geometry. **Body translation** has a constant field (u and v independent of initial position). All other fields show vector variations with values of the initial point. The movements of all points in a surface can be expressed in a pair of equations, the **coordinate transformation equations**

$$\left.\begin{array}{l} x' = f_1(x, y) \\ y' = f_2(x, y) \end{array}\right\} \tag{B.1}$$

or

$$\left.\begin{array}{l} x = f_3(x', y') \\ y = f_4(x', y') \end{array}\right\} \tag{B.2}$$

These two pairs of equations express the same changes. The first is a **Lagrangian specification** in which the input data (on the right hand side) refers to the *original* positions of the points, whereas the second set, known as **Eulerian equations** have input which refers to the *final* positions.

The simplest general forms of the coordinate transformation equations are linear:

$$\left.\begin{array}{l} x' = ax + by \\ y' = cx + dy \end{array}\right\} \tag{B.3}$$

$$\left.\begin{array}{l} x = Ax' + By' \\ y = Cx' + Dy' \end{array}\right\} \tag{B.4}$$

where a, b, c, d, A, B, C and D are constants. There is a connection between these coefficients:

$$A = \frac{d}{ad - bc} \qquad B = \frac{-b}{ad - bc} \qquad C = \frac{-c}{ad - bc} \qquad D = \frac{a}{ad - bc}$$

The geometrical significance of these coefficients is illustrated in Figure B.2.

The equations can also be represented by a 2×2 matrix, the **strain matrix**

$$\begin{array}{l} x' \\ y' \end{array} = \begin{bmatrix} a & b \\ c & d \end{bmatrix} \begin{array}{l} x \\ y \end{array} \tag{B.5}$$

and the **reciprocal strain matrix**

$$\begin{array}{l} x \\ y \end{array} = \begin{bmatrix} A & B \\ C & D \end{bmatrix} \begin{array}{l} x' \\ y' \end{array} \tag{B.6}$$

The **displacement vector field** is given by

$$\left.\begin{array}{l} u = x' - x = (a - 1)x + by \\ v = y' - y = cx + (d - 1)y \end{array}\right\} \tag{B.7}$$

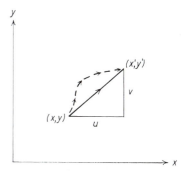

Figure B.1. The displacement vector joins the initial point (x, y) and final point (x', y') and has two components u and v. The movement path of the point (x, y) is shown by the dashed line.

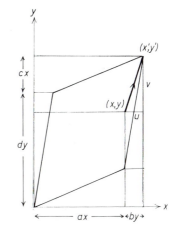

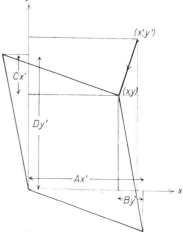

Figure B.2. Geometrical significance of the components a, b, c and d, and A, B, C and D of the coordinate transformation equations.

and the **displacement gradient matrix** is constant

$$\begin{bmatrix} a-1 & b \\ c & d-1 \end{bmatrix}$$

(B.8)

2. Homogeneous strain

Linear coordinate transformation equations have a constant displacement gradient matrix and give rise to a homogeneous state of strain. Homogeneous strain is sometimes defined by the condition that all sets of initially equally spaced parallel straight lines before deformation remain straight, parallel and equally spaced after displacement. Only the orientation and the actual spacing distance are changed. This condition implies, for example that the square units produced by the intersection of any originally orthogonal grid become transformed into identically shaped and identically oriented parallelograms (Figure B.3).

The result of transforming a set of parallel lines given by

$$y = mx + k$$

in the undeformed state is found by transforming all points with initial coordinates (x, y) to new positions using Equation B.4. The new lines are given by:

$$-\frac{cx' + ay'}{ad - bc} = m\left(\frac{dx' - by'}{ad - bc}\right) + k$$

or

$$y' = \left(\frac{c + dm}{a + bm}\right)x' + k\left(\frac{ad - bc}{a + bm}\right)$$

That is the equation of another set of straight lines (see Figure B.3)

$$y' = Mx' + K$$

By varying the values of k to vary the distance of the y-separation of the initial lines to values $2k$, $3k$, etc., then, because K is related to k only by a constant multiplier, the new y-separation of the lines is K, $2K$, $3K$, etc. It therefore follows that the general linear coordinate equations do produce a general homogeneous strain.

3. The strain ellipse concept

One especially important property of homogeneous strain is derived from finding the geometric effect of distorting a circle of unit radius whose centre lies at the origin by the linear displacement equations. The circle

$$x^2 + y^2 = 1$$

is transformed into another shape given by displacing all the points on it according to Equations (B.4):

$$\left(\frac{dx' - by'}{ad - bc}\right)^2 + \left(\frac{-cx' + ay'}{ad - bc}\right)^2 = 1$$

or

$$x'^2\frac{(d^2 + c^2)}{(ad - bc)^2} - 2x'y'\frac{(bd + ac)}{(ad - bc)^2} + y'^2\frac{(a^2 + b^2)}{(ad - bc)^2} = 1$$

(B.9)

an ellipse centred at the origin known as the **strain ellipse** (Figure B.3B). The properties of homogeneous strain may therefore be simply represented by the geometrical properties of this ellipse. As a result of homogeneous straining there are two perpendicular directions which coincide with the major and minor axes of the strain ellipse where extensions reach maximum and minimum values (Figure B.3). These are termed the **principal strains** and the values of the **principal extensions** are e_1 and e_2 ($e_1 \geq e_2$). From the symmetry properties of an ellipse it follows that the properties of strain in a direction making an angle α' with one of the principal strains is the mirror image of that making an angle $-\alpha'$ with the same strain direction, i.e. elongations are identical and shear strains are equal in value but opposite in sign.

In the development of the strain ellipse concept no restrictions were imposed

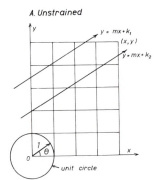

A. Unstrained

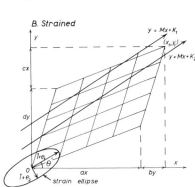

B. Strained

Figure B.3. *Geometrical features of homogeneous strain.*

on the magnitude of the displacement or the internal deformation, the only restrictions were on the homogeneity of the distortion over the area investigated. The concept is therefore valid for both large and small strains, for large permanent flow as well as for small elastic deformations.

4. The reciprocal strain ellipse

The strain ellipse concept is a Lagrangian concept in that its reference is a circle in the initial undeformed state. We can also investigate the form of a reference circle in the deformed material and determine its form before deformation. Taking a circle with unit radius centred at the origin

$$x'^2 + y'^2 = 1$$

and transforming it according to Equation B.3:

$$(ax + by)^2 + (cx + dy)^2 = 1$$

$$(a^2 + c^2)x^2 + 2(ab + cd)xy + (b^2 + d^2)y^2 = 1 \qquad (B.10)$$

which is an ellipse centred at the origin known as the **reciprocal strain ellipse**. If we wish to remove a strain, then it is the reciprocal strain ellipse that must be superposed on the deformed material.

5. Longitudinal strain (extension e) along any line making an initial angle of α with the x-axis

In Figure B.4 the line joining $(0, 0)$ and (x, y) makes an angle α with the x-direction and is of unit length. After displacement (x, y) is positioned at (x', y') and the line now makes an angle of α' with the axis and its length is now $1 + e$ units (where e is the extension). Then

$$x = \cos \alpha$$

$$y = \sin \alpha$$

substituting these values in Equation B.3:

$$x' = a \cos \alpha + b \sin \alpha$$

$$y' = c \cos \alpha + d \sin \alpha$$

From Pythagoras' Theorem:

$$(1 + e)^2 = x'^2 + y'^2 = (a \cos \alpha + b \sin \alpha)^2 + (c \cos \alpha + d \sin \alpha)^2$$

expanding and converting α to 2α

$$[\cos^2 \alpha = \tfrac{1}{2}(1 + \cos 2\alpha); \quad \sin^2 \alpha = \tfrac{1}{2}(1 - \cos 2\alpha); \quad \sin \alpha \cos \alpha = \tfrac{1}{2} \sin 2\alpha]$$

$$\lambda = \tfrac{1}{2}(a^2 - b^2 + c^2 - d^2) \cos 2\alpha + (ab + cd) \sin 2\alpha + \tfrac{1}{2}(a^2 + b^2 + c^2 + d^2) \quad (B.11)$$

where λ is the quadratic extension $(\lambda = (1 + e)^2)$.

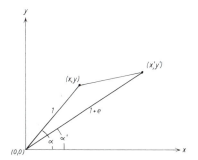

Figure B.4. Change of length of line of initial unit length (join (0, 0) to (x, y) as a result of displacement of (x, y) to (x', y'). The extension is e.

6. Longitudinal strain along any line making a final angle α' with the x-axis

From trigonometrical relationships in Figure B.4:

$$x' = (1 + e) \cos \alpha' \qquad y' = (1 + e) \sin \alpha'$$

substituting these into Equation B.4, using the relationship $x^2 + y^2 = 1$ and simplifying we obtain an equation for the reciprocal quadratic extension $(\lambda' = 1/(1 + e)^2)$,

$$\lambda' = \frac{1}{(ad - bc)^2} [\tfrac{1}{2}(d^2 + c^2 - a^2 - b^2) \cos 2\alpha' - (ac + bd) \sin 2\alpha' + \tfrac{1}{2}(a^2 + b^2 + c^2 + d^2)] \quad (B.12)$$

7. Relationship between α and α'

$$\frac{y'}{x'} = \tan \alpha' = \frac{c \cos \alpha + d \sin \alpha}{a \cos \alpha + b \sin \alpha}$$

$$\tan \alpha' = \frac{c + d \tan \alpha}{a + b \tan \alpha} \tag{B.13a}$$

or

$$\tan \alpha = \frac{c - a \tan \alpha'}{b \tan \alpha' - d} \tag{B.13b}$$

Note that **line rotation w** $= \alpha' - \alpha$.

8. Directions of the principal strains of the strain ellipse after deformation (θ')

Equation B.12 expresses the value of the reciprocal quadratic extension λ' for *any* line after deformation. The orientations of the principal strains can be found by finding its maximum and minimum values. Equation B.12 is differentiated with respect to α' and equated to zero.

$$\frac{d\lambda'}{d\alpha'} = 0 = \frac{1}{(ad - bc)^2}[(a^2 + b^2 - c^2 - d^2)\sin 2\alpha' - 2(ac + bd)\cos 2\alpha']$$

which gives

$$\frac{\sin 2\theta'}{\cos 2\theta'} = \tan 2\theta' = \frac{2(ac + bd)}{a^2 + b^2 - c^2 - d^2} \tag{B.14}$$

where θ' now refers to the orientation of the principal strains (Figure B.3). Solutions in this equation with any 360° range of $2\theta'$ give two values 180° apart. These clearly relate to the major and minor axes of the strain ellipse oriented such that θ' is 90° apart.

9. Initial directions of the lines which will become the principal strains (θ)

Differentiating equation B.11 with respect to α and equating to zero we obtain maxima and minima.

$$\frac{\partial \lambda}{\partial \alpha} = 0 = -(a^2 - b^2 + c^2 - d^2)\sin 2\alpha + 2(ab + cd)\cos 2\alpha$$

which gives

$$\frac{\sin 2\theta}{\cos 2\theta} = \tan 2\theta = \frac{2(ab + cd)}{a^2 - b^2 + c^2 - d^2} \tag{B.15}$$

where θ refers to the initial direction of the line which becomes a principal strain axis (Figure B.3). It is clear from the discussion above that they are perpendicular.

10. Rotation ω

In general θ' and θ will not be equal. The difference defines the rotation ω component of the strain:

$$\omega = \theta' - \theta$$

$$\tan 2\omega = \tan(2\theta' - 2\theta) = \frac{\tan 2\theta' - \tan 2\theta}{1 - \tan 2\theta' \tan 2\theta}$$

Replacing values for $\tan 2\theta'$ (Equation B.14 and 2θ, Equation B.15) and simplifying using the identity

$$\tan 2\omega = \frac{2 \tan \omega}{1 - \tan^2 \omega}$$

we find that

$$\tan \omega = \frac{c - b}{a + d} \tag{B.16}$$

Generally strains are **rotational** strains, but where $c = b$ the rotation is zero and the strain is **irrotational**. This can also be seen by inspection, from a comparison of Equations B.14 and B.15 where $\theta = \theta'$.

11. Values of the principal strains λ_1, λ_2

The values of the principal strains are obtained by substituting the conditions expressed in Equation B.15 (for principal strain directions θ) into equation for the values of quadratic extension (B.11). This then gives the principal quadratic extensions λ_1 and λ_2.

$$\tan 2\theta = \frac{2(ab + cd)}{a^2 - b^2 + c^2 - d^2}$$

Using the identity $\sec^2 2\theta = 1 + \tan^2 2\theta$,

$$\cos 2\theta = \frac{1}{(1 + \tan^2 2\theta)^{1/2}} = \frac{a^2 - b^2 + c^2 - d^2}{[(a^2 - b^2 + c^2 - d^2)^2 + 4(ab + cd)^2]^{1/2}} \qquad (B.17)$$

Using the identity $\operatorname{cosec}^2 2\theta = 1 + \cot^2 2\theta$,

$$\sin 2\theta = \frac{\tan 2\theta}{(1 + \tan^2 2\theta)^{1/2}} = \frac{2(ab + cd)}{[(a^2 - b^2 + c^2 + d^2)^2 + 4(ab + cd)^2]^{1/2}} \qquad (B.18)$$

Substituting Equations B.17 and B.18 in B.11 and simplifying,

$$\lambda_1 \text{ (or } \lambda_2) = \frac{a^2 + b^2 + c^2 + d^2}{2} \pm \tfrac{1}{2}[(a^2 - b^2 + c^2 - d^2)^2 + 4(ab + cd)^2]^{1/2}$$

$$\lambda_1 \text{ (or } \lambda_2) = \tfrac{1}{2}\{a^2 + b^2 + c^2 + d^2 \pm [(a^2 + b^2 + c^2 + d^2)^2 - 4(ad - bc)^2]^{1/2}\} \qquad (B.19)$$

The **ellipticity** R of the strain ellipse is given from (B.19):

$$R = \left(\frac{\lambda_1}{\lambda_2}\right)^{1/2}$$

$$= \left(\frac{a^2 + b^2 + c^2 + d^2 + [(a^2 + b^2 + c^2 + d^2)^2 - 4(ad - bc)^2]^{1/2}}{a^2 + b^2 + c^2 + d^2 - [(a^2 + b^2 + c^2 + d^2)^2 - 4(ad - bc)^2]^{1/2}}\right)^{1/2} \qquad (B.20)$$

12. Area change Δ_A

Any change in area which accompanies deformation is termed **dilatation**, and is represented by the Greek capital "delta" Δ_A. The area of an initial circle is πr^2 and when this is the unit circle its area is π. The strain ellipse derived from this circle has an area $\pi(1 + e_1)(1 + e_2)$, and the dilatation is the proportion given by the equation

$$1 + \Delta_A = \frac{\pi(1 + e_1)(1 + e_2)}{\pi} = (1 + e_2)(1 + e_2) \qquad (B.21)$$

Substitution of the values of the principal strains in terms of the components of the coordinate transformation matrix (using Equation B.19) leads to an expression for the dilatation in terms of these components:

$$1 + \Delta_A = ad - bc \qquad (B.22)$$

This is the determinant of the matrix (leading diagonal product minus the secondary diagonal product).

13. The shearing strain developed along any initial direction α from the x-axis

Figure B.5 shows a line $y = mx$ making an angle α from the x-axis and the line $y = -x/m$ which is perpendicular to it and at an angle of $\alpha + 90°$ to the x-direction ($\tan \alpha = m$). After displacement these two lines are deflected to new positions making angles of α' and α'' to the x axis given from Equation B.4.

$$\frac{-cx' + ay'}{ad - bc} = \frac{m(dx' - by')}{ad - bc}$$

or

$$y' = \frac{x'(md + c)}{mb + a} \qquad (B.23)$$

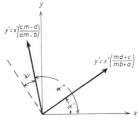

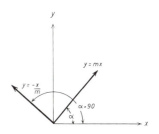

Figure B.5. Relations of angular shear strain ψ to orientation of a line (α before displacement, α' after displacement).

and

$$\frac{m(-cx' + ay')}{ad - bc} = -\frac{(dx' - by')}{ad - bc}$$

or

$$y' = \frac{x'(cm - d)}{am - b} \tag{B.24}$$

The tangent of the angle between these lines is $\tan(90 - \psi)$ where ψ is the angular shear strain

$$\tan(90 - \psi) = \tan(\alpha'' - \alpha') = \frac{\tan \alpha'' - \tan \alpha'}{1 - \tan \alpha'' \tan \alpha'}$$

replacing $\tan \alpha''$ by the slope of the line of Equation B.24 and $\tan \alpha'$ by the slope of (B.23) and simplifying, we obtain:

$$\tan (90 - \psi) = \frac{-[m + (1/m)](ad - bc)}{[m - (1/m)](ab + cd) + (a^2 - b^2 + c^2 - d^2)}$$

using the identities

$$m + \frac{1}{m} = \frac{2}{\sin 2\alpha} \quad \text{and} \quad m - \frac{1}{m} = \frac{-2 \cos 2\alpha}{\sin 2\alpha}$$

$$\gamma_\alpha = \tan \psi = \frac{1}{\tan(90 - \psi)}$$

$$= \frac{2 \cos 2\alpha(ab + cd) - \sin 2\alpha(a^2 - b^2 + c^2 - d^2)}{2(ad - bc)} \tag{B.25}$$

If γ_α is 0, the solution gives a value for the principal strain orientations before deformation (identical to Equation B.15).

14. The shearing strain development along a final direction α' from the x-axis

After substitution of the relations between α and α' expressed by Equation B.13, we obtain:

$$\gamma_{\alpha'} = \frac{[(bm' - d)^2 - (c - am')^2](ab + cd) - (a^2 - b^2 + c^2 - d^2)(c - am')(bm' - d)}{[(c - am')^2 + bm' - d)^2](ad - bc)} \tag{B.26}$$

where $m' = \tan \alpha'$.

It should be clear that *four* components are necessary to define completely the distortional and rotational effects of a finite strain, and that these four terms can be expressed as functions of the coordinate transformation matrix. These four terms could be, for example:

$1 + e_1$		$1 + e_1$		R		R	
$1 + e_2$	or	$1 + e_2$	or	$1 + \Delta_A$	or	$1 + \Delta_A$	etc.
θ		θ'		θ		θ'	
θ'		ω		θ'		ω	

In analysis of strain measurements resulting from geological data we often obtain an *incomplete specification* of these components. Probably the commonest components obtainable in practice are:

R		$1 + e_1$
θ'	or	$1 + e_2$
		θ'

These limitations should be realized when it is necessary to analyse heterogeneous strain fields, where, for a complete mathematical formulation of the strain and displacement field, all four components are necessary throughout the investigated area.

Appendix C
Displacements from Strain

In Appendix B we set out the fundamental relationships between the nature of the coordinate transformation equations and the features of the strain that are set up by the displacement. In Appendix C we will look at the reverse process, that is, if we know the characteristic features of the strain how we may derive the displacements and the coordinate transformations which set up the strain.

Figure C.1 illustrates the geometry of points (x, y) and (x_1, y_1) situated on a unit circle such that they are located at the ends of the two perpendicular radii that will become the two principal axes of longitudinal strain after deformation. They are therefore located at angles of $\theta' - \omega$ from the x-axis and y-axis respectively. After displacement they are located at new positions (x', y') and (x_1', y_1') on the strain ellipse.

Then the following relationships are apparent:

$$x = \cos(\theta' - \omega)$$

$$y = \sin(\theta' - \omega)$$

$$x_1 = -\sin(\theta' - \omega)$$

$$y_1 = \cos(\theta' - \omega)$$

$$x' = (1 + e_1) \cos \theta'$$

$$y' = (1 + e_1) \sin \theta'$$

$$x_1' = -(1 + e_2) \sin \theta'$$

$$y_1' = (1 + e_2) \cos \theta'$$

It follows from the general coordinate transformation giving a homogeneous strain (Equation B.3) that:

$$(1 + e_1) \cos \theta' = a \cos(\theta' - \omega) + b \sin(\theta' - \omega) \tag{C.1}$$

$$(1 + e_1) \sin \theta' = c \cos(\theta' - \omega) + d \sin(\theta' - \omega) \tag{C.2}$$

$$-(1 + e_2) \sin \theta' = -a \sin(\theta' - \omega) + b \cos(\theta' - \omega) \tag{C.3}$$

$$(1 + e_2) \cos \theta' = -c \sin(\theta' - \omega) + d \sin(\theta' - \omega) \tag{C.4}$$

By taking pairs of these equations we can obtain values for a, b, c and d. For example, eliminating b from (C.1) and (C.3):

$$(1 + e_1) \cos \theta' \cos(\theta' - \omega) = a \cos^2(\theta' - \omega) + b \sin(\theta' - \omega) \cos(\theta' - \omega)$$

$$-(1 + e_2) \sin \theta' \sin(\theta' - \omega) = -a \sin^2(\theta' - \omega) + b \cos(\theta' - \omega) \sin(\theta' - \omega)$$

by subtraction

$$(1 + e_1) \cos \theta' \cos(\theta' - \omega) + (1 + e_2) \sin \theta' \sin(\theta' - \omega)$$

$$= a \cos^2(\theta' - \omega) + \sin^2(\theta' - \omega) = a$$

The four equations below are the basic transformations whereby the four components of the coordinate transformation equations may be obtained from four components of the strain.

$$a = (1 + e_1) \cos \theta' \cos(\theta' - \omega) + (1 + e_2) \sin \theta' \sin(\theta' - \omega) \tag{C.5}$$

$$b = (1 + e_1) \cos \theta' \sin(\theta' - \omega) - (1 + e_2) \sin \theta' \cos(\theta' - \omega) \tag{C.6}$$

$$c = (1 + e_1) \sin \theta' \cos(\theta' - \omega) - (1 + e_2) \cos \theta' \sin(\theta' - \omega) \tag{C.7}$$

$$d = (1 + e_1) \sin \theta' \sin(\theta' - \omega) + (1 + e_2) \cos \theta' \cos(\theta' - \omega) \tag{C.8}$$

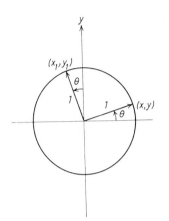

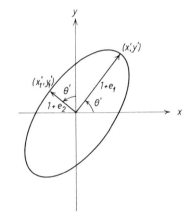

Figure C.1. Relationships between two points (x, y) and (x_1, y_1) situated on the initial unit circle and their positions (x', y') and (x_1', y_1') at the ends of the major and minor diameters of the strain ellipse.

289

Coordinate transformations and displacement gradient matrices for specific types of strain

1. Displacement without rotation or distortion

Coordinate transformations which give only a **body translation** are given by two constant terms A and B specifying the displacement vector component of the body movement, i.e.

$$x' = x + A \qquad \text{strain matrix} \quad \begin{bmatrix} 1 & 0 \\ 0 & 1 \end{bmatrix} \qquad (C.9)$$
$$y' = y + B$$

2. Body rotation ω without distortion

From the geometric features illustrated in Figure (C.2)

$$x = y \tan \omega + \frac{x'}{\cos \omega}$$

$$y' = x' \tan \omega + \frac{y}{\cos \omega}$$

which gives

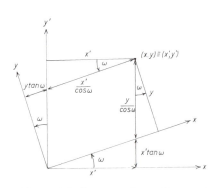

Figure C.2. *Changes of coordinates of a point (x, y) to (x', y') as a result of a body rotation through an angle ω.*

$$x' = x \cos \omega - y \sin \omega \qquad \text{or in matrix form} \quad \begin{bmatrix} \cos \omega & -\sin \omega \\ \sin \omega & \cos \omega \end{bmatrix} \qquad (C.10)$$
$$y' = x \sin \omega + y \cos \omega$$

This matrix is known as a skew-symmetric matrix, the terms b and c—making the secondary diagonal of the matrix—are equal in value but opposite in sign. Only those skew symmetric matrixes with the coefficients showing the special features of Equation C.11 will have no distortion.

$$\begin{bmatrix} (1 - b^2)^{1/2} & -b \\ b & (1 - b^2)^{1/2} \end{bmatrix} \qquad (C.11)$$

3. Simple shear

A simple shear of shear strain γ parallel to the x-axis is given by the matrix

$$\begin{bmatrix} 1 & \gamma \\ 0 & 1 \end{bmatrix} \qquad (C.12)$$

and a simple shear parallel to the y-axis is given by

$$\begin{bmatrix} 1 & 0 \\ \gamma & 1 \end{bmatrix} \qquad (C.13)$$

If the shear zone makes an angle of α with the x-axis, then from the geometry shown in Figure C.3

$$p = (y - x \tan \alpha) \cos \alpha$$

the length of the displacement vector joining (x, y) and (x', y') is

$$\gamma(y - x \tan \alpha) \cos \alpha$$

thus, the component u and v of the displacement vector become

$$u = \gamma(y - x \tan \alpha) \cos^2 \alpha$$

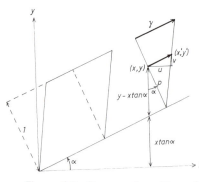

Figure C.3. *Change of position of a point (x, y) to (x', y') as the result of a simple shear with vector oriented at an angle of α to the x-axis.*

$$v = \gamma(y - x \tan \alpha) \cos \alpha \sin \alpha$$

which gives a strain matrix

$$\begin{bmatrix} 1 - \gamma \sin \alpha \cos \alpha & \gamma \cos^2 \alpha \\ \gamma \sin^2 \alpha & 1 + \gamma \sin \alpha \cos \alpha \end{bmatrix} \qquad (C.14)$$

4. General irrotational strain

The general strain matrix is symmetric (see proof, p. 286), the terms of the secondary diagonal of the matrix being equal

$$\begin{bmatrix} a & b = c \\ c = b & d \end{bmatrix} \tag{C.15}$$

If the orientation θ' and values of the principal strains e_1 and e_2 are known this becomes

$$\begin{bmatrix} (1 + e_1) \cos^2 \theta' + (1 + e_2) \sin^2 \theta' & ((1 + e_1) - (1 + e_2)) \sin \theta' \cos \theta' \\ ((1 + e_1) - (1 + e_2)) \sin \theta' \cos \theta' & (1 + e_1) \sin^2 \theta' + (1 + e_2) \cos^2 \theta' \end{bmatrix} \tag{C.16}$$

If only the strain ratio $(1 + e_1)/(1 + e_2) = R$ and the orientation of the principal strains θ' are known this becomes, where dilatation $\Delta_A = 0$:

$$\begin{bmatrix} R^{1/2} \cos^2 \theta' + R^{-1/2} \sin^2 \theta' & \left(\dfrac{R^2 - 1}{R}\right)^{1/2} \sin \theta' \cos \theta' \\ \left(\dfrac{R^2 - 1}{R}\right)^{1/2} \sin \theta' \cos \theta' & R^{1/2} \sin^2 \theta' + R^{-1/2} \cos^2 \theta' \end{bmatrix} \tag{C.17}$$

Strains with no dilatation, and with principal strains $(1 + e_1) = 1/(1 + e_2)$ oriented parallel to the x and y coordinate axes have displacement matrices

$$\begin{bmatrix} 1 + e_1 & 0 \\ 0 & 1/(1 + e_1) \end{bmatrix} \tag{C.18}$$

5. General rotational strain

The general strain matrix has four independent terms (see Figure B.2 for the significance of these terms)

$$\begin{bmatrix} a & b \\ c & d \end{bmatrix} \tag{C.19}$$

whose relationships to the values and orientations of the principal strains, and to the rotation are set out in Equations C.5–C.8. Any general finite strain can be considered as a matrix product between an irrotational (distortion only) component, followed by a body rotation through an angle ω. If the irrotation part is represented by a symmetric matrix

$$\begin{bmatrix} A & B = C \\ C = B & D \end{bmatrix}$$

then the total matrix after anticlockwise body rotation through an angle ω is:

$$\begin{bmatrix} a & b \\ c & d \end{bmatrix} = \begin{bmatrix} A \cos \omega - B \sin \omega & B \cos \omega - D \sin \omega \\ A \sin \omega + B \cos \omega & B \sin \omega + D \cos \omega \end{bmatrix} \tag{C.20}$$

Note that it follows from this matrix that the rotational part of any strain is obtained from

$$\tan \omega = \frac{c - b}{a + d}$$

a result obtained by other arguments in Appendix B (Equation B.16).

6. Superposition of two displacements

The superposition of two displacements is the equivalent of superposing two general strains. Any initial point (x, y) is transformed to another position (x', y') according to the coordinate transformation equations

$$x' = a_1 x + b_1 y$$

$$y' = c_1 x + d_1 y$$

If the points (x', y') are then displaced a second time to new positions (x'', y'') according to the relationships

$$x'' = a_2 x' + b_2 y'$$

$$y'' = c_2 x' + b_2 y'$$

then the total coordinate transformation from (x, y) to (x'', y'') is

$$x'' = a_2(a_1 x + b_1 y) + b_2(c_1 x + d_1 y)$$

$$y'' = c_2(a_1 x + b_1 y) + d_2(c_1 x + d_1 y)$$

i.e.

$$x'' = (a_1 a_2 + c_1 b_2)x + (b_1 a_2 + d_1 b_2)y \qquad (C.21)$$

$$y'' = (a_1 c_2 + c_1 d_2)x + (b_1 c_2 + d_1 d_2)y$$

In matrix form (representative of the matrix product of two strain matrices, this may be written:

$$\begin{bmatrix} a_2 & b_2 \\ c_2 & d_2 \end{bmatrix} \times \begin{bmatrix} a_1 & b_1 \\ c_1 & d_1 \end{bmatrix} = \begin{bmatrix} a_1 a_2 + c_1 b_2 & b_1 a_2 + d_1 b_2 \\ a_1 c_2 + c_1 d_2 & b_1 c_2 + d_1 d_2 \end{bmatrix} \qquad (C.22)$$

The principal strain features can be obtained from the new strain matrix using Equations B.14, B.15 and B.19.

Appendix D
Changes of Lengths and Angles in Strained Bodies

As a result of a finite strain the lengths of lines are generally altered, and the angles between intersecting lines are usually modified. These changes depend only on the distortional effects of the displacement processes, and are independent of body rotation. In addition, angular modifications are independent of dilatation. Some of the methods used to determine the principal strains and their orientations in naturally deformed rocks (Sessions 6, 7 and 8) frequently employ measurements of changes of length and of angles within objects embedded in the rock. It is therefore important to determine how these changes are likely to vary in different directions in a strained material. In order to make the necessary computations we select orthogonal axes x and y in the most convenient position for our analysis; this will not alter the actual geometric properties of the deformed body but it will simplify the mathematical expressions for them. The simplest position for these reference axes coincides with the principal strain directions, that is with the major and minor axes of the strain ellipse. We will now investigate the changes in lengths and modifications in angles by observing the change that has taken place when an initial unit circle (Figure D.1A) is transformed into a strain ellipse (Figure D.1B).

Changes in length

P is located on the undeformed unit circle $x^2 + y^2 = 1$, and has coordinate positions (x, y). After straining it comes to lie on the strain ellipse $x^2/\lambda_1 + y^2/\lambda_2 = 1$ at point $P'(x', y')$. The angle POx was initially ϕ, and as a result of the internal distortions it is modified to $P'Ox = \phi'$. The initial unit length of OP is elongated by an extension e and OP' has quadratic extension λ. Because the x and y axes coincide with the principal strains the ordinate and abscissa of P are modified in proportion to the principal elongations e_1 and e_2

$$x' = x\lambda_1^{1/2} \text{ and } y' = y\lambda_2^{1/2} \tag{D.1}$$

also, from trigonometric relationships

$$x = \cos \phi, \ y = \sin \phi \tag{D.2}$$

combining (D.1) and (D.2)

$$x' = \lambda_1^{1/2} \cos \phi, \ y' = \lambda_2^{1/2} \sin \phi$$

From the application of Pythagoras' theorem in the triangle with hypotenuse OP' and sides x' and y'

$$\lambda = \lambda_1 \cos^2 \phi + \lambda_2 \sin^2 \phi \tag{D.3}$$

This expresses changes in length in terms of the angle ϕ measured in the unstrained state. In most geological computations we make measurements on deformed material and we do not generally know this angle. We therefore transform this equation to express length changes in terms of ϕ'.

$$\left. \begin{array}{l} x' = \lambda^{1/2} \cos \phi' \\ y' = \lambda^{1/2} \sin \phi' \end{array} \right\} \tag{D.4}$$

$$\left. \begin{array}{l} \cos \phi = x'/\lambda_1^{1/2} \\ \sin \phi = y'/\lambda_2^{1/2} \end{array} \right\} \tag{D.5}$$

combining (D.4) and (D.5)

$$\left. \begin{array}{l} \sin \phi = \lambda^{1/2} \sin \phi'/\lambda_2^{1/2} \\ \cos \phi = \lambda^{1/2} \cos \phi'/\lambda_1^{1/2} \end{array} \right\} \tag{D.6}$$

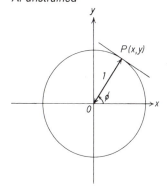

A. unstrained

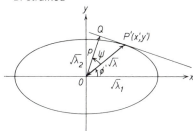

B. strained

Figure D.1. *Change of direction, length and angular shear strain of a line OP to a new position OP' as a result of a homogeneous strain.*

substituting these values in the identity $\cos^2\phi + \sin^2\phi = 1$ using reciprocal quadratic extensions $\lambda' = 1/\lambda$ etc.

$$\lambda' = \lambda_1' \cos^2\phi' + \lambda_2' \sin^2\phi' \qquad \text{(D.7)}$$

For rapid graphic solutions employing the Mohr diagram it is most useful to express (D.7) as a function of $2\phi'$ using the double angle substitutions, $\cos^2\phi = (1 + \cos 2\phi)/2$, $\sin^2\phi = (1 - \cos 2\phi)/2$

$$\lambda' = \frac{\lambda_1' + \lambda_2'}{2} - \frac{(\lambda_2' - \lambda_1')}{2}\cos 2\phi' \qquad \text{(D.8)}$$

Strain generally leads to both extensions and contraction in a body depending upon the values of λ_1, λ_2 and ϕ'. Of the three types of strain ellipse that are possible (Session 4) only one (Field 2) intersects the unit circle. In this ellipse there are two directions, known as directions of **no finite longitudinal strain**, where lines have a finite unchanged length. Their location before and after deformation can be found by substituting $\lambda = \lambda' = 1$ in (D.3) and (D.7) and solving for ϕ and ϕ' respectively.

$$\tan^2\phi = \frac{\lambda_1 - 1}{1 - \lambda_2} \qquad \text{(D.9)}$$

$$\tan^2\phi' = \frac{\lambda_2(\lambda_1 - 1)}{\lambda_1(1 - \lambda_2)} \qquad \text{(D.10)}$$

It must be emphasized that, although these lines show no elongation as a result of the finite strain, it is most unlikely that they retained this property at all stages during the deformation history. In general these directions will have suffered a history of elongation and contraction which have compensated one for the other. Two types of strain ellipse have values of λ_1 and λ_2 which lead to negative quantities for $\tan^2\phi$ and $\tan^2\phi'$ with no real solutions for these angles. These ellipses are illustrated in Figure 4.10, (Fields 1 and 3), and it will be seen that they lie either entirely inside or entirely outside the unit circle from which they were derived.

Changes in angles

A change which takes place in the angles between intersecting lines as a result of deformation depend upon the relative differences in the principal strains expressed in terms of the *ratio of the principal strains* or **ellipticity** $R = (1 + e_1)/(1 + e_2)$. This is because any uniform change in extension in all directions (a dilatation) has no effect on the angular relationships of intersecting lines. The principal strains may always be expressed in terms of the ratio of the two principal strains and the dilatation Δ_A

$$\lambda_1 = R(1 + \Delta_A) \qquad \text{(D.11)}$$

$$\lambda_2 = (1 + \Delta_A)/R \qquad \text{(D.12)}$$

Because the dilatation effect does not lead to change of angular relationships the initial angle α between two lines is modified to the same angle α' providing R is constant.

As a result of strain, the angle (Figure D.1A, ϕ) a line makes with the direction which becomes the principal extension is changed to ϕ'

$$\tan\phi' = y'/x'$$

using Equation D.1

$$\tan\phi' = y\lambda_2^{1/2}/x\lambda_1^{1/2}$$

Because $y/x = \tan\phi$ we can derive the relationship between ϕ' and ϕ, in a form used extensively by Wettstein (1886) and sometimes known as **Wettstein's equation**

$$\tan\phi' = \tan\phi/R \qquad \text{(D.13)}$$

As R is always greater than unity $\phi' < \phi$, and line elements therefore always move to make a smaller angle with the principal extension.

Shear strain

The two directions which are initially perpendicular and which become the two principal strains are clearly always directions of zero shear strain. All other initially perpendicular lines loose their orthogonal relationships during straining by an angle (ψ) which depends on their intitial orientation (Figure D.1, ϕ) and the strain ratio R. We will now compute the value of the angular shear strain ψ and derive formulae of great practical use in geological strain analysis, particularly in determinations of strain from deformed fossils. In Figure D.1A, the tangent to the unit circle has been drawn at the point P, and this line is perpendicular to OP. In the deformed state (Figure D.1B) the circle $x^2 + y^2 = 1$ is transformed into the strain ellipse $x^2/\lambda_1 + y^2/\lambda_2 = 1$, and the original tangent to the unit circle becomes a tangent to the ellipse at the point $P'(x', y')$ given by

$$\frac{xx'}{\lambda_1} + \frac{yy'}{\lambda_2} = 1$$

replacing for x' and y' (D.1)

$$x \cos \phi/\lambda_1^{1/2} + y \sin \phi/\lambda_2^{1/2} = 1$$

This tangent is no longer perpendicular to OP; if a line OQ is drawn perpendicular to this tangent the angle $P'OQ$ is a measure of the angular shear strain ψ. We now compute this angle ψ in triangle $P'OQ$ knowing OP' is of length $\lambda^{1/2}$, and OQ is given from the standard coordinate geometry formula that the length of a perpendicular from the origin on to $ax + by = c$ is $c/(a^2 + b^2)^{1/2}$

$$p = \left(\frac{1}{\cos^2 \phi/\lambda_1 + \sin^2 \phi/\lambda_2} \right)^{1/2}$$

In triangle $P'OQ$

$$\sec \psi = \lambda^{1/2}/p = \lambda(\cos^2 \phi/\lambda_1 + \sin^2 \phi/\lambda_2)^{1/2}$$

From the definition of shear strain $\gamma = \tan \psi$

$$\gamma^2 = \tan^2 \psi = \sec^2 \psi - 1 = \lambda(\cos^2 \phi/\lambda_1 + \sin^2 \phi/\lambda_2) - 1$$

substituting for λ (Equation D.3) and simplifying the result using the identity $(\sin^2 \phi + \cos^2 \phi)^2 = 1$

$$\gamma = \frac{\lambda_1 - \lambda_2}{(\lambda_1 \lambda_2)^{1/2}} \cos \phi \sin \phi \tag{D.14}$$

or using Equations D.11 and D.12

$$\gamma = (R - 1/R)\cos \phi \sin \phi \tag{D.15}$$

These expressions are in terms of the angle ϕ measured in the unstrained state, and for geological computations it is generally more convenient to change them so that the angles are measured in the deformed state by substituting Equation D.6

$$\gamma = \frac{\lambda_1 - \lambda_2}{(\lambda_2 \lambda_2)^{1/2}} \frac{\lambda^{1/2} \cos \phi'}{\lambda_1^{1/2}} \frac{\lambda^{1/2} \sin \phi'}{\lambda_2^{1/2}}$$

$$\frac{\gamma}{\lambda} = \left(\frac{1}{\lambda_2} - \frac{1}{\lambda_1} \right) \sin \phi' \cos \phi' \tag{D.16}$$

This is extremely important for rapid analysis of strain using graphical techniques (Session 6) and it is generally expressed in a slightly different form by defining a new strain parameter (gamma prime, γ') by combining shear strain and quadratic extension $\gamma' = \gamma/\lambda$. Converting the single angles to double angles using $\sin \phi' \cos \phi' = (\sin 2\phi')/2$, and using the reciprocals of the quadratic extensions $\lambda_1' = 1/\lambda_1$

$$\gamma' = \frac{(\lambda_2' - \lambda_1')}{2} \sin 2\phi' \tag{D.17}$$

Another useful expression for shear strain may be derived by substituting for λ (D.3) in (D.16), dividing throughout by $\lambda_2 \cos^2 \phi$ and simplifying using $R = (\lambda_1/\lambda_2)^{1/2}$

$$\gamma = \frac{(R^2 - 1) \tan \phi'}{1 + R^2 \tan^2 \phi'} \tag{D.18}$$

Maximum shear strain

There are always two directions in a strained body symmetrically arranged to the principal extensions where the shear strain attains maximum and minimum values. These may be found by differentiating (D.14) with respect to ϕ and equating to zero

$$\frac{d\gamma}{d\phi} = \frac{\lambda_1 - \lambda_2}{(\lambda_1 \lambda_2)^{1/2}} \cos 2\phi = 0$$

a condition which occurs where $(\lambda_1 - \lambda_2)/(\lambda_1 \lambda_2)^{1/2} = 0$, or when $\cos 2\phi = 0$. The first condition only refers to specific types of strain ($\lambda_1 = \lambda_2$, i.e. uniform dilation where $\gamma = 0$ for all values of ϕ; where $1/\lambda_1 = 0$, i.e. infinitely large strain where γ is infinity for all values of ϕ except $\phi = 0$), the general condition for all strains is where $\cos 2\phi = 0$ or $\phi = \pm 45°$. Using (D.13) maximum shear strain occurs in the deformed body in direction where $\phi' = \tan^{-1}(1/R)$. The maximum and minimum values of the shear strain may be found by putting $\phi = \pm 45°$ in (D.15)

$$\gamma_{max} = \frac{\lambda_1 - \lambda_2}{2(\lambda_1 \lambda_2)^{1/2}} = \tfrac{1}{2}(R - 2 + 1/R) \tag{D.19}$$

$$\gamma_{min} = \frac{\lambda_2 - \lambda_1}{2(\lambda_1 \lambda_2)^{1/2}} = \tfrac{1}{2}(2 - R - 1/R) \tag{D.20}$$

These positions of maximum and minimum shearing strain do not generally coincide with the directions of no finite longitudinal strain.

Source List

This list contains titles of papers, books and articles which are recommended for further study of the subjects covered in this volume. Space limitations have not permitted anything like a complete list for each topic so we have selected the titles in three ways:

1. general books and special reports of conferences, especially those giving a wide variety of ideas and with extensive reference lists;
2. individual papers which, in our opinion, provide particularly useful background reading for the topics of this volume;
3. papers of historical interest which formulated new ideas or which, because of their clarity of expression and breadth of cover, provided strong impetus to advances in structural geology.

Strain

General

Ramsay, J. G. and Wood, D. S. (1976): A discussion on natural strain and geological structure. *Phil. Trans. R. Soc. Lond. A* **283**.

Anhaeusser, C. R. (1969). A comparison of pebble and fold deformation in the Nelspruit granite aureole, Barberton Mountain Land. *Trans. Geol. Soc. S. Afr.* **72**, 49–60.

Badoux, H. (1963). Les bélemnites tronçonées de Leytron (Valais). *Bull. Lab. Géol. Min. Géoph. Musée Géol. Univ. Lausanne* **138**, 1–7.

Barr, M. and Coward, M. P. (1974). A method for the measurement of volume change. *Geol. Mag.* **111**, 293–296.

Beach, A. (1979). The analysis of deformed belemnites. *J. Struct. Geol.* **1**, 127–135.

Bell, A. M. (1979). Factorization of finite strains in three dimensions—a computer method. *J. Struct. Geol.* **1**, 163–167.

Bell, A. M. (1981). Strain factorizations from lapilli tuff, English Lake District. *J. Geol. Soc. Lond.* **138**, 463–474.

Bell, T. H. (1981). Foliation development—the contribution, geometry and significance of progressive, bulk, inhomogeneous shortening. *Tectonophysics* **75**, 273–296.

Bilby, B. A., Eshelby, J. D. and Kundu, A. K. (1975). The change of shape of a viscous ellipsoidal region embedded in a slowly deforming matrix having a different viscosity. *Tectonophysics* **28**, 265–274.

Blake, J. F. (1878). On the measurements of the curves formed by cephalopods and other mollusks. *Phil. Mag.* **5**, 241–262.

Borradaile, G. J. (1977). On cleavage and strain: results of a study in West Germany using tectonically deformed sand dykes. *J. Geol. Soc. Lond.* **133**, 146–164.

Borradaile, G. J. (1979). Strain study of the Caledonides in the Islay region, SW Scotland: implication for strain histories and deformation mechanisms in green-schists. *J. Geol. Soc. Lond.* **136**, 77–88.

Borradaile, G. J. (1981). Minimum strain from conglomerates with ductility contrast. *J. Struct. Geol.* **3**, 295–297.

Borradaile, G. J. and Poulsen, K. H. (1981). Tectonic deformation of pillow lava. *Tectonophysics* **79**, T17–T26.

Boulter, C. A. (1976). Sedimentary fabrics and their relation to strain-analysis methods. *Geology* **4**, 141–146.

Brace, W. F. (1960). Analysis of large two dimensional strain in deformed rock. *Rep. 21st Int. Geol. Congr.*, Copenhagen, Part 18, 261–269.

Brace, W. F. (1961). Mohr construction in the analysis of large geologic strain. *Geol. Soc. Am. Bull.* **72**, 1059–1080.

Breddin, H. (1956). Die tektonische Deformation der Fossilien im Rheinischen Schiefergebirge. *Z. deutsch. geol. Ges.* **106**, 227–305.

Breddin, H. (1957). Tektonische Fossil- und Gesteinsdeformation im Gebiet von St. Goarshausen (Rheinisches Schiefergebirge). *Decheniana* **110**, 289–350.

Breddin, H. (1964). Die tektonische Deformation der Fossilien und Gesteine der Molasse von St. Gallen (Schweiz) *Geol. Mitt.* **4**, 1–68.

Breddin, H. (1967). Quantitative Tektonik. 1. Teil. *Geol. Mitt.* **7**, 205–238.

Brun, J. P. and Pons, J. (1981). Strain patterns of pluton emplacement in a crust undergoing non-coaxial deformation, Sierra Morena, Southern Spain. *J. Struct. Geol.* **3**, 219–229.

Bucher, W. H. (1944). The stereographic projection, a handy tool for the practical geologist. *J. Geol.* **52**, 191–212.

Burns, K. L. and Spry, A. H. (1969). Analysis of the shape of deformed pebbles. *Tectonophysics* **7**, 177–196.

Chapman, T. J., Milton, N. J. and Williams, G. D. (1979). Shape fabric variations in deformed conglomerates at the base of the Laksefjord Nappe, Norway. *J. Geol. Soc. Lond.* **136**, 683–691.

Chapple, W. M. (1968). The analysis of strain in deformed rocks: a discussion. *J. Geol.* **76**, 491–494.

Cloos, E. (1943). Distortion of stratigraphic thicknesses due to flowage and folding. *Am. Geoph. Un. Tr.*, 273–280.

Cloos, E. (1947). Oolite deformation in the South Mountain Fold, Maryland. *Geol. Soc. Am. Bull.* **58**, 843–918.

Cloos, E. (1971). "Microtectonics", 234 pp. Johns Hopkins University, Baltimore.

Cobbold, P. R. (1979). Removal of finite deformation using strain trajectories. *J. Str. Geol.* **1**, 67–72.

Cobbold, P. R. (1980). Compatibility of two-dimensional strains and rotations along strain trajectories. *J. Str. Geol.* **2** (1980), 379–382.

Coward, M. P. (1980). The analysis of flow profiles in a basaltic dyke using strained vesicles. *J. Geol. Soc. Lond.* **137**, 605–615.

Coward, M. P. and James, P. R. (1974). The deformation patterns of two Archean greenstone belts in Rhodesia and Botswana. *Prec. Res.* **1**, 235–258.

Daubrée, G. A. (1876). Expériences sur la schistosité des roches et sur les déformations des fossiles, corrélatives de ce phénomène. *Compt. Rend. Acad. Sci. Paris* **82**, 710, 798.

Debat, P., Sirieys, P., Déramond J. and Soula, J. C. (1975). Paleodéformations d'un massif orthogneissique. *Tectonophysics* **28**, 159–183.

De Paor, D. G. (1980). Some limitations of the R_f/ϕ technique of strain analysis. *Tectonophysics* **64**, T29–T31.

Da Paor, D. G. (1981). Strain analysis using deformed line distributions. *Tectonophysics* **73**, T9–T14.

Da Paor, D. G. (1981). A new technique of strain analysis using three-dimensional distributions of passively deformed linear markers. *Tectonophysics* **76**, T13–T16.

Déramond, J. and Litaize, D. (1976). Méthode informatique pour la détermination du taux de déformation. Applications. *Bull. Soc. géol. France*, série 7, **18**, 1423–1433.

Déramond, J. and Rambach J.-M. (1979). Mesure de la déformation dans la nappe de Gavarnie (Pyrénées centrales): interprétation cinématique. *Bull. Soc. géol. France*, série 7, **21**, 201–211.

De Wit, M. J. (1974). On the origin and deformation of the Fleur de Lys metaconglomerate, Appalachian fold belt, Northwest Newfoundland. *Can. J. Earth Sci.* **11**, 1168–1180.

Dunnet, D. (1969). A technique of finite strain analysis using elliptical particles. *Tectonophysics* **7**, 117–136.

Dunnet, D. and Siddans, A. W. B. (1971). Non random sedimentary fabrics and their modification by strain. *Tectonophysics* **12**, 307–325.

Ehlers, C. (1976). Homogeneous deformation in Precambrian supracrustal rocks of Kumlinge area, Southwest Finland. *Prec. Res.* **3**, 481–504.

Elliott, D. (1970). Determination of finite strain and initial shape from deformed elliptical objects. *Geol. Soc. Am. Bull.* **81**, 2221–2236.

Elliott, D. (1972). Deformation paths in structural geology. *Geol. Soc. Am. Bull.* **83**, 2621–2638.

Engelder, T. and Engelder, R. (1977). Fossil distortion and décollement tectonics of the Appalachian Plateau *Geology* **5**, 457–460.

Fanck, A. (1929). Die bruchlose Deformation von Fossilien durch tektonischen Druck und ihr Einfluss auf die Bestimmung der Arten. *Diss. Zürich*, 59 pp.

Ferguson, C. C. (1981). A strain reversal method for estimating extension from fragmented rigid inclusions. *Tectonophysics* **79**, T43–T52.

Flinn, D. (1956). On the deformation of the Funzie conglomerate, Fetlar, Shetland. *J. Geol.* **64**, 480–505.

Flinn, D. (1962). On folding during three dimensional progressive deformation. *Q. J. Geol. Soc. Lond.* **118**, 385–428.

Flinn, D. (1978). Construction and computation of three-dimensional progressive deformations. *J. Geol. Soc. Lond.* **135**, 291–305.

Flinn, D. (1979). The deformation matrix and the deformation ellipsoid. *J. Str. Geol.* **1**, 299–307.

Fry, N. (1979). Density distribution techniques and strained length methods for determination of finite strains. *J. Str. Geol.* **1**, 221–229.

Fry, N. (1979). Random point distributions and strain measurement in rocks. *Tectonophysics* **60**, 89–105.

Furtak, H. (1962). Die "Brechung" der Schiefrigkeit. *Geol. Mitt.* **2**, 177–196.

Furtak, H. and Hellermann, E. (1961). Die tektonische Verformung von pflanzlichen Fossilien des Karbons. *Geol. Mitt.* **2**, 49–69.

Gay, N. C. (1968a). The motion of rigid particles embedded in a viscous fluid during pure shear deformation of the fluid. *Tectonophysics* **5**, 81–88.

Gay, N. C. (1968b). Pure shear and simple shear deformation of inhomogeneous viscous fluids. 1. Theory— *Tectonophysics* **5**, 211–234.

Gay, N. C. (1968c). 2. The determination of the total finite strain in a rock from objects such as deformed pebbles. *Tectonophysics* **5**, 295–302.

Gay, N. C. (1969). The analysis of strain in the Barberton Mountain Land, Eastern Transvaal, using deformed pebbles. *J. Geol.* **77**, 377–396.

Ghosh, S. K. (1973). Compression and simple shear of test models with rigid and deformable inclusions. *Tectonophysics* **17**, 133–175.

Graham, R. H. (1978). Quantitative deformation studies in the Permian rocks of the Alpes Maritimes. *Proc. Goguel Symp.* (*Bur. Rech. Géol. Mines, France*), 220–238.

Gratier, J. P. and Vialon, P. (1980). Deformation pattern in a heterogeneous material: Folded and cleaved sedimentary cover immediately overlying a crystalline basement (Oisans, French Alps). *Tectonophysics* **65**, 151–180.

Hanna, S. S. and Fry, N. (1979). A comparison of methods of strain determination in rocks from Southwest Dyfed (Pembrokeshire) and adjacent areas. *J. Str. Geol.* **1**, 155–162.

Harker, A. (1885). On slaty cleavage and allied rock structures with special reference to the mechanical theories of their origin. *Rep. Br. Ass.* 55th meeting, 1–40.

Harvey, P. K. and Ferguson, C. C. (1981). Directional properties of polygons and their application to finite strain estimations. *Tectonophysics* **74**, T33–T42.

Haughton, S. (1856). On slaty cleavage and the distortion of fossils. *Phil. Mag.* Ser. 4 **12**, 1–13.

Heim, A. (1878). Untersuchungen über den Mechanismus der Gebirgsbildung. Schwabe, Basel 85p.

Helm, D. G. and Siddans, A. W. B. (1971). Deformation of a slaty, lapillar tuff in the English Lake District: Discussion. *Geol. Soc. Am. Bull.* **82**, 523–531.

Hobbs, B. E., Means, W. D. and Williams, P. F. (1976). "An Outline of Structural Geology", 571 pp. Wiley International, New York.

Hobbs, B. E. and Talbot, J. L. (1966). The analysis of strain in deformed rocks. *J. Geol.* **74**, 500–513.

Holst, T. B. (1982). The role of initial fabric on strain determination from deformed ellipsoidal objects. *Tectonophysics* **82**, 329–350.

Hossack, J. R. (1968). Pebble deformation and thrusting in the Bygdin area (S. Norway). *Tectonophysics* **5**, 315–339.

Hossain, K. M. (1979). Determination of strain from stretched belemnites. *Tectonophysics* **60**, 279–288.

Hsu, T. C. (1966). The characteristics of coaxial and non-coaxial strain paths. *J. Strain Anal.* **1**, 216–222.

Jaeger, J. C. (1956). "Elasticity, Fracture and Flow", 208 pp. Methuen, London.

Kligfield, R., Carmignani, L. and Owens, W. H. (1981). Strain analysis of a Northern Apennine shear zone using deformed marble breccias. *J. Str. Geol.* **3**, 421–436.

Kligfield, R., Owens W. H. and Lowrie, W. (1981). Magnetic susceptibility anisotropy, strain and progressive deformation in Permian sediments from the Maritime Alps (France). *Earth Planet. Sci. Lett.* **5**, 181–189.

Kneen, S. (1976). The relationship between the magnetic and strain fabrics of some haematite bearing Welsh slates. *Earth Planet. Sci. Lett.* **31**, 413–416.

Lisle, R. J. (1977). Clastic grain shape and orientation in relation to cleavage from the Aberystwyth Grits, Wales. *Tectonophysics* **39**, 381–395.

Lisle, R. J. (1977). Estimation of tectonic strain ratio from the mean shape of deformed elliptical markers. Geol. Mijnb **56**, 140–144.

Lisle, R. J. (1979). Strain analysis using deformed pebbles: the influence of initial pebble shape. *Tectonophysics* **60**, 263–277.

March, A. (1932). Mathematische Theorie der Regelung nach der Korngestalt. *Z. Krist.* **81**, 285–297.

Matthews, P. E., Bond, R. A. B. and Van den Berg. J. J. (1974). An algebraic method of strain analysis using elliptical markers. *Tectonophysics* **24**, 31–67.

McKenzie, D. (1979). Finite deformation and fluid flow. *Geophys. J. R. Astr. Soc.* **58**, 689–715.

McLeish, A. J. (1971). Strain analysis of deformed Pipe Rock in the Moine Thrust zone, Northwest Scotland. *Tectonophysics* **12**, 469–504.

Means, W. D. (1976). "Stress and Strain", 339 pp. Springer-Verlag, Heidelberg.

Means, W. D., Hobbs, B. E., Lister, G. S. and Williams, P. F. (1981). Vorticity and non-coaxiality in progressive deformations. *J. Struct. Geol.* **2**, 371–378.

Mimran, Y. (1976). Strain determination using a density-distribution technique and its application to deformed Upper Cretaceous Dorset chalks. *Tectonophysics* **31**, 175–192.

Mitra, S. (1978). Microscopic deformation mechanisms and flow laws in quartzites within the South Mountain anticline. *J. Geol.* **86**, 129–152.

Mitra, S. and Tullis, J. (1979). A comparison of intracrystalline deformation in naturally and experimentally deformed quartzites. *Tectonophysics* **53**, T21–T27.

Mosely, H. (1838). On the geometric form of turbinated and discoid shells. *Phil. Trans. R. Soc. Lond.* **1**, 351–370.

Mosher, S. (1980). Pressure solution deformation of conglomerates in shear zones, Narragansett Basin, Rhode Island. *J. Str. Geol.* **2**, 219–225.

Mukhopadhyay, D. (1973). Strain measurements from deformed quartz grains in the slaty rocks from the Ardennes and the Northern Eifel. *Tectonophysics* **16**, 279–296.

Nadai, A. (1963). "Theory of Flow and Fracture of Solids", 705 pp. McGraw-Hill, New York.

Nickelsen, R. P. (1966). Fossil distortion and penetrative rock deformation in the Appalachian Plateau Pennsylvania *J. Geol.* **74**, 924–931.

Oertel, G. (1970). Deformation of slaty, lapillar tuff in the Lake District, England. *Geol. Soc. Am. Bull.* **81**, 1173–1188.

Owens, W. H. (1973). Strain modification of angular density distributions. *Tectonophysics* **16**, 249–261.

Owens, W. H. and Bamford D. (1976). Magnetic, seismic, and other anisotropic properties of rocks. *Phil. Trans. R. Soc. Lond.* A **283**, 55–68.

Percevault, M. N. and Cobbold, P. R. (1982). Mathematical removal of regional ductile strains in the Central Bittany: evidence for wrench tectonics. *Tectonophysics* **82**, 317–328.

Pfiffner, O. A. and Ramsay, J. G. (1982). Constraints on geological strain rates, arguments from finite strain states of naturally deformed rocks. *J. Geophys. Res.* **87**, 311–321.

Phillips, F. C. (1954). "The Use of Stereographic Projection in Structural Geology." Edward Arnold, London.

Phillips, J. (1857). Report on cleavage and foliation in rocks. *Rept. Brit. Assoc. Sci.* **269**, 60–61.

Plessmann, W. (1966). Diagenetische und kompressive Verformung in der Oberkreide des Harz—Nordrandes sowie im Flysch von San Remo. *N. Jb. Geol. Pal. Mh.* 480–493.

Ragan, D. M. (1973). "Structural Geology", 2nd edn. Wiley, New York.

Ramberg, H. (1975). Particle paths, displacement and progressive strain applicable to rocks. *Tectonophysics* **28**, 1–37.

Ramberg, H. and Ghosh, S. K. (1977). Rotation and strain of linear and planar structures in three-dimensional progressive deformation. *Tectonophysics* **40**, 309–337.

Ramsay, J. G. (1967). "Folding and Fracturing of Rocks", 568 pp. McGraw-Hill, New York.

Ramsay, J. G. (1969). The measurement of strain and displacement in orogenic belts. *In* "Time and Place in Orogeny" (P. E. Kent *et al.*, eds) 43–79. Special Publication 3. Geological Society, London.

Ramsay, J. G. (1976). Displacement and strain. *Phil. Trans. R. Soc. Lond.* A **283**, 3–25.

Ramsay, J. G. and Wood, D. S. (1973). The geometric effects of volume change during deformation processes. *Tectonophysics* **16**, 263–277.

Roberts, B. and Siddans, A. W. B. (1971). Fabric studies in the Llwyd Mawr ignimbrite, Caernarvonshire, North Wales. *Tectonophysics* **12**, 283–306.

Roder, G. H. (1977). Adaption of polygonal strain markers. *Tectonophysics* **43**, T1–T10.

Sanderson, D. J. (1976). The superposition of compaction and plane strain. *Tectonophysics* **30**, 35–54.

Sanderson, D. J. (1977). The analysis of finite strain using lines with an initial random orientation. *Tectonophysics* **43**, 199–211.

Sanderson, D. J. and Meneilly, A. W. (1981). Analysis of three dimensional strain modified uniform distributions: ardalusite fabrics from a granite aureole. *J. Struct. Geol.* **3**, 109–116.

Schwerdtner, W. M. (1973). A scale problem in paleostrain analysis. *Tectonophysics* **16**, 47–54.

Schwerdtner, W. M. (1977). Geometric interpretation of regional strain analysis. *Tectonophysics* **39**, 515–531.

Seymour, D. B. and Boulter, C. A. (1979). Tests of computerised strain analysis methods by the analysis of simulated deformation of natural unstrained sedimentary fabrics. *Tectonophysics* **58**, 221–235.

Shimamoto, T. and Ikeda, Y. (1976). A simple algebraic method for strain estimation from deformed ellipsoidal objects. *Tectonophysics* **36**, 315–337.

Siddans, A. W. B. (1980). Analysis of three-dimensional homogeneous, finite strain using ellipsoidal objects. *Tectonophysics* **64**, 1–16.

Siddans, A. W. B. (1980). Elliptical markers and non-coaxial strain increments. *Tectonophysics* **67**, T21–T25.

Sorby, H. C. (1853). On the origin of slaty cleavage. *Edin. New. Philos. J.* **55**, 137–148.

Sorby, H. C. (1856). On the theory of the origin of slaty cleavage. *Phil. Mag.* **12**, 127–135.

Sorby, H. C. (1858). On some facts connected with slaty cleavage. Rep. 27th. meeting British Assoc., 92–93.

Sorby, H. C. (1908). On the application of quantitative methods to the study of the structure and history of rocks. *Q. J. Geol. Soc. Lond.* **64**, 171–232.

Stauffer, M. R. and Burnett, A. I. (1979). Down-plunge viewing: a rapid method for estimating the strain ellipsoid for large clasts in deformed rocks. *Can. J. Earth Sci.* **16**, 290–304.

Stringer, P. and Treagus, J. E. (1980). Non axial planar S_1 cleavage in the Hawick Rocks of the Galloway area, Southern Uplands, Scotland. *J. Struct. Geol.* **2**, 317–331.

Talbot, C. J. (1970). The minimum strain ellipsoid using deformed quartz veins. *Tectonophysics* **9**, 47–76.

Tan, B. K. (1973). Determination of strain ellipses from deformed ammonites. *Tectonophysics* **16**, 89–101.

Tan, B. K. (1974). Deformation of particles developed around rigid and deformable nuclei. *Tectonophysics* **24**, 243–257.

Thakur, V. C. (1972). Computation of the values of the finite strains in the Molare region, Ticino, Switzerland, using stretched tourmaline crystals. *Geol. Mag.* **109**, 445–450.

Thompson, D. W. (1942). "On Growth and Form." Univ. Press Cambridge, 1–1116.

Thompson, W. and Tait, P. G. (1879). "Principles of Mechanics and Dynamics", Part 1 (paperback version published 1962) 508 pp. Dover.

Tobisch, O. T., Fiske, R. S., Sacks, S. and Taniguchi, D. (1977). Strain in metamorphosed volcanoclastic rocks and its bearing on the evolution of orogenic belts. *Geol. Soc. Am. Bull.* **88**, 23–40.

Truesdell, C. and Toupin, R. A. (1960). The classical field theories. *In* "The Encyclopaedia of Physics" (S. Flugge, ed) 226–793. Springer-Verlag, Heidelberg.

Tuck, G. J. and Stacey, F. D. (1978). Dielectric anisotropy as a petrofabric indicator. *Tectonophysics* **50**, 1–11.

Vialon, P., Ruhland, M. and Grolier, J. (1976). "Eléments de Tectonique Analytique." Masson, Paris.

Watterson, J. (1968). Homogeneous deformation of the gneisses of Vesterland, SW Greenland. *Medd. Grønland.* **175**, 72 p.

Wickham, J. and Anthony, M. (1977). Strain paths and folding of carbonate rocks near Blue Ridge, Central Appalachians. *Geol. Soc. Am. Bull.* **88**, 920–924.

Wilkinson, P., Soper, N. J. and Bell, A. M. (1975). Skolithos pipes as strain markers in mylonites. *Tectonophysics* **28**, 143–157.

Wilson, G. (1982). "Introduction to Small-scale Geological Structures" (in collaboration with J. W. Cosgrove). George Allen and Unwin, London.

Wood, D. S. (1973). Patterns and magnitudes of natural strain in rocks. *Phil. Trans. R. Soc. Lond.* **A274**, 373–382.

Wood, D. S., Oertel, G., Singh, J. and Bennett, H. F. (1976). Strain and anisotropy in rocks. *Phil. Trans. R. Soc. Lond.* A **283**, 27–42.

Wright, T. O. and Platt, L. B. (1982). Pressure dissolution and cleavage in the Martinsburg shale. *Am. J. Sci.* **282**, 122–135.

Zingg, T. (1935). Beitrag zur Schotteranalyse. *Schweiz. Mineral. Petrog. Mitt.* **15**, 39–140.

Ductile shear zones

General

Carreras, J., Cobbold P. R., Ramsay, J. G., White, S. H. (eds) (1980). Shear zone in rocks. *J. Struct. Geol.* **2**, No. 1/2.

Bak, J., Korstgård, J. and Sørensen, K. (1975). A major shear zone within the Nagssugtoqidian of West Greenland. *Tectonophysics* **27**, 191–209.

Beach, A. (1974). The measurement and significance of displacement on Laxfordian shear zones, North-West Scotland. *Proc. Geol. Ass.* **85**, 13–21.

Beach, A. (1976). The inter-relations of fluid transport, deformation, geochemistry and heat flow in early Proterozoic shear zones in the Lewisian complex. *Phil. Trans. R. Soc. Lond.* **A280**, 569–604.

Berthé, D., Choukroune, P. and Jegouzo, P. (1979). Orthogneiss, mylonite and non-coaxial deformation of granites: the example of the South Armorican Shear Zone. *J. Str. Geol.* **1**, 31–42.

Burg, J. P. and Laurent, Ph. (1978). Strain analysis of a shear zone in a granodiorite. *Tectonophysics* **47**, 15–42.

Burg, J. P., Iglesias, M., Laurent, Ph., Matte, Ph. and Ribeiro, A. (1981). Variscan intracontinential deformation: the Coimbra–Cordoba shear zone (SW Iberian Peninsula). *Tectonophysics* **78**, 161–177.

Cobbold, P. R. (1977a). Description and origin of banded deformation structures: I. Regional strain, local perturbations, and deformation bands. *Can. J. Earth Sci.* **14**, 1721–1731.

Cobbold, P. R. (1977b): II. Rheology and the growth of banded perturbations. *Can. J. Earth Sci.* **14**, 2510–2523.

Coward, M. P. (1976). Strain within ductile shear zones. *Tectonophysics* **34**, 181–197.

Escher, A., Escher, J. C. and Waterson, J. (1975). The reorientation of the Kangâmuit dike swarm, West Greenland. *Can. J. Earth Sci.* **12**, 158–173.

Grocott, J. (1979). Shape fabrics and superimposed simple shear strain in a Precambrian shear belt, W Greenland. *J. Geol. Soc. Lond.* **136**, 471–488.

Knipe, R. J. and White, S. H. (1979). Deformation in low grade shear zones in the Old Red Sandstone, S.W. Wales. *J. Str. Geol.* **1**, 53–66.

Lister, G. S. and Williams, P. F. (1979). Fabric development in shear zones: theoretical controls and observed phenomena. *J. Str. Geol.* **1** (1979), 283–297.

Mitra, G. (1978). Ductile deformation zones and mylonites: The mechanical processes involved in the deformation of crystalline basement rocks. *Am. J. Sci.* **278**, 1057–1084.

Mitra, G. (1979). Ductile deformation zones in Blue Ridge basement rocks and estimation of finite strains. *Geol. Soc. Am. Bull.* **90**, 935–951.

Ramsay, J. G. (1980). Shear zone geometry: a review. *J. Struct. Geol.* **2**, 83–89.

Ramsay, J. G. and Graham, R. H. (1970). Strain variation in shear belts. *Can. J. Earth Sci.* **7**, 786–813.

Schwerdtner, W. M. (1982). Calculation of volume change in ductile band structures. *J. Str. Geol.* **4**, 57–62.

Watts, M. J. and Williams, G. D. (1979). Fault rocks as indicators of progressive shear deformation in the Guingamp region, Brittany. *J. Str. Geol.* **1**, 323–332.

Ptygmatic folding and boudinage

Agostino, P. N. (1971). Theoretical and experimental investigations on ptygmatic structures. *Geol. Soc. Am. Bull.* **82**, 2651–2660.

Brühl, H. (1969). Boudinage in den Ardennen und in der Nordeifel als Ergebnis der inneren Deformation. *Geol. Mitt.* **8**, 263–308.

Burg, J. P. and Harris, L. B. (1982). Tension fractures and boudinage oblique to the maximum extension direction: an analogy with Lüders' bands. *Tectonophysics* **83**, 347–363.

Cloos, E. (1947). Boudinage. *Am. Geophys. Union Trans.* **28**, 626–632.

Coe, K. (1959). Boudinage structure in West Cork, Ireland. *Geol. Mag.* **116**, 191–201.

Corin, F. (1932). A propos du boudinage en Ardenne. *Bull. Soc. belge. Géol. Pal. Hydr.* **42**, 101–117.

De Sitter, L. U. (1958). Boudins and parasitic folds in relation to cleavage and folding. *Geol. Mijnbouw.* **37**, 277–286.

Dietrich, R. V. (1960). Genesis of ptygmatic features. Rep. 21st. Int. Geol. Cong. Copenhagen, Part 14, 138–148.

Fletcher, R. C. (1982). Analysis of the flow in layered fluids at small, but finite, amplitude with application to mullion structures. *Tectonophysics* **81**, 51–66.

Griggs, D. and Handin, J. (1960). Rock deformation. *Geol. Soc. Am. Mem.* **79**, 382 pp.

Hambrey, M. J. and Milnes, A. G. (1975). Boudinage in glacier ice—some examples. *J. Glac.* **14**, 383–393.

Holmquist, P. J. (1931). On the relations of the boudinage structure. *Geol. Fören. Stockh. Förh.* **53**, 193–208.

Kuenen, Ph. H. (1938). Observations and experiments on ptygmatic folding. *Bull. Comm. Geol. Finlande.* **123**, 11–27.

Kuenen, Ph. H. (1968). Orgin of ptygmatic features. *Tectonophysics* **6**, 143–158.

Lloyd, G. E. and Ferguson, C. C. (1981). Boudinage structure: Some new interpretations based on elastic-plastic finite element simulations. *J. Str. Geol.* **3**, 117–128.

Lohest, M., Stainier, X. and Fourmarier, P. (1909). Compte rendu de la Session extraordinaire de la Société géologique de Belgique, Tenue à Eupen et à Bastogne. *Ann. Soc. géol. Bélgique* **35**, B351–434.

Mitra, S. and Datta, J. (1978). Ptygmatic structures: an analysis and review. *Geol. Rdsch.* **67**, 880–895.

Platt, J. P. and Vissers, R. L. M. (1980). Extensional structures in anisotropic rocks. *J. Str. Geol.* **2**, 397–410.

Quirke, T. T. (1923). Boudinage, an unusual structural phenomenon. *Geol. Soc. Am. Bull.* **34**, 649–660.

Ramberg, H. (1955). Natural and experimental boudinage and pinch and swell structure. *J. Geol.* **63**, 512–526.

Ramberg, H. (1959). Evolution of ptygmatic folding. *Norsk, Geol. Tidsskr.* **39**, 99–151.

Rast, N. (1956). The origin and significance of boudinage. *Geol. Mag.* **93**, 401–408.

Sanderson, D. J. (1974). Patterns of boudinage and apparent stretching lineation developed in folded rocks. *J. Geol.* **82**, 651–661.

Sederholm, J. J. (1913). Über ptygmatische Faltungen. *N. Jahrbuch Min. Geol. Paläont.* **36**, 491–512.

Sen, R. and Mukherjee, A. D. (1975). Comparison of experimental and natural boudinage. *Geol. Mag.* **112**, 191–196.

Smith, R. B. (1975). Unified theory of the onset of folding, boudinage, and mullion structure. *Geol. Soc. Am. Bull.* **86**, 1601–1609.

Stephansson, O. and Berner, H. (1971). The finite element method in tectonic processes. *Phys. Earth Pl. Int.* **4**, 301–321.

Strömgård, K.-E. (1973). Stress distribution during formation of boudinage and pressure shadows. *Tectonophysics* **16**, 215–248.

Wegmann, C. E. (1932). Note sur le boudinage. *Bull. Soc. Géol. France* **2**, 477–489.

Foliation, lineation and strain

General

Borradaile, G. J., Bayly, M. B. and Powell, C. McA. (eds) (1982). Atlas of deformational and metamorphic rock fabrics. Springer-Verlag.

Beutner, E. C., Jancin, M. D. and Simon, R. W. (1977). Dewatering origin of cleavage in light of deformed calcite veins and clastic dykes in Martinsburg slate, Delaware Water Gap, New Jersey. *Geology* **5**, 118–122.

Borradaile, G. J. (1978). Transected folds: A study illustrated with examples from Canada and Scotland. *Geol. Soc. Am. Bull.* **89**, 481–493.

De Sitter, L. U. (1954). Schistosity and shear in micro and macrofabrics. *Geol. Mijnbouw* **16**, 429–439.

Geiser, P. A. (1974). Cleavage in some sedimentary rocks of the Central Valley and Ridge Province, Maryland. *Geol. Soc. Am. Bull.* **85**, 1399–1412.

Geiser, P. A. (1975). Slaty cleavage and the dewatering hypothesis—An examination of some critical evidence. *Geology* **3**, 717–720.

Goguel, J. (1945). Sur l'origine mécanique de la schistosité. *Bull. Soc. géol. France*, série 5, **15**, 509–522.

Groshong Jr, R. H. (1976). Strain and pressure solution in the Martinsburg slate, Delaware Water Gap, New Jersey. *Am. J. Sci.* **276**, 1131–1146.

Maxwell, J. C. (1962). Origin of slaty and fracture cleavage in the Delaware Water Gap Area, New Jersey and Pennsylvania. *Geol. Soc. Am. Buddington volume*, 281–311.

Means, W. D. (1975). Natural and experimental microstructures in deformed micaceous sandstones. *Geol. Soc. Am. Bull.* **86**, 1221–1229.

Means, W. D. (1977). Experimental contributions to the study of foliations in rocks: a review of research since 1960. *Tectonophysics* **39**, 329–354.

Powell, C. McA. (1972). Tectonically dewatered slates in the Ludlovian of the Lake District, England. *Geol. J.* **8**, 95–110.

Roberts, D. (1971). Abnormal cleavage patterns in fold hinge zones from Varanger Peninsula, Northern Norway. *Am. J. Sci.* **271**, 170–180.

Sharpe, D. (1847). On slaty cleavage. *Q. J. Geol. Soc.* **3**, 74–105.

Sharpe, D. (1849). On slaty cleavage. *Q. J. Geol. Soc.* **5**, 111–129.

Siddans, A. W. B. (1972). Slaty cleavage, a review of research since 1815. *Earth Sci. Rev.* **8**, 205–232.

Siddans, A. W. B. (1977). The development of slaty cleavage in a part of the French Alps. *Tectonophysics* **39**, 533–557.

Sorby, H. C. (1879). The structure and origin of limestones. *J. Geol. Soc. Lond. Proc.* **35**, 56–93.

Tullis, T. E. (1976). Experiments on the origin of slaty cleavage and schistosity. *Geol. Soc. Am. Bull.* **87**, 745–753.

Weber, K. (1976). Gefügeuntersuchungen an transversal-geschieferten Gesteinen aus dem östlichen Rheinischen Schiefergebirge (Ein Beitrag zur Genese der transversalen Schieferung). *Geol. Jb. Reihe D.* **15**, 3–98.

Williams, P. F. (1977). Foliation: a review and discussion. *Tectonophysics* **39**, 305–328.

Wood, D. S. (1974). Current views of the development of slaty cleavage. *A. Rev. Earth Plant Sci.* **2**, 369–401.

Veins and pressure shadows

Adams, S. F. (1920). A microscopic study of vein quartz. *Econ. Geol.* **15**, 623–664.

Beach, A. (1975). The geometry of en-echelon vein arrays. *Tectonophysics* **28**, 245–263.

Beach, A. (1977). Vein arrays, hydraulic fractures and pressure-solution structures in a deformed flysch sequence, S.W. England. *Tectonophysics* **40**, 201–225.

Beach, A. and Jack, S. (1982). Syntectonic vein development in a thrust sheet from the external French Alps. *Tectonophysics* **81**, 67–84.

Casey, M., Dietrich, D. and Ramsay, J. G. (1982). Methods for determining deformation history for chocolate tablet boudinage with fibrous crystals. *Tectonophysics* **92**, 211–239.

Choukroune, P. (1971). Contribution à l'étude des mécanismes de la déformation avec schistosité grâce aux cristallisations syncinématiques dans les "zones abritées". *Bull. Soc. géol. France* **13**, 257–271.

Durney, D. W. and Ramsay, J. G. (1973). Incremental strains measured by syntectonic crystal growths. *In* "Gravity and Tectonics" (K. A. De Jong and R. Scholten, eds) 67–96. Wiley, New York.

Hancock, P. L. (1972). The analysis of en-echelon veins. *Geol. Mag.* **109**, 269–276.

Hancock, P. L. and Atiya, M. S. (1975). The development of en-echelon vein segments by the pressure solution of formerly continuous veins. *Proc. Geol. Assoc. Lond.* **86**, 281–286.

Nicholson, R. (1978). Folding and pressure solution in a laminated calcite-quartz vein from the Silurian slates of the Llangollen region of N Wales. *Geol. Mag.* **115**, 47–54.

Mugge, O. (1930). Bewegungen von Porphyroblasten in Phylliten und ihre Messung. *Neues Jahrb. Mineral. Geol. Palaeont.* **61**, 469–520.

Pabst, A. (1931). "Pressure shadows" and the measurements of the orientation of minerals in rocks. *Am. Mineral.* **16**, 55–61.

Phillips, W. J. (1974). The development of vein and rock textures by tensile strain crystallization. *J. Geol. Soc. Lond.* **130**, 441–448.

Ramsay, J. G. (1980). The crack-seal mechanism of rock deformation. *Nature* **284**, 135–139.

Roering, C. (1968). The geometrical significance of natural en-echelon crack arrays. *Tectonophysics* **5**, 107–123.

White, S. H. and Wilson, C. J. L. (1978). Microstructure of some quartz pressure fringes. *Neues Jahrb. Mineral.* **134**, 33–51.

Wickham, J. S. (1973). An estimate of strain increments in a naturally deformed carbonate rock. *Am. J. Sci.* **273**, 23–47.

Index

Page numbers in bold refer to citations where the term is defined or explained; page numbers in italic refer to figures.

303